# Selected Papers from the 9th World Congress on Industrial Process Tomography

# Selected Papers from the 9th World Congress on Industrial Process Tomography

Special Issue Editors

**Manuchehr Soleimani**
**Thomas Wondrak**
**Chao Tan**


MDPI • Basel • Beijing • Wuhan • Barcelona • Belgrade

*Special Issue Editors*
Manuchehr Soleimani
University of Bath
UK

Thomas Wondrak
Institut für Fluiddynamik
Germany

Chao Tan
Tianjin University
China

*Editorial Office*
MDPI
St. Alban-Anlage 66
4052 Basel, Switzerland

This is a reprint of articles from the Special Issue published online in the open access journal *Sensors* (ISSN 1424-8220) in 2019 (available at: https://www.mdpi.com/journal/sensors/special_issues/WCIPT9).

For citation purposes, cite each article independently as indicated on the article page online and as indicated below:

LastName, A.A.; LastName, B.B.; LastName, C.C. Article Title. *Journal Name* **Year**, *Article Number*, Page Range.

**ISBN 978-3-03928-248-7 (Pbk)**
**ISBN 978-3-03928-249-4 (PDF)**

# Contents

# About the Special Issue Editors

**Manuchehr Soleimani** received his B.S. degree in Electrical Engineering, M.S. degree in Biomedical Engineering, and finally in 2005, his Ph.D. degree in Inverse Problems and Electromagnetic Tomography from University of Manchester, Manchester, UK. From 2005 to 2007, he was Research Associate with the School of Materials, University of Manchester. In 2007, he joined the Department of Electronic and Electrical Engineering, University of Bath, Bath, UK, where he is currently Professor and works in the area of intelligent tomographic imaging for medical and industrial applications.

**Thomas Wondrak** works at HZDR in Germany. He is a Research Scientist specializing in magnetohyrodynamics (MHD) and inverse problems in magnetic imaging methods. Dr Wonderak leads the Liquid Metal section on Tomocon, an EU training network project, and is one of the pioneers of CIFT, a contactless inductive method for imaging metal flow in continuous casting.

**Chao Tan** Chao Tan (M'09–SM'15) received his B.S., M.S., and Ph.D. degrees from Tianjin University, Tianjin, China, in 2003, 2006, and 2009, respectively, all in Control Science and Engineering, where he is currently Professor with the School of Electrical and Information Engineering. His current research interests include process parameter detection and control systems, multiphase flow measurement and instrumentation, industrial process tomography, and multisensor/data fusion.

*Editorial*

# Selected Papers from the 9th World Congress on Industrial Process Tomography

**Manuchehr Soleimani [1,*], Thomas Wondrak [2] and Chao Tan [3]**

[1] Engineering tomography lab (ETL), Department of Electronic & Electrical Engineering, University of Bath, Bath BA2 7AY, UK

[2] Helmholtz-Zentrum Dresden-Rossendorf, Institut für Fluiddynamik, Abt. Magnetohydrodynamik, 01328 Dresden, Germany

[3] School of Electrical and Information Engineering, Tianjin University, Tianjin 300072, China

* Correspondence: M.Soleimani@bath.ac.uk

Received: 2 September 2019; Accepted: 2 September 2019; Published: 3 September 2019

---

Industrial process tomography (IPT) is a set of multi-dimensional sensor technologies and methods that aim to provide unparalleled internal information on industrial processes used in many sectors. The World Congress on Industrial Process Tomography (WCIPT) is the flagship conference of the International Society for Industrial Process Tomography (ISIPT. https://www.isipt.org/), which is held every two years. After successful previous events: UK (1999), Germany (2001), Canada (2003), Japan (2005), Norway (2007), China (2010), Poland (2013), and Brazil (2016), WCIPT-2018 took place in University of Bath, UK and WCIPT-2020 is planned to be hosted in Tianjin University in China. The field of IPT is gathering momentum and several large scale international projects such as EU Tomocon (https://www.tomocon.eu/) are well underway.

A selection of papers was invited for this Special Issue. Of the 22 extended papers submitted to this Special Issue, 14 papers were accepted for publication. The authors in [1] introduce a new method for one-shot multi-frequency electrical tomography (ET). Promising early stage experimental results demonstrated that the method could indeed be useful in spectral imaging. In [2] the authors propose a fully automated software and design workflow for 3D electrical capacitance tomography (ECT) sensor, field modelling, and image reconstruction aided by 3D printers. Multilayer ECT sensors can be fabricated and linked to their corresponding computational and experimental system. In [3] the authors introduce a crowdsourcing approach with the example of highly complex silo flow imaging based on the X-ray tomography method. This provides a major step in the direction of human computer interfacing for IPT. In [4] a novel waste water metre is developed based on the ET method. The low cost ET system was verified with several experimental data with potential applications in environmental sustainable wastewater area. Authors in [5] show the use of electrical resistance tomography (ERT) for monitoring silica nanoparticles for sand-pack flooding experiments for oil recovery. The new ERT method has the potential to be used instead of X-ray CT or MRI for such an application offering a robust, online, and low cost solution. In [6] the authors demonstrate a novel image reconstruction method applied to magnetic induction tomography (MIT) for surface defect detection in metallic samples; the high resolution achieved in this approach makes MIT a unique candidate for such an application. The method was tested in lab-based experiments and was demonstrated for the first time using MIT data from a high temperature environment in continues casting in real steel manufacturing process.

In [7] the author discusses the use of EIT for granular or solid materials. A correlation was demonstrated between the absolute impedance data and moisture content, an important parameter in many of these processes. Grouting ducts with steel strands are widely used to increase the structural strengths of infrastructures. The determination of the steel strand's integrity inside of the ducts and the grouting quality are important for a strength evaluation of the structure. In [8], a capacitive sensing technique was applied to identify the cross-sectional distribution of the steel strands.

In [9] the authors present a new method of using a simulated inductor to help with capacitive coupled electrical impedance tomography (CCEIT). The new method of CCEIT offers great promise in the contactless imaging of complex dielectric properties, in particular, electrical conductivity, and this inductive compensation method is critical to enhancing the quality of the measured signals. In [10] a conformal mapping approach was developed for converting the open EIT geometry to the traditional circular EIT system. The method was verified using both simulated and real experimental data. The results greatly enhance our understanding of open EIT and planar array EIT. Calderon's method has been successfully used for direct image reconstruction in ECT. In the method proposed in [11], the truncation radius adopted in the numerical integral greatly influences the reconstruction results. As our understanding of the complex value of ECT makes ECT a more versatile technology, the direct method proposed by the authors in [11], which directly reconstructs the admittivity, will become more important in near future. In [12], the authors introduce a new method to combine EIT and ultrasound tomography (UT). A Lagrange-Newton Method to combine EIT and UT reflection imaging a great step forward in realization of such a great multi-modality. In [13] the authors provide a comprehensive comparison of different machine learning (ML) methods for ET data. Similar to other IPT modalities the ET system are producing very large number of data in form of 2, 3, or 4D images. ML approaches will gain more and more traction as these imaging methods make their way into real industrial process applications. Wire-mesh sensors are used to determine the phase fraction of gas–liquid two-phase flow in many industrial applications. In [14], the authors report the use of a sensor to study the flow behaviour inside an offshore oil and gas industry device for subsea phase separation.

**Funding:** This research received no external funding.

**Acknowledgments:** The editors of this Special Issue would like to thank all the authors for their great contribution, the anonymous reviewers for their very valuable work, and the *Sensors* editorial team for their cooperation, suggestions and advice for making this Special Issue possible.

**Conflicts of Interest:** The authors declare no conflict of interest.

## References

1.  Darnajou, M.; Dupré, A.; Dang, C.; Ricciardi, G.; Bourennane, S.; Bellis, C. On the Implementation of Simultaneous Multi-Frequency Excitations and Measurements for Electrical Impedance Tomography. *Sensors* **2019**, *19*, 3679. [CrossRef] [PubMed]

2.  Kowalska, A.; Banasiak, R.; Romanowski, A.; Sankowski, D. 3D-Printed Multilayer Sensor Structure for Electrical Capacitance Tomography. *Sensors* **2019**, *19*, 3416. [CrossRef] [PubMed]

3.  Romanowski, A.; Łuczak, P.; Grudzień, K. X-ray Imaging Analysis of Silo Flow Parameters Based on Trace Particles Using Targeted Crowdsourcing. *Sensors* **2019**, *19*, 3317. [CrossRef] [PubMed]

4.  Wu, C.; Hutton, M.; Soleimani, M. Smart Water Meter Using Electrical Resistance Tomography. *Sensors* **2019**, *19*, 3043. [CrossRef]

5.  Nwufoh, P.; Hu, Z.; Wen, D.; Wang, M. Nanoparticle Assisted EOR during Sand-Pack Flooding: Electrical Tomography to Assess Flow Dynamics and Oil Recovery. *Sensors* **2019**, *19*, 3036. [CrossRef]

6.  Li, F.; Spagnul, S.; Odedo, V.; Soleimani, M. Monitoring Surface Defects Deformations and Displacements in Hot Steel Using Magnetic Induction Tomography. *Sensors* **2019**, *19*, 3005. [CrossRef] [PubMed]

7.  Porzuczek, J. Assessment of the Spatial Distribution of Moisture Content in Granular Material Using Electrical Impedance Tomography. *Sensors* **2019**, *19*, 2807. [CrossRef] [PubMed]

8.  Li, N.; Cao, M.; Du, H.; He, C.; Wu, B. Detection of Single Steel Strand Distribution in Grouting Duct Based on Capacitive Sensing Technique. *Sensors* **2019**, *19*, 2564. [CrossRef]

9.  Ye, X.; Wang, Y.; Tang, X.-Y.; Ji, H.; Wang, B.; Huang, Z. On the Design of a New Simulated Inductor Using a Contactless Electrical Tomography System as an Example. *Sensors* **2019**, *19*, 2463. [CrossRef]

10. Wang, Y.; Ren, S.; Dong, F. Focusing Sensor Design for Open Electrical Impedance Tomography Based on Shape Conformal Transformation. *Sensors* **2019**, *19*, 2060. [CrossRef] [PubMed]

11. Sun, S.; Xu, L.; Cao, Z.; Sun, J.; Tian, W. Adaptive Selection of Truncation Radius in Calderon's Method for Direct Image Reconstruction in Electrical Capacitance Tomography. *Sensors* **2019**, *19*, 2014. [CrossRef] [PubMed]

12. Liang, G.; Ren, S.; Zhao, S.; Dong, F. A Lagrange-Newton Method for EIT/UT Dual-Modality Image Reconstruction. *Sensors* **2019**, *19*, 1966. [CrossRef] [PubMed]

13. Rymarczyk, T.; Kłosowski, G.; Kozłowski, E.; Tchórzewski, P. Comparison of Selected Machine Learning Algorithms for Industrial Electrical Tomography. *Sensors* **2019**, *19*, 1521. [CrossRef] [PubMed]

14. Ofuchi, C.Y.; Eidt, H.K.; Rodrigues, C.C.; Dos Santos, E.N.; Dos Santos, P.H.D.; Da Silva, M.J.; Neves, F., Jr.; Domingos, P.V.S.R.; Morales, R.E.M. Multiple Wire-Mesh Sensors Applied to the Characterization of Two-Phase Flow inside a Cyclonic Flow Distribution System. *Sensors* **2019**, *19*, 193. [CrossRef] [PubMed]

 *sensors* 

*Article*

# On the Implementation of Simultaneous Multi-Frequency Excitations and Measurements for Electrical Impedance Tomography [†]


**Mathieu Darnajou [1,2,*], Antoine Dupré [3], Chunhui Dang [1,2], Guillaume Ricciardi [1], Salah Bourennane [2] and Cédric Bellis [4]**

[1]   CEA Cadarache, 13115 Saint-Paul-lez-Durance, France
[2]   École Centrale de Marseille, 38 Rue Frédéric Joliot Curie, 13013 Marseille, France
[3]   Department of Paediatrics, University of Geneva, 1205 Geneva, Switzerland
[4]   Aix Marseille Univ, CNRS, Centrale Marseille, LMA UMR 7031, Marseille, France
[*]   Correspondence: mathieu.darnajou@centrale-marseille.fr
[†]   This paper is an extended version of our paper published in The Design of Electrical Impedance Tomography Detectors in Nuclear Industry. In Proceedings of the 9th World Congress on Industrial Process Tomography, Bath, UK, 2–6 September 2018.



Received: 18 July 2019; Accepted: 20 August 2019; Published: 24 August 2019



**Abstract:** The investigation of quickly-evolving flow patterns in high-pressure and high-temperature flow rigs requires the use of a high-speed and non-intrusive imaging technique. Electrical Impedance Tomography (EIT) allows reconstructing the admittivity distribution characterizing a flow from the knowledge of currents and voltages on its periphery. The need for images at high frame rates leads to the strategy of simultaneous multi-frequency voltage excitations and simultaneous current measurements, which are discriminated using fast Fourier transforms. The present study introduces the theory for a 16-electrode simultaneous EIT system, which is then built based on a field programmable gate array data acquisition system. An analysis of the propagation of uncertainties through the measurement process is investigated, and experimental results with fifteen simultaneous signals are presented. It is shown that the signals are successfully retrieved experimentally at a rate of 1953 frames per second. The associated signal-to-noise ratio varies from 59.6–69.1 dB, depending on the generated frequency. These preliminary results confirm the relevance and the feasibility of simultaneous multi-frequency excitations and measurements in EIT as a means to significantly increase the imaging rate.



**Keywords:** FPGA; high-speed EIT; frequency division multiplexing; ONE-SHOT; EIDORS


---

## 1. Introduction

In the context of high pressure and high temperature flow rigs, flow patterns are investigated. Tomography is a suitable technique for its applicability and non-intrusiveness. This imaging technique is used to reconstruct the volume of an object from measurements of penetrating waves or particles on its external boundary. Tomography is based on measurements of the boundary conditions of a system, where both the signal emitter and receiver have to be well known in space. Various tomography techniques exist using electromagnetic, acoustic waves, or electron and muon particles. Among the use of tomography in applications, the most well established is medical imaging using X-ray-based CT scans or electromagnetic waves in magnetic resonance imaging. Nevertheless, these techniques are not robust in this context, complex to implement, and expensive.

Electrical Impedance Tomography (EIT) [1–4] is another tomographic method that consists of imposing (resp. measuring) an electrical current passing through a set of electrodes at the surface of a body, while measuring (resp. imposing) the electrical potential over another set of electrodes. This technique is preferred for its applicability at high pressure and temperature, robustness, and relatively low cost. The standard approach of EIT is based on the concept of Time-Division Multiplexing (TDM), where the excitation signal is routed to a single pair of electrodes at any given time. The sequential selection of these pairs of electrodes with multiplexers or electronic switching [5] creates an excitation strategy and builds an EIT data frame, which includes the measurement data for all excitation pairs. The EIT data frame defines the Neumann and Dirichlet boundary conditions to be used in the associated inverse problem of reconstructing the electrical conductivity field inside the body.

EIT necessitates potentially a high data-frame acquisition rate, for instance in flow study involving rapidly-evolving flow regimes [6,7]. The challenge is that increasing the number of frames per seconds (fps) reduces the measurement time, while a large number of measurements is required to limit the ill-posedness of the inverse problem. In addition, the sequential excitation of the electrodes in contact with ionic solutions results in significant difficulties caused by transient voltages due to the presence of capacitive components in the electrode–electrolyte interface [8]. Therefore, the TDM sequential measurement procedure requires dead time, complex signal processing techniques, or hardware improvement to limit the effects of contact impedance [9,10].

Several high-speed EIT systems of hundreds of fps have been proposed in the literature. In [11], a multichannel architecture system was presented, containing a control module that manages and synchronizes 64 channels capable of generating and measuring voltages and currents. The system was able to collect 182 frames per seconds when acquiring 15 spatial patterns with a Signal-to-Noise Ratio (SNR) from 65.5–96 dB when averaging the signal from 32 samples. Another EIT system proposed in [12] allows capturing over 100 frames per seconds with an SNR greater than 90 dB when averaging the signal. Concerning Electrical Capacitance Tomography (ECT), another soft field imaging technique, several systems were also reported in [13–15] with high frame rate. Furthermore, an EIT system was developed in our group [16] to reach the Data Acquisition (DAQ) rate of 833 fps when acquiring 120 spatial patterns with TDM. Nevertheless, in any case, high speed is reached at the price of a small number of electrodes, a partial scanning strategy, or short measurement times.

Equivalent to TDM, in the early stages of X-ray tomography, a single source/detector pair was rotated around a body to be imaged, resulting in a low acquisition rate. The improvement of X-ray tomography systems led to multiple emitter/receiver pairs that do not interact with each other as X-rays pass straight through the body. The sources and receivers are activated simultaneously, improving significantly the acquisition rate. By analogy with the evolution of X-ray tomography, a solution for increasing the speed of EIT is to perform simultaneous measurements. However, the sensitivity regions within the body depend on the conductivity of the material, which is a priori unknown since it depends on the conductivity distribution inside the body, which is unknown. Therefore, simultaneous electrical emissions and measurements through surface electrodes result in a superposition of the signals in the measurement channels.

The ONe Excitation for Simultaneous High-speed Operation Tomography (ONE-SHOT) is an innovative method developed by our group [17] to give a solution for the problem of the superposition of signals for simultaneous excitation in EIT systems. Based on simultaneous multi-frequency stimulations, the overlapping of signals is discriminated using Frequency-Division Multiplexing (FDM) techniques. FDM, used in telecommunications, consists of dividing the total bandwidth available into a series of non-overlapping frequency bands, each of which is used to carry a separate signal. FDM is used in the context of EIT [18,19], and simultaneous excitations at different frequencies can be discriminated by demodulation, resulting in simultaneous excitations and measurements.

The ONE-SHOT approach brings several novelties in the field of FDM EIT. Firstly, in the excitation strategy of the previous systems, half of the electrodes are tagged with an excitation frequency, and the other half is used for measurement: the electrodes are either used for excitation or measurement. The ONE-SHOT method provides current measurement in the excitation circuit, and each electrode is simultaneously used for both excitation and measurement. Secondly, in the previous systems, the current excitation electrodes are tagged with a single frequency. On the other hand, ONE-SHOT introduces in this article an experiment where 15 excitation signals at 15 frequencies are imposed on a single electrode. It aims at demonstrating the applicability of FDM in this situation. Thirdly, a central argument in using FDM is the absence of transients at the electrode–electrolyte, interface resulting in a high data frame rate and low noise. The association of continuous excitations with a point-by-point synchronous Fourier transform is discussed to optimize the data acquisition speed. In addition, an innovative hardware method leads to the implementation of every excitation and measurement, including real-time fast Fourier transform on a single FPGA chip.

The present article proposes a practical implementation of the ONE-SHOT in a physical experiment to prove the feasibility of the method. The method was introduced in [17] for four electrodes; however, in practice, EIT systems usually contain eight electrodes or more for better performances. After an overview of the mathematical aspect of EIT in Section 2, a first step is to determine the adequate number of electrodes for satisfying performances and to adapt the theory of ONE-SHOT consequently (Section 3). Secondly, the high frame rate generates a large amount of data to process. In the past few years the development of electronics made it possible to handle systems allowing large measurement data to transfer [20]. The estimation of the optimal data transfer rate and online computation speed is needed to choose the adequate hardware components, an issue that is discussed in Section 4. Thirdly, the error propagation through the FDM is estimated in this context based on two quantitative numerical simulations (Section 5). Finally, the article presents preliminary results of the demodulation of 15 simultaneous excitation frequencies in Section 6.

## 2. Electrical Impedance Tomography

EIT is a type of non-intrusive technique widely used in medical, geo-physical, and industrial imaging. From the information of electrical current and potential on the boundary of an object, the Calderón's inverse problem aims at determining the conductivity distribution inside the object. This section introduces the mathematical aspects of EIT as generally reported in the literature.

The system $\Omega \subset \mathbb{R}^3$ in which the image reconstruction is considered is cylindrical (Figure 1). EIT is a technique to solve the inverse problem of determining the conductivity function $\gamma(x, \omega) = \sigma(x) + i\omega\epsilon(x)$ within the interior of $\Omega$ [2–4]. Here, $\sigma(x)$ is the isotropic electrical conductivity field, $\epsilon(x)$ the permittivity, and $\omega$ the angular frequency. EIT operates at $\omega \ll 1$ GHz, so the imaginary part of $\gamma$ can be neglected $\gamma(x, \omega) \to \sigma(x)$. The closure of $\Omega$ is denoted as $\bar{\Omega}$, and its boundary is $\partial\Omega$.

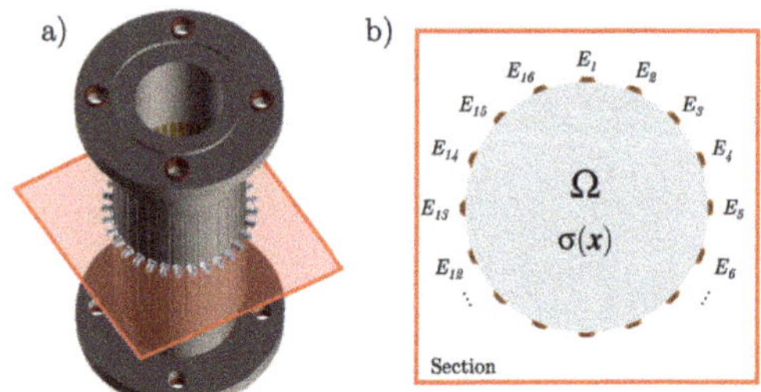


**Figure 1.** (a) Scheme of the EIT sensor prototype of height 336 mm, external diameter 120 mm, and inner diameter 80 mm containing 32 electrodes of length 150 mm and width 6 mm. In the current experiment, one over two, 16 electrodes are connected to coaxial cables with the SubMiniature Version A (SMA) connectors, also shown in the figure. (**b**) Representation of the 16 electrodes in the cross-section of the EIT detector.

In EIT, the conductivity function $\sigma(x)$ is unknown and has to be determined from simultaneous excitations and measurements of the voltage $u(x)$ and the current $j(x) = \sigma(x)\nabla u$, on the boundary. The Maxwell equations define $j(x)$ to be divergence free. Thus, Ohm's law gives the partial differential equation:

$$\nabla \cdot j(x) = 0 \quad \text{for } x \in \Omega. \tag{1}$$

The EIT detector contains a set of $n_e$ electrodes. On each electrode $E_n \subset \partial\Omega$ for $0 \leq n \leq n_e$ (Figure 1b), the potential $V_n$ and the current $I_n$ are either given or measured, defining the boundary conditions. Equation (1) is taken with either the *Dirichlet boundary conditions*:

$$u(x) = \begin{cases} V_n(x) & \text{for } x \in E_n, \\ 0 & \text{for } x \in \partial\Omega \text{ and } x \notin E_n, \end{cases} \tag{2}$$

or the *Neumann boundary conditions*:

$$j \cdot n(x) = \begin{cases} I_n(x) & \text{for } x \in E_n, \\ 0 & \text{for } x \in \partial\Omega \text{ and } x \notin E_n, \end{cases} \tag{3}$$

where $n(x)$ is the unitary vector on the surface pointing outward at $x \in \partial\Omega$. The Neumann boundary conditions have the additional requirement that:

$$\sum_{n=1}^{n_e} I_n(x) = 0, \tag{4}$$

as required by Equation (1) for well-posedness. In addition, in the case of imposed potentials, the Dirichlet boundary conditions are chosen with the average-free boundary voltage requirement:

$$\sum_{n=1}^{n_e} V_n(x) = 0, \tag{5}$$

in order to define the ground value.

The so-called Dirichlet-to-Neumann (DtN) map $\Lambda_\sigma : u|_{\partial\Omega} \longmapsto \sigma(x)\nabla u \cdot n|_{\partial\Omega}$ is the operator that links the imposed potential (i.e., Equation (2)) to the current on the boundary. Equivalently, the Neumann-to-Dirichlet (NdD) map $\mathcal{R}_\sigma : \sigma\nabla u \cdot n|_{\partial\Omega} \longmapsto u|_{\partial\Omega}$ yields the voltage distribution from any given current-density distribution on the boundary (i.e., Equation (3)). The DtN or NtD map contains the information to determine $\sigma(x)$ from the set of $V_n(x)$ (or $I_n(x)$). The aim of EIT is to reconstruct the conductivity distribution $\sigma(x)$ while **imposing** either the *Dirichlet* or the *Neumann* boundary conditions while **measuring** respectively the *currents* or the *potentials* over the electrodes [21,22]. Finally, the measurements relative to different excitation patterns constitute the data frame, which is used to solve the associated inverse problem to reconstruct $\sigma(x)$.

## 3. Specifications of the Proposed EIT System

The ONE-SHOT method was introduced in [17], which contains motivating results that predict the feasibility of the demodulation of simultaneous excitations with respect to their frequencies. Nevertheless, the proof-of-principle experiment is based on a four-electrode EIT system. The number of independent pairs of excitation for such systems is six, implying six simultaneous excitations at six different frequencies. In practical EIT, the number of electrodes is an important parameter since it allows more measurements and a better conditioning of the inverse problem. Usually, EIT systems contain more than eight electrodes.

The need for more simultaneous measurements is the motivation to evolve the ONE-SHOT method with the increased number of electrodes and, consequently, the number of excitation frequencies. This section discusses the choice of developing the ONE-SHOT excitation strategy for 16 electrodes in Section 3.1. Secondly, the ONE-SHOT excitation strategy is generalized to any number of electrodes in Section 3.2 and then adapted for 16 electrodes in Section 3.3. This process increases significantly the number of independent pairs and the complexity of the corresponding excitation patterns. Finally, choosing the frequencies is a challenging task. A choice is proposed in Section 3.4 to ensure discriminability, a high rate, and adaptivity to high-speed hardware systems.

### 3.1. Adequate Number of Electrodes

A large number of measurements is required to well condition the inverse problem for an accurate and stable solution. Consider an EIT device containing a set of $n_e$ electrodes. The maximum number of measurements is:

$$M = N \times n_e, \tag{6}$$

where $N$ is the total number of independent excitation pairs, i.e.:

$$N = \frac{n_e\,(n_e - 1)}{2}. \tag{7}$$

As $M \sim n_e^3$, the dataset size increases rapidly as the number of electrodes increases. One can expect that larger values of the parameter $n_e$ lead to images of better quality.

To evaluate this, the open access code EIDORS (Electrical Impedance Tomography and Diffuse Optical Tomography Reconstruction Software) [23] was used. The code provides software algorithms for forward and inverse modeling for EIT and diffusion-based optical tomography, in medical and industrial settings. Synthetic data of several flow patterns with EIT detectors containing $n_e = 8$, 16, and 32 electrodes are computed. The data are associated with an image based on the linear back projection reconstruction algorithm. This non-iterative method projects the set of voltage variations $[\delta V_n]$ between homogenous and an inhomogeneous conductivity data onto the maps of conductivity change $\delta\sigma(x)$ with a set of sensitivity coefficients $S$ calculated with a linearized version of the inverse problem [24].

In the simulations of Figure 2, the domain is filled with homogenous water of conductivity $0.5$ S·m$^{-1}$ with the addition of steam bubbles of conductivity $1 \times 10^{-5}$ S·m$^{-1}$. Three patterns are represented in the figure. The number of measurements is $M = 224$ for $n_e = 8$, $M = 1\,920$ for $n_e = 16$, and $M = 15\,872$ for $n_e = 32$. For every pattern, increasing $n_e$ improves the image quality, as expected. However, the situation $n_e = 32$ does not give much more accuracy than $n_e = 16$, even if the number of measurements is multiplied by more than eight. This result is a known matter in the field of EIT.

Any system with a high frame rate brings an important amount of data. The balance between high image accuracy and low data size led us to develop an EIT system with a specific number of 16 electrodes. According to Equation (7), the full set of excitations for an EIT system of 16 electrodes is $N = 120$.

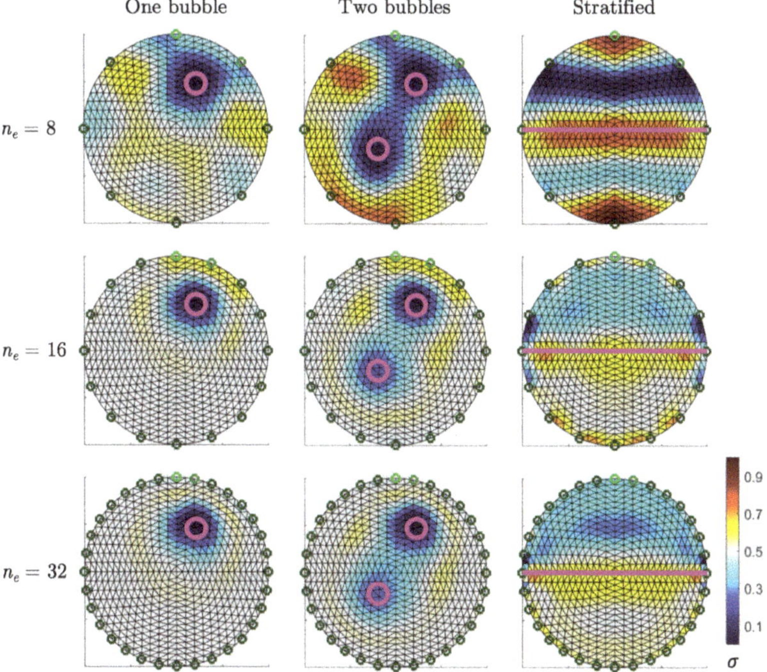

**Figure 2.** Image reconstruction using linear back projection from simulated data of bubbles in liquid for EIT detectors containing 8, 16, and 32 electrodes, represented with the green circles. On the left, one bubble of diameter 0.1 $\mathcal{D}$, with $\mathcal{D}$ being the diameter of the pipe. In the middle, two bubbles of the same diameter. On the right, stratified flow. The gas–liquid interface is shown with the purple line.

### 3.2. General Simultaneous Excitation Pattern for $n_e$ Electrodes

ONE-SHOT has to be adapted for a larger number of electrodes. As in Equation (7), the total number of independent measurements $N$ defines the number of frequencies that have to be generated simultaneously to maximize the number of measurements for a given system. The excitation is a set of voltages imposed on the electrodes and is generated from a basis of $N$ sines. Moreover, as discussed in Section 2 the excitation pattern ensures that the sum of boundary voltages is zero at any time.

One defines the signal $\Psi_i$ as:

$$\Psi_i(t) = A\sin(2\pi f_i t). \tag{8}$$

Then, given an arbitrary number $n_e$ of electrodes, one defines the excitation voltage $V_n(t)$ at an arbitrary electrode $n \in \{1, \ldots, n_e\}$ using the following recurrence relation:

$$V_n(t) = -\sum_{j=1}^{n-1} \left[V_j(t)\right]_{n-1} + \sum_{\ell=\ell_n^{\min}}^{\ell_n^{\max}} \Psi_\ell(t) \tag{9}$$

where:

$$\ell_n^{\min} = (n-1)n_e - \frac{n(n-1)}{2} + 1 \quad \text{and} \quad \ell_n^{\max} = n\,n_e - \frac{(n+1)n}{2} \tag{10}$$

and with $\left[V_j(t)\right]_{n-1}$ designating the $(n-1)^{\text{th}}$ element of the identity defining the voltage $V_j(t)$, under the convention that the terms $\Psi_i$ are always ordered with increasing index $i$ in such an identity. Moreover, in (9), we adopt the convention that a sum is identically zero if the value of the starting index is larger than that of the ending one.

### 3.3. Simultaneous Excitation Pattern for 16 Electrodes

We are now interested in applying Equation (9) in the situation of $n_e = 16$ electrodes. In this situation, $N = 120$ and the set of excitation voltages $[V_n(t)]$ is:

$$
\left\{
\begin{array}{lllllllll}
V_1(t) & = & +\Psi_1 & +\Psi_2 & +\Psi_3 & + & ... & +\Psi_{14} & +\Psi_{15} \\
V_2(t) & = & -\Psi_1 & +\Psi_{16} & +\Psi_{17} & + & ... & +\Psi_{28} & +\Psi_{29} \\
V_3(t) & = & -\Psi_2 & -\Psi_{16} & +\Psi_{30} & + & ... & +\Psi_{41} & +\Psi_{42} \\
V_4(t) & = & -\Psi_3 & -\Psi_{17} & -\Psi_{31} & + & ... & +\Psi_{53} & +\Psi_{54} \\
& & & \vdots & & & & & \\
V_{15}(t) & = & -\Psi_{14} & -\Psi_{28} & -\Psi_{41} & - & ... & -\Psi_{118} & +\Psi_{120} \\
V_{16}(t) & = & -\Psi_{15} & -\Psi_{29} & -\Psi_{42} & - & ... & -\Psi_{119} & -\Psi_{120}
\end{array}
\right.
\tag{11}
$$

where the frequency $f_{120}$ is the $N^{\text{th}}$ frequency for $n_e = 16$. Finally, the verification of Equation (5) from the excitation pattern of Equation (11) is straightforward.

### 3.4. Determination of the Excitation Frequencies Based on the Measurement Time Window

The main advantage in the use of a continuous multi-frequency excitation method is the absence of transients between successive projections, resulting in a diminution of the measurement error from the absence of the residual voltage in the electrode–electrolyte contact impedance. Concerning ONE-SHOT, the values of the frequencies remain as free parameters. In this section, we suggest suitable values based on the performances of the DAQ system.

The first observation is that in order to ensure continuous excitations and measurements in parallel, the frequencies of the excitation signals have to be harmonics of a fundamental frequency $f_0$. A wise choice is to define $f_0$ as the frequency of the Discrete Fourier Transform (DFT) of a $P$-point measurement sequence $\{V_n(p)\}$ and $p$ the discrete time. The discretisation of the time is due to the sampling rate of the DAQ system at the frequency $f_{DAQ}$, with the discretisation time interval $\Delta p = 1/f_{DAQ}$, implying:

$$
f_0 = \frac{f_{DAQ}}{P}.
\tag{12}
$$

The DFT of the real-valued sequence $\{V_n(p)\}$ in the $1 \leq n \leq n_e$ measurement channel is:

$$
\hat{V}_n(k) = \frac{1}{P} \sum_{p=0}^{P-1} V_n(p) e^{-ik\beta_p}, \quad k = 0, ..., P-1,
\tag{13}
$$

with $\beta_p = 2\pi p/P$, and a synchronous sampling is assumed. The real and imaginary parts of $\hat{V}_n(k)$:

$$
R_n(k) = \frac{1}{P} \sum_{p=0}^{P-1} V_n(p) \cos(k\beta_p)
\tag{14}
$$

and:

$$
I_n(k) = -\frac{1}{P} \sum_{p=0}^{P-1} V_n(p) \sin(k\beta_p)
\tag{15}
$$

define the module $M_n(k)$ and the phase $\phi_n(k)$ of each frequency domain sample $\hat{V}_n(k)$:

$$
M_n(k) = \sqrt{R_n^2(k) + I_n^2(k)}
\tag{16}
$$

and:

$$
\phi_n(k) = \arctan\left(\frac{I_n(k)}{R_n(k)}\right).
\tag{17}
$$

The adequate situation where the frequencies $f_i$ of the generated sines $\Psi_i$ are multiples of $f_0$ leads to the following observation: To each $k$-coefficient is associated a frequency $f_k$, which is linked to the DFT computation frequency $f_0$ by $f_k = kf_0$, for $k = 0, ..., P - 1$, and the following difference, which defines the resolution in the Fourier space:

$$\Delta f_k = f_{k+1} - f_k = f_0. \tag{18}$$

Choosing the generated frequencies as harmonics of $f_0$ leads to a match between $f_i$ and $f_k$ such that $f_i = f_k$, for $i = k$ and $i = 0, ..., P - 1$.

This observation is particularly interesting as with a given magnitude $M_n(k)$ is associated a frequency $f_k$, which corresponds to one and only one generated signal frequency $f_i$ on the electrode $n$. Therefore, the set of $[M_n(k)]$ for all $n$ and $k$ is one frame of the EIT data. The frame is acquired at the frequency $f_0$ and contains the full set of independent excitations and measurements to define the Dirichlet or the Neumann boundary conditions.

A first remark is that the highest frequency $f_N$ is constrained below the Nyquist limit: $f_N \leq f_{DAQ}/2$. A second remark concerns the phase shift, which describes the difference in radians when two or more alternating quantities reach their maximum or zero values. The phase shift of an electric signal passing through a material depends on the frequency [16]. In ONE-SHOT, the continuous generation of signals makes the phase shift independent of the Fourier magnitude. This is true because of the choice of the generated signals that are harmonics of the Fourier computation time window.

In standard EIT, the measurement associated with one excitation configuration corresponds to a voltage or current input over one or several periods of an alternative excitation at a given fixed frequency. The corresponding data are a set of tens of points per measurement, which multiplied by $M$ (Equation (6)) gives the total number of samples per frame. On the other hand, in the Fourier space, the corresponding data for the same measurement becomes a single element $M_n(k)$ with $n$ and $k$ defined. Building a data frame with the Fourier elements reduces significantly the data size.

Finally, the determination of the frequencies for the application of a 16-electrode ONE-SHOT excitation strategy strongly depends on the acquisition rate $f_{DAQ}$ of the DAQ system and the choice for the number of measurement points $P$ used to compute the DFT.

## 4. The Data Acquisition System

The concretization of an experiment to implement ONE-SHOT is based on the proposed hardware system presented in Figure 3. It is composed of three elements: the EIT sensor that contains the electrodes, the Printed Circuit Board (PCB) to distribute the signals, and the DAQ controller to manage the excitations and measurements.

Section 4.1 contains a discussion on the requirements for each component that led to the choice of this particular system. Then, in Section 4.2, the excitation and measurement strategy is applied to this system.

### 4.1. Selection of the Hardware

The starting point in creating a physical experiment for simultaneous EIT excitations and measurements is to establish the requirements on the DAQ system. The following details the issues and solutions found for the conception of its different components.

### 4.1.1. The EIT Sensor

The EIT sensor is composed of a ring of electrodes in contact with the fluid and is non-intrusive. The electrodes must be insulated from each other and made in a conductive and robust material. In addition, the excitation and measurement signals at the electrodes are possibly of low amplitude.

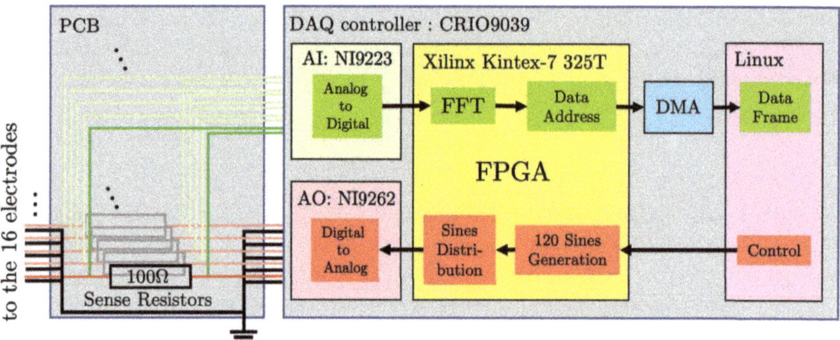

**Figure 3.** Layout of the DAQ system. On the left, details of the PCB that includes 16 independent circuits for voltage measurements over 100 Ω resistors. On the right, the FPGA-based CRIO-9039 configuration containing four NI-9223 AI modules, three NI-9262 AO modules, and a Direct Memory Access (DMA) buffer all controlled by the host, operated by Linux. In red, the excitation circuit. In green, the measurement circuit.

The EIT sensor prototype used in our experiment is shown in Figure 1a. It was a 336 mm-long cylinder PMMA prototype with an internal diameter of 80 mm containing a set of 32 electrodes. The electrodes were chosen to be made of stainless steel. They were 150 mm long and 6 mm wide, and their surface was tangential to the inner pipe surface and collinear to the axis of the cylinder. In the experimental setting considered, only one over two, 16 electrodes were connected. The characteristics of the detector were chosen for future tests on static and dynamic flow measurements at the Laboratory of analytical Thermohydraulics and Hydromechanics of Core and Circuits (LTHC). More details on the design of the EIT for flow measurement applications can be found in [25].

The signal was routed to the electrodes using shielded coaxial cables. The EIT sensor provided Sub Miniature version A (SMA) adaptors to ensure the connection of the electrodes with the inner wire of the coax and the insulation of the shielding.

### 4.1.2. The Printed Circuit Board

The main motivation in using the ONE-SHOT strategy is its high data acquisition rate. The natural choice for the hardware is to provide a very high sampling rate for excitation and measurement. As in Section 2, EIT relies on imposing or measuring either a voltage or a current on the boundary. However, the sampling rate for generation and acquisition of current is usually much slower than for the voltage.

A one-layer PCB (Figure 3) was designed to manage the voltage excitations and, in parallel, the current measurements. The currents passing through the electrodes (Neumann boundary conditions) were reconstructed from voltage measurements over 16 sense resistors of 100 Ω in the 16-electrode circuits on the PCB. This allowed current measurement using an analog voltage sampling at the DAQ level for a high sampling rate. Moreover, the PCB provided coaxial connectors and ensured the signal shielding with a ground connection.

### 4.1.3. The DAQ Controller

The DAQ controller was required to perform 16 voltage excitations and 16 voltage measurements in parallel at high rate with great accuracy. The output configuration contained the generation of 120 sines at 120 different frequencies and a distribution onto the 16 electrodes under Equation (11). The input configuration computed the DFT online for each electrode signal at a high rate, based on the Fast Fourier Transform (FFT) algorithm. Finally, the DAQ controller was required to perform fast data storage.

Apart from the performance requirements, the DAQ controller must be robust and light weight for transportation purposes in future prospects of experiments on several hydraulic loops. It also has to be controllable at a distance for an application to high pressure and high temperature flow experiments.

The strict requirements of fast computations in both output and input for multiple channel operations, including FFT, suggest the use of a Field Programmable Gate Array (FPGA). In recent years, the number of logic blocks contained in an FPGA chip increased, and the generation of a large number of arbitrary signals became feasible.

A suitable device for the DAQ controller in the ONE-SHOT hardware system is the National Instruments CRIO-9039 (NI CRIO-9039: http://www.ni.com/pdf/manuals/375697d_02.pdf) controller, which includes a 1.91-GHz Quad-Core CPU, 2 GB of DRAM, and the Xilinx's FPGA Kintex-7 325T. The FPGA includes 326,080 logic cells, 840 digital signal processing slices, and 16,020 block RAM elements. The main advantage of the FPGA is that it includes a large number of logic cells and large block RAM, which are required in our experiment to perform the operations described in Section 3. The adequate use of the FPGA allows fast DFT for a minimal data frame size, which allows efficient data transfer and reduces storage.

The CRIO-9039 DAQ controller includes a Linux host computer that operates ONE-SHOT with LabView 2017 Real-Time and FPGA. The monitoring of ONE-SHOT results in a large amount of data over numerous channels. The fast data monitoring is based on Direct Memory Access (DMA), a buffer to send the data to the host computer. In addition, the CRIO-9039 provides 8 slots in which Analog Output (AO) and Analog Input (AI) modules are connected.

### 4.1.4. Output and Input Configuration

Regarding the necessity for fast and accurate operations, the AO was managed with the NI-9262 (NI-9262: http://www.ni.com/pdf/manuals/377139a_02.pdf) module to provide the voltage outputs. The six-channel module had a typical output voltage range of $\pm10.742$ V, including an internal noise of 150 μV RMS per channel. In the system, three NI-9262 modules were connected to the 16-output excitations.

Concerning the AI, the NI-9223 (NI-9223: http://www.ni.com/pdf/manuals/374223a_02.pdf) module is suitable to provide voltage inputs. The NI-9223 includes four channels with a typical input voltage range of $\pm10.6$ V with a noise of 229 μV RMS per channel. A total of four NI-9223 modules were required to connect the 16 input measurements.

Finally, the digitalized data format for both AO and AI modules were (20,5) fixed points: 20 bits allocated to the number including 5 precision digits. The data were operated at a sampling rate of 1 Mega Sample per second (MS/s) for both AO and AI, which is very competitive in terms of the performance requirements.

### 4.2. Practical Excitation and Measurement Strategy

In Figure 3, the excitation circuit is detailed in red. The host computer controls the FPGA to provide a fast generation of 120 sine signals. A second function inside the FPGA distributes the sines to the electrodes as in Equation (11). The 16 signals are transformed from 1 MS/s digital to analog signals using the NI-9262 modules and sent with coaxial cables to the electrodes through the resistors of the PCB.

In parallel, the measurement circuit, shown in green in Figure 3, contains AI modules to digitalize the voltages taken over the 16 resistances on the PCB. The FPGA-based DFT computation provides the signal magnitude and phase for each Fourier coefficient at the high rate of 1 MS/s. The electrode number $n$ and Fourier coefficient $k$ are associated with the magnitude value to provide its address: the host computer uses the address as $x$ and $y$ coordinates for the magnitude to build the data matrix. The FPGA sends the data with a Direct Memory Access (DMA) to ensure fast operations and reliability.

Considering the sampling rate of 1 MS/s of the DAQ system, one solution for the frequencies of the generated signals (Equation (11)) in the situation of a 16-electrode EIT sensor consists of $f_i = i f_0$

for $1 \leq i \leq N$ and $N = 120$, as in Equation (7). Furthermore, the DFT computation can be chosen to be over $P = 512$ points and results in a data frame acquisition rate of $1 \times 10^6/512 = 1953$ fps. This choice implies the lowest sine frequency $f_1$ to be the same as the DFT computation frequency. The highest frequency $f_N = f_{120}$ is 234.375 kHz, below the Nyquist limit of 500 kHz for the considered system.

## 5. Analysis of the Error Propagation in DFT

Besides the hardware configuration, the measurement performances strongly depend on the DFT algorithm, which may have an influence on the uncertainty of the output data [26]. Consequently, there is a great interest in evaluating the systematic uncertainty in the DFT reconstruction. Two simulations are proposed, firstly in Section 5.1 without noise to assess the systematic uncertainty due to the discretisation of the DFT into $P = 512$ points and, secondly, in Section 5.2, another simulation with a noisy signal to assess the propagation of noise in the DFT by quantifying the magnitude change in the Fourier space.

### 5.1. Noise-Free Sine Simulation

In the first step, the experiment focused on a limited number of frequencies. Let us impose 16 voltage excitation signals over the electrodes: 15 sines $V_n$ of frequencies $f_i$ and amplitude $A$ such that:

$$V_n = A \sin(2\pi f_i p), \quad \text{for } n = 1, \dots, 15 \text{ and } i = n \tag{19}$$

over the electrodes $E_n$. In addition, the definition of the ground (Equation (5)) suggests a drain voltage:

$$V_{Drain} = -A \sum_i \sin(2\pi f_i p) \tag{20}$$

imposed on the remaining electrode $E_{16}$ (Figure 4).

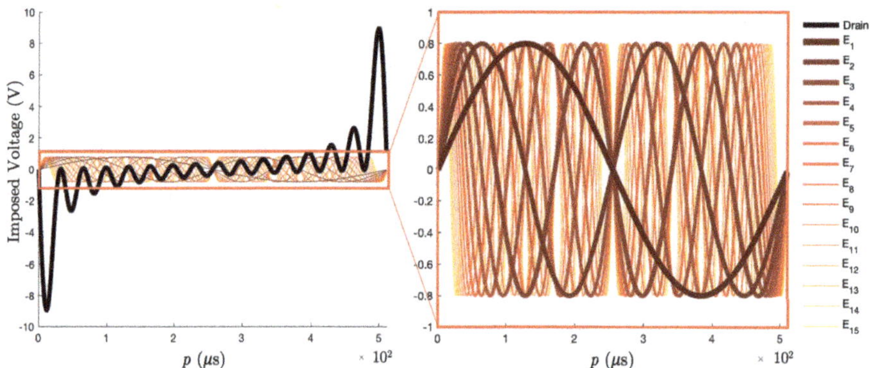


**Figure 4.** The 15 Excitation signals $V_n$ imposed on electrodes $E_n$ of frequencies $i f_1$ for $1 \leq i \leq 15$, and $f_1 = (512 \, \mu s)^{-1}$. The drain excitation signal is shown in black on the left plot.

The lowest frequency is defined such that it corresponds to the frequency of the DFT computation time window (Section 3.4). The DAQ system described in Section 4 has a sampling frequency of 1 MS/s, resulting in $f_1 = (512 \, \mu s)^{-1} = 1953.125$ Hz, which is also the resolution in the frequency space. Furthermore, the frequencies $f_i$ are the harmonics of $f_1$ such that $f_i = i f_1$. The amplitude $A$ is 0.8 V, so the drain signal remains in the NI-9262 AO module range limit of $\pm 10.742$ V.

The reconstruction of the magnitudes is shown in Figure 5. The systematic uncertainty due to the 512 points discretisation of the DFT was of the order of $O(1 \times 10^{-15})$ V.

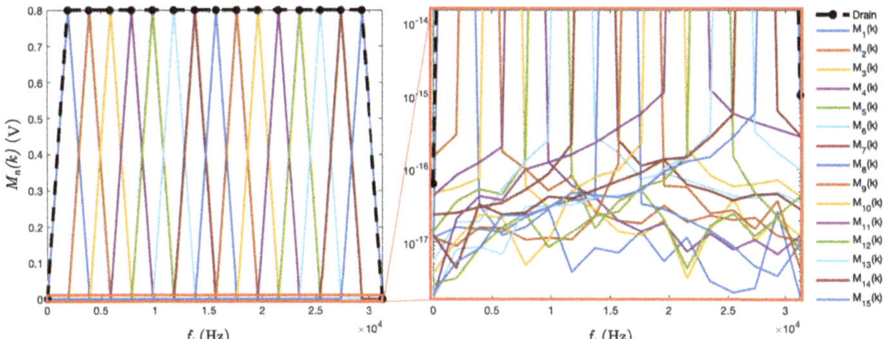

**Figure 5.** Simulation of the magnitudes of the 15 frequencies plus the drain at the 16 channels. On the right, rescaled zoom in the bottom part in the logarithmic scale of the DFT plot to show the tails of the peaks due to the discretisation onto 512 points.

### 5.2. Noisy Sines Simulation

The second objective was to assess the magnitude of the distortion of the $k^{\text{th}}$ harmonic due to the noise in the measured signal. In the time domain, the noisy signal $\tilde{V}_n(p)$ in the circuit of electrode $E_n$ contained a true signal $V_n$ and a Gaussian additive noise $\delta$ such that:

$$\tilde{V}_n(p) \;=\; V_n(p) + \delta(p). \tag{21}$$

Since the Fourier transform is linear, the addition of noisy signal gives the following magnitude in the $k^{\text{th}}$ harmonic of the measurement on the $n^{\text{th}}$ electrode, as in Equations (13)–(16):

$$\tilde{M}_n(k) = \left[ \left( \frac{1}{P} \sum_{p=0}^{P-1} (V_n(p) + \delta(p)) \cos(k\beta_p) \right)^2 + \left( \frac{1}{P} \sum_{p=0}^{P-1} (V_n(p) + \delta(p)) \sin(k\beta_p) \right)^2 \right]^{1/2}. \tag{22}$$

A simulation was proposed to assess the effects of a noisy signal by quantifying the magnitude change $\Delta M_n(k) = \tilde{M}_n(k) - M_n(k)$. The probability function, $\delta$, of the Gaussian-distributed Gaussian noise pattern reads:

$$\delta(p) = \frac{1}{s\sqrt{2\pi}} e^{-\frac{1}{2}\left(\frac{p}{s}\right)^2}. \tag{23}$$

Several noise amplitudes with various standard deviation $s$ were generated and added to the signal $V_n$ of Figure 4. The DFT coefficients of four different frequencies of low and high harmonic range are shown in Figure 6. On the abscissa, the Root Mean Square (RMS) of the generated noise $\delta_{RMS}$ is compared with the amplitude $A$ of the generated sine as in Equation (8).

Due to signal filtering in the DFT computation, the error of the DFT magnitude for a given coefficient remains small by comparison with the original signal. However, the above simulation is based on white noise, which is well filtered by the DFT computation. Nevertheless, in the experimental case, it is important to detect possible constructing interference patterns in the noise, which could correspond to a generated frequency. The lengths of the cables in the DAQ system for instance have to be taken into account to not generate electromagnetic noise at one of the generated frequencies.

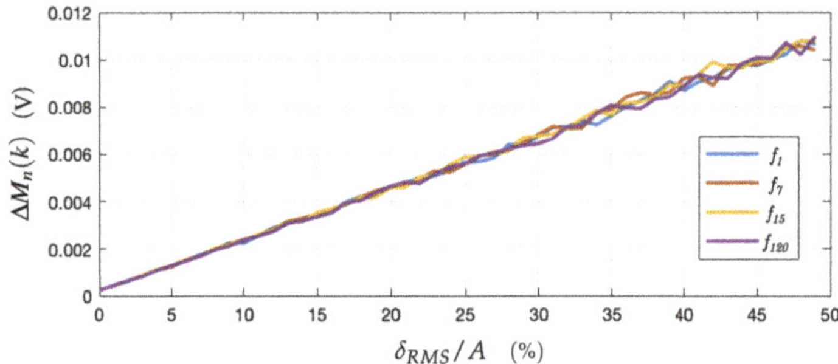

**Figure 6.** Propagation of the Gaussian white noise through DFT computation for four generated frequencies. The Noise RMS was computed from 1000 Gaussian white noises and averaged out. The four curves follow the same linear interpolation: $y = 2.14 \times 10^{-4}\, x + 1.81 \times 10^{-4}$.

## 6. Preliminary Experimental Results with 15 Different Frequencies

As discussed in the Introduction, the imaging rate of the well-established X-ray tomography suddenly increased when considering several pairs of emitters and receptors for simultaneous measurements. The novel idea, which consists of applying simultaneous excitations and measurements in EIT, brings new challenges due to the fundamental differences existing between such hard field and soft field systems. Apart from the radical difference in solving the inverse problem (Section 2), multifrequency excitations and measurements have to be considered with the association of TDM in the ONE-SHOT strategy.

The discriminability of the raw data is discussed in Section 6.1 in two harmonic ranges. In Section 6.2, the noise measurement is shown with a discussion of the SNR of the measured data. Section 6.3 introduces the images reconstructed from a set of measurement data.

### 6.1. Raw Data

The ONE-SHOT excitation strategy has as its goal one single excitation for all independent pairs of electrodes, resulting in the generation of 240 positive and negatives sines (Equation (11)) of 120 different frequencies (Equation (7)) for 16 electrodes. Implementing ONE-SHOT is a challenging task as it requires fast voltage excitations of arbitrary signals and FFT computation of current measurements, all in parallel over numerous channels. Nevertheless, the simulation results in Section 5 are a great incentive to build the experiment as described in Section 4.

The preliminary results in this experiment are presented to prove the feasibility of using TDM in EIT, based on the simultaneous excitation of 30 positive and negative sines at 15 frequencies and the experimental DFT reconstruction of the same voltage excitation pattern as in Figure 4. Two frequency domains are investigated: the low harmonic range where the 15 positives sines are the 15 first harmonics of the DFT frequency $f_0$ (Section 3.4). Furthermore, the high harmonic range contains the harmonics 106–120: as discussed above, these higher frequencies are planned to be generated in the future development of a full simultaneous excitation set.

#### 6.1.1. Experimental Results in the Low Harmonic Range

An experiment was set up to measure a homogenous field of conductivity 300 μS/cm. The results in the low harmonic range (Figure 7) are the DFT magnitudes $M_n(k)/R$ of the 16 experimental measurement signals computed from $P = 512$ data points, at a frequency of 1953.125 Hz for a data sampling of 1 MS/s. The results are shown as the current by including $R = 100\ \Omega$, the resistance in the measurement channel on the PCB. The authors observed a good accordance with the simulations of Section 5. The comparison of the experimental results (Figure 7) with the simulated DFT reconstruction of pure sines (Figure 4) led to the following observations:

- The readers familiar with EIT may expect the following phenomenon. In the experimental data, at the bottom of Figures 7 and 8, non zero currents were measured over electrodes that were not excited at the corresponding frequency. Let us consider an electrode $E_n$ with $1 \leq n \leq 15$ excited by a single sine of frequency $f_i$. In the Fourier space (Figures 7 and 8), the measured current passing through an electrode gives non-zero values for other frequencies. The reason comes from the electric potential difference that appears between the excited electrode $E_n$ with an imposed potential $V_n$ with $-A \leq V_n \leq +A$ and a non-excited electrode $E_m$ whose potential is $V_m = 0$ for a given frequency $f_i$ in the Fourier space (Figure 5). This effect is particularly large for the current in the adjacent measurement of the drain electrode in $E_1$ and $E_{15}$. These data depend on the electrical conductivity distribution inside the body and have to be considered as well.
- The magnitudes of the Fourier coefficients in the experimental results were lower for the mid-range harmonics than $f_1$ and $f_{15}$: Considering the finite electrical conductivity inside the system, the current resulting from the potential imposed between two neighboring electrodes was larger than the current resulting from the same potential imposed between two opposite electrodes in the circular shape of the EIT sensor [16].

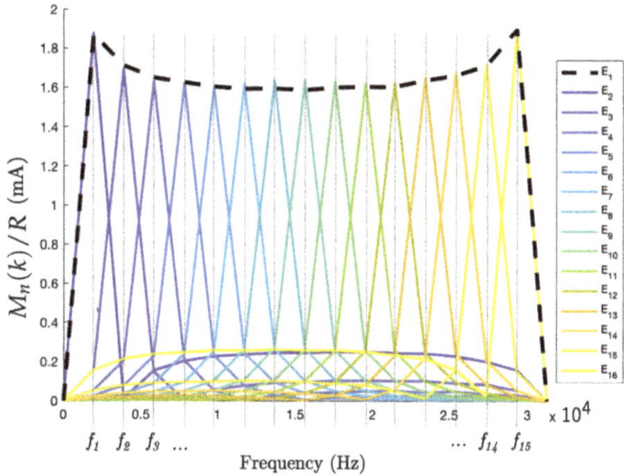

**Figure 7.** Measured current, represented as the magnitude of the DFT of 16 voltages over resistances in the excitation circuit of the 16 electrodes. This result corresponds to excitations in the low harmonic domain.

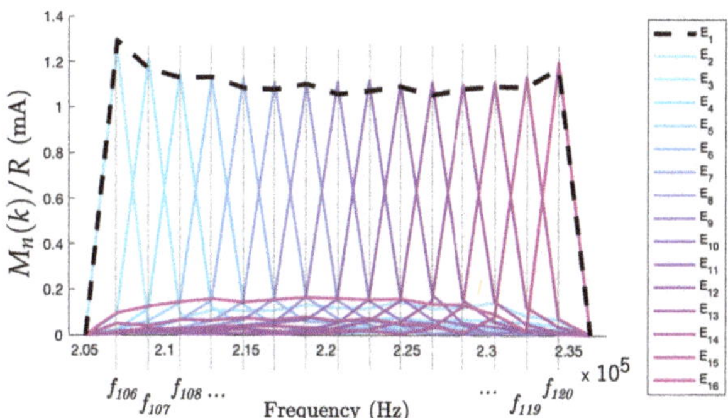

**Figure 8.** Magnitudes of the DFT from excitations in the high harmonic domain.

### 6.1.2. Experimental Results in the High Harmonic Range

The main prospect of the ONE-SHOT excitation strategy is to consider simultaneous excitations at 120 different frequencies for the EIT sensor of 16 electrodes. As in Section 3.4, the frequencies were chosen to be harmonics of the fundamental frequency $f_0$. The 120th harmonic $f_{120}$ was then 234.357 kHz. We investigated the performance of ONE-SHOT in this higher harmonic range.

The experimental measurement resulted in the high harmonic range from $f_{106}$–$f_{120}$ is shown in Figure 8. The discrimination of the signals by frequency offered equivalent performances to the low harmonic case. However, an important remark is that the signal amplitudes in the high harmonic range were lower than previously due to the impedance of water, which depends on the excitation frequency. The result was a weaker response in amplitude in the high harmonic range for the same conductivity change in the system. Investigations are ongoing for calibrating the amplitude of the generated sines to have a similar response in every harmonic for a future generation of a full set of 120 excitations.

### 6.2. Noise Measurement

A set of voltage magnitudes $M_n^0(k)$, defined as in Equation (16), was experimentally measured without any excitation to identify the noise spectrum. The results in Figure 9 show localized peaks in the noise amplitude beyond the high harmonic frequency domain.

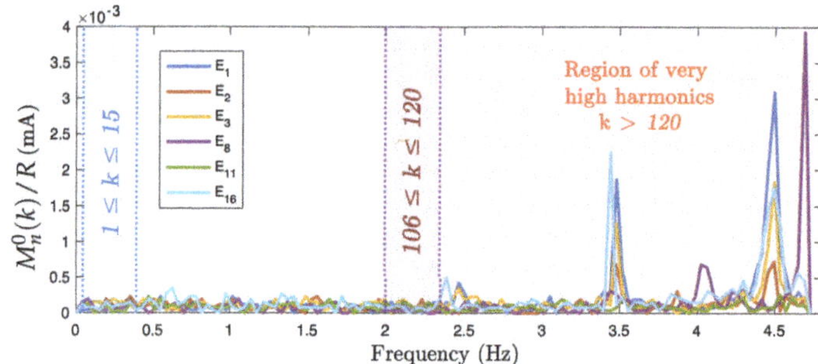


**Figure 9.** Noise measured over the full spectrum available for the sampling rate of 1 MS/s. The results show five of the 16 electrode channels. The low, high, and very high harmonic ranges are also shown.

The SNR of a signal measured at a given electrode $n$ and tagged with a given frequency $k$ is defined as follows:

$$SNR_n(k) = log_{10}\left[\frac{\sum_\lambda [M_n(k)]_\lambda / R}{\sum_\lambda ([M_n(k)]_\lambda / R)^2 + \sum_\lambda ([M_n^0(k)]_\lambda / R)^2}\right] \tag{24}$$

where $\lambda$ is the number of data taken for averaging. The noise in the low harmonic range resulted in an SNR of the raw signal of 69.1 dB in the lowest frequency magnitude in the excitation electrode channels when averaging the signal from $\lambda = 20$ samples. In a non-excited channel, the signal was O(0.1) weaker, resulting in an SNR of about 60 dB depending on the signal magnitude. In the high harmonic range, the SNR was 59.6 dB for the highest harmonic.

The very high frequencies $250 \leq 500$ kHz, bounded above by the Nyquist limit, could be considered as excitation frequencies. However, a proper shielding of the DAQ system must be investigated. A large part of the noise was caused by electrical noise in the PCB. For this reason, a new PCB is being manufactured. Finally, the results showed that the noise completely covered the DFT systematic uncertainty (Section 5.1), which is completely negligible.

### 6.3. Image Reconstruction from 16 Datasets

The main objective of the present manuscript is the implementation of a method to obtain a very fast EIT data frame acquisition rate. From such data, the user can then reconstruct an image with any given reconstruction algorithm in post-processing, a step that we consider to be distinct from our main goal. As an example, we are interested in the image reconstruction of the above data. However, the datasets are incomplete by comparison with the ONE-SHOT excitation prospects of Equation (11). A solution consists of reconstructing the full dataset in the post-processing.

The results described above contained a complete set of excitations for the drain electrode. It is possible to consider another drain electrode and another frequency set to obtain another set of measurements, independent of location and frequency.

The rotation of the drain over the 16 electrodes resulted in $16 \times 15 = 240$ pairs of excitations. The symmetry suggested to consider only half of the data. The full dataset was deduced and contained the decomposition of 240 signals from 120 excitation pairs, measured over the 16 resistances of the PCB. Therefore, the full dataset included a total of 1920 elements. To sum up, the sequential excitations with the drain located on each of the 16 electrodes resulted in a dataset equivalent to the one expected by the full implementation of the ONE-SHOT method.

Several datasets were measured from the inclusion of non-conducting PMMA rods in the test section of Figure 1 filled with tap water with a conductivity of $\sigma = 635$ µS·m$^{-1}$. The one-step least-squares iterative reconstruction method [27,28] was implemented, and the results are shown in Figure 10.

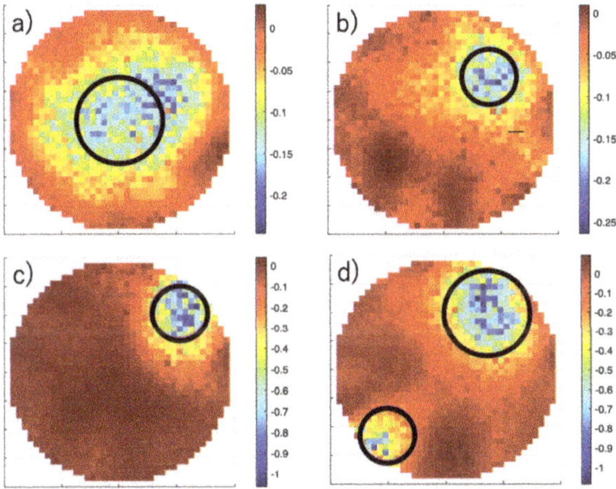

**Figure 10.** Electrical conductivity indicator function. Two non-conductive rods of diameter 20 and 30 mm, shown with black circles, are inserted in the test section filled with salt water. (**a**) A 30-mm rod at the center. (**b**) A 20-mm rod half way to the edge. (**c**) A 20-mm rod on the edge. (**d**) Two rods on the edge.

The results showed similar image reconstruction performances as another EIT system [16] based on TDM and using the same reconstruction algorithm. As usual in EIT, the sensitivity was much higher close to the electrodes. This effect resulted in sharp gradients when imaging the inclusions inserted at the edge.

Finally, the frame rate of the ONE-SHOT system can be changed by integrating the FFT over more or less data points. For instance, choosing the FFT to be computed over 512 points as considered in this article resulted in the frame rate of 1953 fps. In this case, the 120 frequencies can fill all the discrete values from 1953 Hz to the Nyquist limit of 500 kHz with $\Delta f = 1/512$ µs = 1953 Hz the minimal gap between two neighboring values. The high frame rate was at the cost of a large frequency bandwidth.

## 7. Conclusions

This article assessed the feasibility of simultaneous excitations and measurements using a multi-frequency strategy for high-speed electrical impedance tomography, based on the ONE-SHOT excitation method. The motivations for increasing the data frame rate for EIT measurement followed by an overview of the mathematical aspect of EIT were presented in Sections 1 and 2, respectively. The requirements for the hardware system based on the excitation strategy and measurement process, as well as the motivations to determine the number of electrodes were discussed in Section 3. An efficient hardware solution for implementing the simultaneous EIT excitation strategy was proposed in Section 4 with an analysis of the error and noise propagation through the measurement process in Section 5. Finally, experimental results were shown in Section 6 in the low and high harmonic ranges of the excitation signals.

Table 1 compares the performances of ONE-SHOT with a number of existing high-rate EIT systems. The description of the different hardware systems contains several characteristics that have to be considered while comparing the data frame rates, including the Number of Measurements per Seconds (NMS). Firstly, the AI sampling frequency directly impacted the data frame rate. Secondly, the data size was an important factor to maximize for a good conditioning of the inverse problem. Thirdly, the noise also directly impacted the image quality. The noise of the ONE-SHOT system was the value currently measured in a partially-shielded DAQ prototype. Future tests, especially including the new PCB, will result in better performances in terms of SNR.

**Table 1.** Comparison of the performances of the ONE-SHOT simultaneous excitation difference with several high-rate EIT systems with single-frequency multiplexing strategies.

| System | $f_{DAQ}$ (MHz) | $n_e$ | Data Size | NMS | Data Frame Acquisition Rate (fps) | Excitation Frequency (kHz) | SNR (dB) |
|---|---|---|---|---|---|---|---|
| Phantom [11], 2008 | 1 | 64 | $15 \times 63$ | 172,652 | 182.7 | 1–10,000 | 65.5–98.6 |
| Phantom [12], 2015 | 10 | 16 | $16 \times 16$ | 28,160 | 110 | 97.65 | 90 |
| ProME-T[16], 2017 | 2 | 16 | $16 \times 120$ | 1,599,360 | 833 | 15 | 60 |
| ONE-SHOT, 2019 | | | | | | | |
| (preliminary results) | 1 | 16 | $16 \times 15$ | 468,744 | 1953.1 | 2–250 | 59.6–69.1 |
| (prospects) | 1 | 16 | $16 \times 120$ | 7,500,000 | 3906.3 | 4–469 | - |

The next experimental step is to measure a full scan by exciting all independent pairs of electrodes. For a 16-electrode EIT device, the number of frequencies becomes 120, bringing several new challenges. Mainly, the limit imposed by the Nyquist frequency of 500 kHz for the current system associated with the resolution in the frequency space from the choice of the DFT computation time window results in a maximum number of harmonics. Adding more frequencies may impose a longer time window, directly impacting the data frame acquisition rate. However, considering frequencies in the very high harmonic range may drastically affect the SNR due to higher resistivity and noise peaks. Furthermore, the limited physical space in the FPGA restricts the number of sine generators. The prospects of the performances of ONE-SHOT are also shown in Table 1. In these prospects, the DFT computed over measurements of $P = 256$ points may increase the data acquisition rate, keeping $f_{120} = 469$ kHz, below the Nyquist limit.

**Author Contributions:** Conceptualization, M.D., A.D., C.D., G.R., and C.B.; data curation, M.D. and A.D.; formal analysis, M.D., A.D., and C.B.; funding acquisition, G.R., S.B., and C.B.; investigation, M.D.; methodology, M.D., A.D., and C.B.; project administration, G.R., S.B., and C.B.; resources, G.R., S.B., and C.B.; software, M.D.; supervision, G.R., S.B., and C.B.; validation, M.D., C.D., and C.B.; visualization, M.D.; writing, original draft, M.D.; writing, review and editing, A.D., C.D., G.R., S.B., and C.B.

**Funding:** This research received no external funding.

**Conflicts of Interest:** The authors declare no conflict of interest.

## Abbreviations

The following abbreviations are used in this manuscript:

| | |
|---|---|
| AI | Analog Input |
| AO | Analog Output |
| DAQ | Data Acquisition |
| DFT | Discrete Fourier Transform |
| DMA | Direct Memory Access |
| DtN | Dirichlet-to-Neumann |
| ECT | Electrical Capacitance Tomography |
| EIDORS | Electrical Impedance Tomography and Diffuse Optical Tomography Reconstruction Software |
| EIT | Electrical Impedance Tomography |
| FDM | Frequency-Division Multiplexing |
| FFT | Fast Fourier Transform |
| FPGA | Field Programmable Gate Array |
| fps | frames per seconds |
| LTHC | Laboratory of analytical Thermohydraulics and Hydromechanics of Core and Circuits |
| MS/s | Mega Samples per second |
| NtD | Neumann-to-Dirichlet |
| NMS | Number of Measurements per Seconds |
| ONE-SHOT | ONe Excitation for Simultaneous High-speed Operation Tomography |
| PCB | Printed Circuit Board |
| RMS | Root Mean Square |
| SMA | SubMiniature version A |
| SNR | Signal-to-Noise Ratio |
| TDM | Time-Division Multiplexing |

## References


1. Breckon, W.R.; Pidcock, M.K. Mathematical aspects of impedance imaging. *Clin. Phys. Physiol. Meas.* **1987**, *8*, 77–84. [CrossRef] [PubMed]
2. Cheney, M.; Isaacson, D.; Newell, J. Electrical Impedance Tomography. *SIAM Rev.* **1999**, *41*, 85–101. [CrossRef]
3. Borcea, L. Electrical impedance tomography. *Inverse Probl.* **2002**, *18*, R99–R136. [CrossRef]
4. Hanke, M.; Bruhl, M. Recent progress in electrical impedance tomography. *Inverse Probl.* **2003**, *19*, S65–S90. [CrossRef]
5. Kauppinen, P.; Hyttinen, J.; Malmivuo, J. Sensitivity distribution visualizations of impedance tomography measurement strategies. *Int. J. Bioelectromagn.* **2006**, *8*, VII/1–VII/9.
6. Dupre, A.; Ricciardi, G.; Bourennane, S.; Mylvaganam, S. Electrical Capacitance-Based Flow Regimes Identification: Multiphase Experiments and Sensor Modeling. *IEEE Sens. J.* **2017**, *17*, 8117–8128. [CrossRef]
7. Ricciardi, G.; Pettigrew, M.J.; Mureithi, N.W. Fluidelastic Instability in a Normal Triangular Tube Bundle Subjected to Air-Water Cross-Flow. *J. Press. Vessel. Technol.* **2011**, *133*, 061301. [CrossRef]
8. Zhang, X.; Chen, S.; Wang, H.; Zhang, Y. FPGA-based multi-frequency excitation and modulation technology in EIT system. In Proceedings of the 2010 3rd International Conference on Biomedical Engineering and Informatics, Yantai, China, 16–18 October 2010; Volume 2, pp. 907–911. [CrossRef]
9. Wang, M.; Ma, Y. Over-zero switching scheme for fast data collection operation in electrical impedance tomography. *Meas. Sci. Technol.* **2006**, *17*, 2078–2082. [CrossRef]
10. Wright, P.; Basu, W.; Lionheart, W.; Crabb, M.; Green, P. Improved amplitude estimation of lung EIT signals in the presence of transients: Experimental validation using discrete phantoms. In Proceedings of the 18th International Conference on Biomedical Applications of Electrical Impedance Tomography, Hanover, NH, USA, 21–24 June 2017; p. 26.

11. Halter, R.J.; Hartov, A.; Paulsen, K.D. A Broadband High-Frequency Electrical Impedance Tomography System for Breast Imaging. *IEEE Trans. Biomed. Eng.* **2008**, *55*, 650–659. [CrossRef]

12. Khan, S.; Manwaring, P.; Borsic, A.; Halter, R. FPGA-Based Voltage and Current Dual Drive System for High Frame Rate Electrical Impedance Tomography. *IEEE Trans. Med. Imaging* **2015**, *34*, 888–901. [CrossRef]

13. Wang, B.; Ji, H.; Huang, Z.; Li, H. A high-speed data acquisition system for ECT based on the differential sampling method. *IEEE Sens. J.* **2005**, *5*, 308–312. [CrossRef]

14. Cui, Z.; Wang, H.; Chen, Z.; Xu, Y.; Yang, W. A high-performance digital system for electrical capacitance tomography. *Meas. Sci. Technol.* **2011**, *22*, 055503. [CrossRef]

15. Morales, D.; Lopez-Ruiz, N.; Castillo, E.; Garcia, A.; Martinez Olmos, A. Adaptive ECT System Based on Reconfigurable Electronics. *Measurement* **2015**, *74*, 238–245. [CrossRef]

16. Dupré, A. Electrical Impedance Tomography for Void Fraction Measurements of Harsh Two-Phase Flows: Prototype Development and Reconstruction Techniques. Ph.D. Thesis, Ecole Centrale Marseille, Marseille, France, 2017.

17. Dupré, A.; Mylvaganam, S. A Simultaneous and Continuous Excitation Method for High-Speed Electrical Impedance Tomography with Reduced Transients and Noise Sensitivity. *Sensors* **2018**, *18*, 13. [CrossRef]

18. Granot, Y.; Ivorra, A.; Rubinsky, B. Frequency-Division Multiplexing for Electrical Impedance Tomography in Biomedical Applications. *Int. J. Biomed. Imaging* **2007**, *2007*, 201 – 232. [CrossRef]

19. Dowrick, T.; Holder, D. Phase division multiplexed EIT for enhanced temporal resolution. *Physiol. Meas.* **2018**, *39*, 034005. [CrossRef]

20. Kryszyn, J.; Wróblewski, P.; Stosio, M.; Wanta, D.; Olszewski, T.; Smolik, W. Architecture of EVT4 data acquisition system for electrical capacitance tomography. *Measurement* **2017**, *101*, 28–39. [CrossRef]

21. Barber, D.C.; Brown, B.H. Applied potential tomography. *J. Phys. E* **1984**, *17*, 723–733. [CrossRef]

22. Kim, Y.; Woo, H.W. A prototype system and reconstruction algorithms for electrical impedance technique in medical body imaging. *Clin. Phys. Physiol. Meas.* **1987**, *8*, 63–70. [CrossRef]

23. Adler, A.; Lionheart, W.R.B. Uses and abuses of EIDORS: An extensible software base for EIT. *IOP Publishing Physiol. Meas.* **2006**, *5*, S25–S42. [CrossRef]

24. Kotre, C.J. Studies of Image Reconstruction Methods for Electrical Impedance tomography. Ph.D. Thesis, Newcastle University, Newcastle upon Tyne, UK, 1993.

25. Darnajou, M.; Dang, C.; Ricciardi, G.; Bourennane, S.; Bellis, C.; Schmidt, H. The Design of Electrical Impedance Tomography Detectors in Nuclear Industry. In Proceedings of the 9th World Congress on Industrial Process Tomography, Bath, UK, 2–6 September 2018; pp. 457–465.

26. Betta, G.; Liguori, C.; Pietrosanto, A. Propagation of uncertainty in a discrete Fourier transform algorithm. *Measurement* **2000**, *27*, 231–239. [CrossRef]

27. Yorkey, T.J.; Webster, J.G.; Tompkins, W.J. Comparing Reconstruction Algorithms for Electrical Impedance Tomography. *IEEE Trans. Biomed. Eng.* **1987**, *BME-34*, 843–852. [CrossRef]

28. Kim, M.C.; Kim, S.; Kim, K.Y.; Lee, Y.J. Regularization methods in electrical impedance tomography technique for the two-phase flow visualization. *Int. Commun. Heat Mass Transf.* **2001**, *28*, 773–782. [CrossRef]

*Article*

# 3D-Printed Multilayer Sensor Structure for Electrical Capacitance Tomography

**Aleksandra Kowalska *, Robert Banasiak, Andrzej Romanowski and Dominik Sankowski**

Institute of Applied Computer Science, Lodz University of Technology, 90-924 Lodz, Poland
* Correspondence: akowalska@iis.p.lodz.pl

Received: 13 June 2019; Accepted: 29 July 2019; Published: 4 August 2019

**Abstract:** Presently, Electrical Capacitance Tomography (ECT) is positioned as a relatively mature and inexpensive tool for the diagnosis of non-conductive industrial processes. For most industrial applications, a hand-made approach for an ECT sensor and its 3D extended structure fabrication is used. Moreover, a hand-made procedure is often inaccurate, complicated, and time-consuming. Another drawback is that a hand-made ECT sensor's geometrical parameters, mounting base profile thickness, and electrode array shape usually depends on the structure of industrial test objects, tanks, and containers available on the market. Most of the traditionally fabricated capacitance tomography sensors offer external measurements only with electrodes localized outside of the test object. Although internal measurement is possible, it is often difficult to implement. This leads to limited in-depth scanning abilities and poor sensitivity distribution of traditionally fabricated ECT sensors. In this work we propose, demonstrate, and validate experimentally a new 3D ECT sensor fabrication process. The proposed solution uses a computational workflow that incorporates both 3D computer modeling and 3D-printing techniques. Such a 3D-printed structure can be of any shape, and the electrode layout can be easily fitted to a broad range of industrial applications. A developed solution offers an internal measurement due to negligible thickness of sensor mount base profile. This paper analyses and compares measurement capabilities of a traditionally fabricated 3D ECT sensor with novel 3D-printed design. The authors compared two types of the 3D ECT sensors using experimental capacitance measurements for a set of low-contrast and high-contrast permittivity distribution phantoms. The comparison demonstrates advantages and benefits of using the new 3D-printed spatial capacitance sensor regarding the significant fabrication time reduction as well as the improvement of overall measurement accuracy and stability.

**Keywords:** 3D; ECT; 3D-printing; sensors; modeling

---

## 1. Introduction

Electrical Capacitance Tomography (ECT) is a non-intrusive and non-invasive imaging modality dedicated to monitoring industrial processes in pipelines and reactors, wherever non-conducting dielectric materials' mixtures are processed. Typical examples may be shown as follows: gas-oil flows [1], solid particle flows [2] or reservoirs [1,3–5]. A typical ECT system measures mutual capacitance changes between pairs of electrodes distributed around the circumference of an industrial process container or pipe [6–9]. The amount of measurement data depends on used hardware (typically 66–496) and its acquisition rate may vary from a few to hundreds of images per second [6,9]. Thereafter, collected data can be processed by a high-performance PC system using mathematical modeling [5,10] and specialized algorithms for reconstruction of images [5,11–13], analyzing raw data [3,8,14,15] and finally making a right diagnostic decision for process control and automation [16–18].

Industrial processes generally have a 3-dimensional nature. Hence, it is an obvious tendency to sense and measure phenomena in real 3D space occupied by these industrial processes. Therefore ECT tomography is evolving from the *z*-axis averaged measurement recorded in the cross-sectional plane to the 3D scanning of the entire volume [8,13,16,19–22]. Classical ECT measurement systems employ a single-layer regular electrode layout for cross-sectional sensing. Image reconstruction processes strongly rely on a quality of the measurement sensitivity analysis [5,16,23–25]. Sensitivity analysis focuses on the modeling of an electric field inside the sensing space volume. Hence, it tends to minimize the reconstruction approximation error for the given measurement data set. However, 3D ECT sensors concepts extend the notion of the reconstruction from the plane cross-sectional imaging to the fully volumetric reconstruction of the given material distribution in the entire sensor space as has been reported in [2,23,24,26,27]. This work reveals the new concept of various ECT sensor 3D structure arrangements. The main components of 3D ECT systems are illustrated in Figure 1.

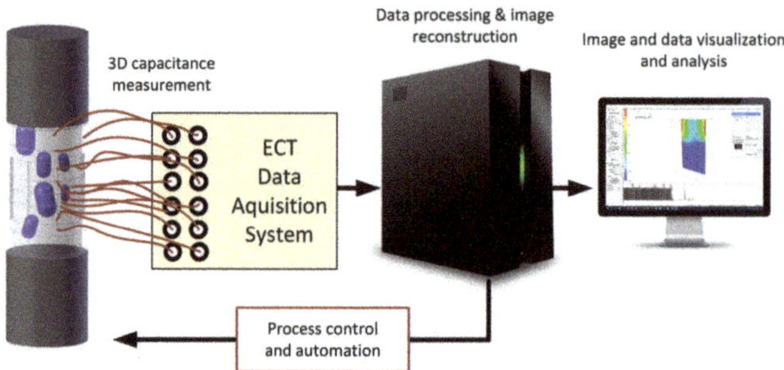


**Figure 1.** 3D ECT system and its components.

This research work is devoted to a new type of 3D ECT sensor structure where its mounting support is completely fabricated using 3D printing technology. A comparison of 3D-printed ECT sensor and traditionally fabricated sensor will be performed in to investigate their detectability for low- and high-contrast dielectric phantoms. The comparison demonstrates differences and similarities of using the new 3D-printed spatial capacitance sensor over a traditionally fabricated device.

## 2. Theoretical Fundamentals of 3D ECT Sensor Modeling

3D capacitance tomography sensor modeling is typically named as a forward problem. Forward problem-solving techniques employ simulation of output measurement data for a given set of excitation values as well as permittivity values characterizing the dielectric material distribution. The inverse problem can be considered to be an imaging result or a raw analysis for a given dataset of measurement records. In 3D ECT tomography data workflow, the forward problem must be solved before the inverse problem solution.

To solve a 3D ECT forward problem, two assumptions can be made. The first one is no wave effect. The second one is a low-frequency approximation approach to Maxwell's equation-solving. Therefore in a simplified mathematical 3D ECT model, the following electrostatic approximation (Equation (1)):

$$0 = \nabla \times E \tag{1}$$

is made. This approach leads to neglecting the effect of wave propagation. Taking (Equation (2)):

$$E = -\nabla \varphi \tag{2}$$

one can assume no internal charges. Hence, the following Equation (3) holds:

$$\nabla \varepsilon(x, y, z) \nabla \varphi = 0 \tag{3}$$

where: $\varphi$ is a spatial distribution of electric potential inducted by electric field $E$, $\varepsilon(x, y, z)$ is a spatial dielectric permittivity distribution. The potential $\varphi$ of each electrode can be derived from Equation (4):

$$\varphi = v_k \quad at \quad elec_k \tag{4}$$

where: $elec_k$ is the $k$th electrode at the potential $v_k$. To solve this equation, one can apply a variety of numerical methods. In this work, the finite element method (FEM) method is employed to derive (Equation (5)):

$$Y(\varepsilon)\varphi = F \tag{5}$$

where: the matrix $Y$ is the discrete representation of the operator $\nabla\epsilon\nabla$ including the Dirichlet-type boundary conditions, the vector $F$ is the Neumann type boundary condition term and $\varphi$ is the approximated electric potential solution. The total electric charge $Q$ on the electrode $e_k$ is given by Equation (6):

$$Q_k = \int_{e_k} \varepsilon \frac{\partial \varphi}{\partial n} dx^2 \tag{6}$$

where: $n$ is the inward normal on the $k$th electrode.

The FEM is recognized as an efficient computational tool that can be applied for a 3D ECT sensor design and modeling process to simulate an interior electric field dispersion [24]. It may significantly help for the capacitance value estimation as well as sensitivity maps to optimize sensor geometry and structure. The sensitivity maps, dependent on the sensor geometry, are essential characteristics of the 3D ECT systems [2]. One of the ECT sensing systems features an extremely small sensing ability in the middle of the sensor, due to the incapitalization capacitance values can be very small for most distant layers, typically in the range of 0.01–1 pF [13] and it differences with low and high-contrast media the dimensions of the electrodes must be a trade-off between their heights and mutual capacitances. It is necessary to achieve a measurement range comparable with measurement range of the used system hardware. Homogeneous sensitivity distribution inside a sensing domain is also an electrical field intensity uniformly inside an investigated volume [2,13,21,28].

The common 3D ECT sensor is an array of metal-plate electrodes arranged around the sensed medium, typically on an outer boundary of a non-conducting pipe. External, grounded potential shield keeps the electric field lines inside the sensor space. In some cases of conducting reactors, such as metal pipes or containers, the electrodes are located on the inside surface of the reactor. Whenever this is the case, the metal walls are then regarded as the electric shielded layer with the other external coupling components such as the radial and guard electrodes, to improve the quality of measurements and hence the reconstruction process.

## 3. 3D Modeling & Printing of ECT Capacitance Sensors

3D printing usually refers to the additive type of manufacturing 3D objects. This process is conducted with use of 3D printers following detailed, computer-aided scripts generated from a digital model prepared for fabrication processes. The resulting object is produced in an iterative process of adding consecutive layers of melted substance on top of another. This work presents a design and outcomes of a spatial capacitance sensor fabricated with the aid of the Fused Filament Fabrication technique. To develop a new mechanical structure of precise 3D ECT sensor, its computational design must be prepared first. The Blender modeling software suite was employed to prepare a complete design of a 3D sensor structure. We used the stereolithography (STL) format output from the Blender software. The mechanical design model for 3D printing was developed under several important constraints, especially in accordance to the mesh structure design guidelines. The internal support mesh structure was strictly validated for thin-wall issues, redundancy required for stiffness, layered production compliance, etc.

The proposed design of the sensor was of substantial size exceeding standard equipment maximum print height in vertical direction. Hence, it was decided to divide the sensor print into 2 distinct elements. The maximum print height for Ultimaker 3, employed in this study, is 200 mm. The final design was rearranged into 2 symmetrical parts, therefore some additional joint elements (insets and holes were added at the contact edges) were required to connect both elements after the printing. Additionally, walls separating electrode layouts at the outside external side were printed at half thickness to ensure tight fit and uniformity of the 3D sensing structure.

The next step was to prepare the 3D printing configuration. In this study, the Ultimaker 3 printer with PLA (polylactic acid) filament and Ultimaker CURA 4.1 printing software were used for the building and controlling of the printing process. It is possible to construct detailed spatial ECT sensor arrangements of variable shapes and in a wide range of dimensions using 3D printing equipment. The maximum resolution of the used 3D printer is 0.4 mm for xy-axes and 0.06 mm for z-axis (print layer thickness) and can be extended by using extended-resolution extruders. A high XY resolution was used to generate a very thin sensor wall for keeping the distance between electrodes and scanned volume as low as possible. In the past it was very hard or even impossible to achieve this in previous traditionally manufactured PMMA (or PVC)-based 3D ECT sensors. In this research, the PLA filament has been used as a widely used popular 3D printing solution with its relative permittivity value 3–5 at 1 MHz [29]. The PLA filament provides a good balance between durability, flexibility, and ease to use. Its dielectric properties are well-fitted to the dielectric properties of tested phantoms. For 3D ECT printed sensor, a PLA extruding temperature is worth raising by 5–10 °C to provide better Z-direction bonding. The one disadvantage is that PLA is characterized by a relatively low temperature, in which it is losing its stiffness and durability (60 °C). This must be taken into consideration when 3D ECT sensor industrial applications are developed. As a much stronger alternative, a nylon filament can be used as a printing material, but its relative permittivity value is slightly lower than PLA–2.5 at 1 MHz. A nylon 3D print offers even better durability and it is more resistant to higher temperatures (>150 °C). However, ABS-based filaments are too breakable to be used for 3D ECT sensor manufacturing, however their temperature resistance is relatively high (>180 °C). True three-dimensional ECT sensor arrangement needs a specific design with additional walls that separate electrodes working in an inter-measurement plane mode. As a result, a dual extruder mode had to be used to build extra support. PVA water-dissolvable material was used to produce these supplementary supports. PVA is a unique filament that works properly with PLA, nylon, and other filaments with similar melting temperature. As was mentioned earlier, main electrode placements have been optimized to be as thin as possible (0.2 mm) to strengthen electric field penetration abilities by using a single layer of filament. However, more layers (2–3) can be applied as well to make a sensor pipe stiffer and more durable. The increased number of filament layers with fixed thickness (1 mm) has been used to build inter-plane separators for horizontal and vertical screening systems. Eletrical components were added to the finished prefabricated structure in the form of wiring, electrodes, and shielding. Figure 2 presents the workflow diagram for 3D printing of ECT sensor structures.

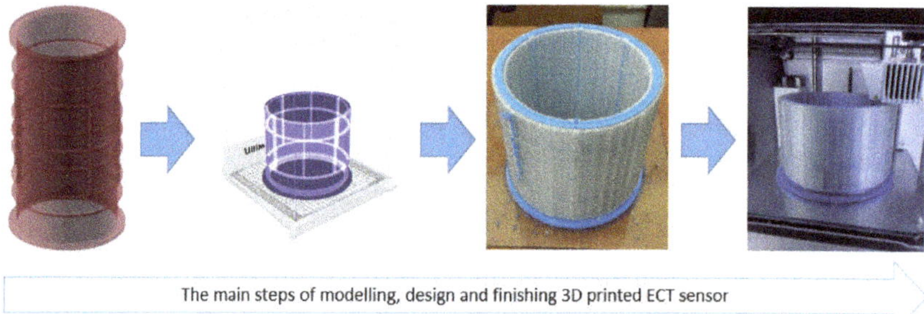


**Figure 2.** 3D printed ECT sensor building workflow.

## 4. Experimental Setup

Experiments and measurement was conducted in Tom Dyakowski Process Tomography Lab at Lodz University of Technology, Poland. Verification of the classical vs. proposed 3D printed ECT sensor was based on two distinct testing concepts. The first strategy was to use the precise Agilent E4980A impedance meter combined with 64-channel computer-controlled multiplexer for low-contrast solid-gas mixtures (objects) in offline mode. The LCR meter-based system can precisely detect signals that can be useful in capturing extremely small capacitance value changes in spatial measurements for distant ECT sensor electrodes. In this paper, an LCR meter-based experimental system was directly used for investigating low-permittivity contrast mixtures that can be challenging objects in terms of accurate measurement of spatial low-amplitude (<1 pF) capacitance value changes. For this approach, a single 496-measurement raw data frame was acquired three times: Step 1—for low-permittivity distribution (empty sensor); Step 2—for high permittivity (a sensor was completely fulfilled by PE granules) and Step 3—for the investigated object. The second approach was to use a specialized charge–discharge principle-based 32-channel ECT measurement system (ET3) with flexible two-way gain amplifier setup. ET3 measurement hardware was applied in online mode for high-contrast dielectric liquid–gas mixtures. This strategy used the internal ET3 calibration procedure and generated normalized capacitance data. For 32-channel, the ET3 system was able to acquire 11 frames per second. As with the LCR-based approach, three steps of data acquisition were performed as well. By contrast, 50 frames at each step were taken and averaged to smooth a signal and reduce a measurement noise. Both measurement systems used in this study are presented in Figure 3.

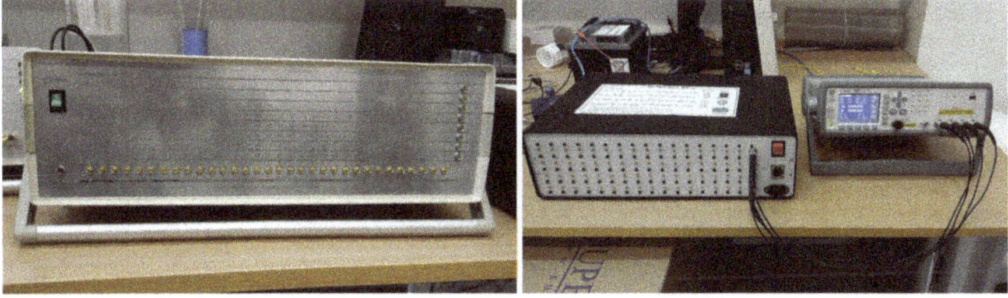

**Figure 3.** Experimental setup hardware: left—32-channel ET3 measurement hardware, right—Agilent E4980A with 64-channel multiplexer device.

The 3D capacitance-measurement sensors used in this test was arranged of 32 electrodes grouped in four cylindrical layers consisting of 8 electrodes in each one [12]—Figure 4.

**Figure 4.** Pictures of two experimental constructions of 3D ECT sensors under study.

Each of those sensors included two border planes (1st and 4th) and two inner planes (2nd and 3rd). Therefore, they have met major requirements of the spatial capacitance-measurement principle which need typically a significant number of 16–32 electrodes to capture an industrial process with reasonable axial and spatial resolution. To obtain performance of studied ECT sensors, the intra-layer and inter-layer electrode excitation strategy was used to provide $m$ = 496 independent capacitance data for $n_{el}$ = 32 excluding repeating mutual measurements (Equation (7)):

$$m = \frac{n_{el}(n_{el} - 1)}{2} \tag{7}$$

The traditionally made 3D ECT sensor (we name it here as S1) has already been developed and verified in the past [10,19]. Thus, in this study we positioned it as a baseline reference capacitance sensor for all tests. S1 was designed and built with the aid of the classical, "hand-made" method. S1 is built out of PMMA (polymethyl methacrylate) off-the-shelf pipe. S1 has the following dimensions: external diameter of 158 mm, the total height of 304 mm and pipe thickness of 4 mm with slight variations over the length (±2 mm) due to the inaccurate PMMA fabrication process. The sensor was asymmetrical in terms of the electrode layers arrangement: outer electrode height equals 70 mm and inner electrode only 30 mm. The electrode layers were placed on the outer surface of the sensor which led to external type of measurement. Sensor shielding was composed of double 25 mm guard screens located on the top and bottom parts of the structure plus extra, outer protective screens for external electric interference.

A novel 3D printed sensor (we name it here as S2) had same dimensions as S1 and was developed using a novel 3D modeler design and 3D-printing technology. PLA (transparent material) was employed for building the internal mechanical support sensor structure. The mounting pipe thickness of S2 was only 0.4 mm and constant over the pipe wall. It simultaneously provided the insulation between the electrode's active surface and industrial process and at the same time preserved deep measurement penetration. Additionally, this arrangement can enhance sensitivity in the central area of a sensor. The new concept of S1 included a copper internal radial screen between adjacent electrodes in the same layer and between adjacent layers. This prevented the electric field uncontrolled leakage for excited electrodes and kept the electric field distribution more uniform. This idea was an extension to some previous research using guard electrodes in ECT [28]. Using a new 3D-printing approach, it was easy to integrate such an advanced screening structure within electrodes—see Figure 5.

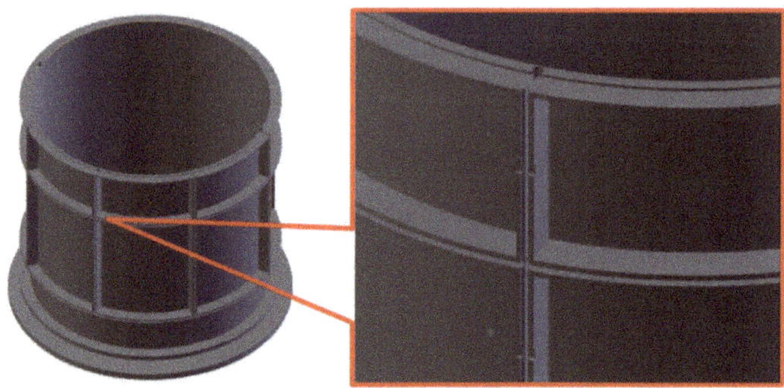

**Figure 5.** A concept of horizontal and vertical internal screening for 3D ECT sensor structure.

To investigate the low-contrast medium, a set of three cylindrical 3D printed phantoms (objects) was developed (TestA, TestB, and TestC)—Figure 6. The test phantoms were constructed using the 3D-printing technique (PLA) and filled with 3 mm PE (polyethylene) granules in two ways—see Table 1.

**Table 1.** Configuration of PE granules filler low-contrast Test*A*, Test*B*, and Test*C* phantoms.

| Config | Test*A* | Test*B* | Test*C* |
|---|---|---|---|
| inside | Phantom Test$A_{in}$ | Phantom Test$B_{in}$ | Phantom Test$C_{in}$ |
| outside | Phantom Test$A_{out}$ | Phantom Test$B_{out}$ | Phantom Test$C_{out}$ |

For capacitance characteristics and standard deviation evaluation, raw capacitance-measurement data taken from the LCR meter were normalized using high $C_{full}$ and low $C_{empty}$ permittivity distributions (PE granulate dielectric constant $\approx$3.2 at room temperature, with the air dielectric constant = 1—as the background).

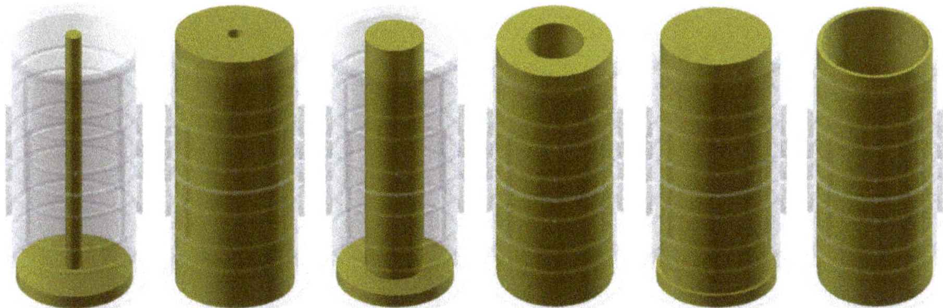

**Figure 6.** Arrangement of tested low-contrast objects according to Table 1—from the leftmost Test*A*, Test*B*, Test*C*.

In this study, the normalization of experimental capacitance data $C_n$ was computed according to formula [4]:

$$C_n = \frac{C - C_{empty}}{C_{full} - C_{empty}} \tag{8}$$

To evaluate performance of both sensors for high-contrast objects (gas–liquid mixtures) a set of ten cylindrical-shaped 3D printed phantoms were developed—Figure 7. The test phantoms were constructed using PLA filament and printed using the "spiral vase" option to avoid liquid leakages. As a test liquid, the propylene glycol (dielectric constant $\approx$32 in room temperature) was used, while the air acted as a background. All the high-contrast phantoms were designed as cylinders with various diameters (10–40 mm). The shapes, diameters, and positions of these test objects were somehow similar to the previous low-contrast set, but its roles were different. In this study, only the high-contrast material "liquid inside a cylinder object" option was considered to investigate small (10 mm, 15 mm, 20 mm) and middle-size (2 × 40 mm) high-contrast cylinders. Small-diameter cylinders were located in three positions (70 mm—geometrical sensor center, 35 mm from geometrical sensor center, 60 mm from geometrical sensor center) on the path along the 3D ECT sensor radius. The middle-size diameter object included two 40 mm cylinders and it was located symmetrically in two positions along the sensor diameter at equal distance from the sensor center (35 mm). The arrangement of the specific electrode area and height asymmetry relation for outer and inner layers was preserved for both sensors to retain the distant measurement signal for the inter-plane electrodes (for instance, the electrode pair: 1–29) at a level possible to be sensed by the ECT measurement unit. The following measurement workflow was applied. 15 V positive potential was excited on the sender electrode with the other electrodes grounded. Then consecutively each electrode was switched to be the sender, one by one.

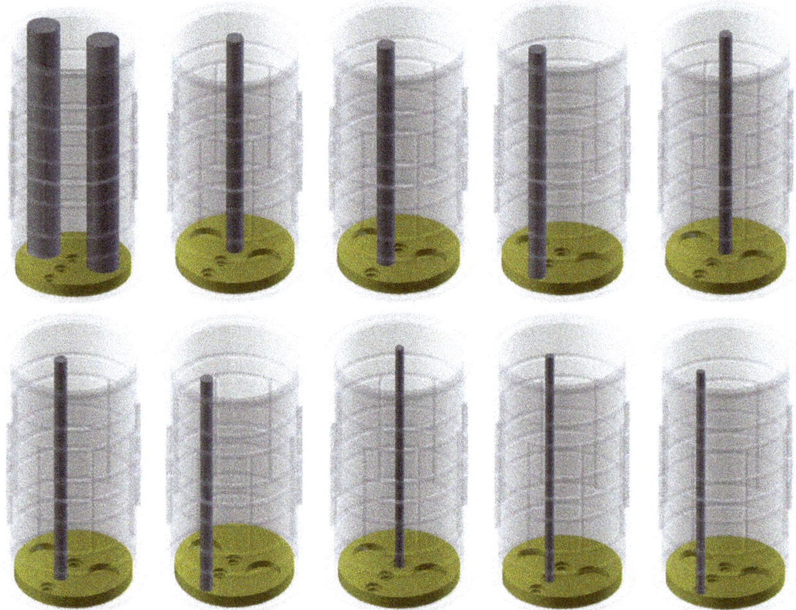

**Figure 7.** A set of ten phantoms used during high-contrast media measurements tests. The mounting stand had 5 holes. Three 10 mm of diameter holes were positioned along the sensor profile diameter at given positions: "P1" at x = 70 mm, y = 70 mm; "P2" at x = 70 mm, y = 40 mm; "P3" at x = 70 mm, y = 10 mm. Two additional 40 mm of diameter holes were positioned symmetrically at x1 = 110 mm and x2 = 40 mm for y = 70 mm. All the rods were parallel to sensor walls.

## 5. Results and Discussion

### 5.1. Low-Contrast Objects Investigation

In this section, the comparison of detection capability for sensors S1 and S2 is discussed in six experimental conditions of low-contrast (low permittivity distribution) solid-gas phantom objects. To analyze and compare the measurement abilities of the tested sensors, four capacitance-measurement inner-layer and inter-layer configurations for acquiring *RCD* (Raw Capacitance Data) have been selected: 1–2 (Layer1–Layer2), 1–9 (Layer1–Layer2), 1–17 (Layer1–Layer3), and 1–25 (Layer1–Layer4) for 6 various tested low-contrast objects—see Table 2.

**Table 2.** Results of capacitance measurement (in pF) for selected electrode pairs and all testing phantoms.

| *RCD* | Test$A_{out}$ | Test$A_{in}$ | Test$B_{out}$ | Test$B_{in}$ | Test$C_{out}$ | Test$C_{in}$ |
|---|---|---|---|---|---|---|
| S1-1-2 | 7.446 | 7.291 | 7.460 | 7.278 | 7.408 | 7.336 |
| S1-1-9 | 6.604 | 6.536 | 6.614 | 6.531 | 6.616 | 6.541 |
| S1-1-17 | 3.926 | 3.914 | 3.928 | 3.911 | 3.918 | 3.930 |
| S1-1-25 | 3.919 | 3.917 | 3.921 | 3.917 | 3.917 | 3.928 |
| S2-1-2 | 4.909 | 4.626 | 4.922 | 4.617 | 4.827 | 4.673 |
| S2-1-9 | 4.577 | 4.414 | 4.591 | 4.410 | 4.568 | 4.427 |
| S2-1-17 | 3.920 | 3.912 | 3.921 | 3.911 | 3.913 | 3.916 |
| S2-1-25 | 3.922 | 3.920 | 3.922 | 3.920 | 3.919 | 3.922 |

A capacitance measurement has been conducted using precise LCR impedance analyzer and multiplexer hardware and a set of 6 capacitance characteristics for full 496 measurements cycle has been captured and normalized using low (for the air) and high (for PE granules) reference measurements. A standard deviation value of all normalized capacitance characteristics was calculated as a "signal stability" indicator for evaluating the performance of the S1 and S2 sensor.

For most tests sensor S2 with configuration S2-1-17 achieved similar capacitance-measurement value to S2 for both close-to-wall and central area. For most distant configurations 1–25 an advantage S2 over S1 could be observed. S1-1-2 and S2-1-2 configurations resulted in the maximum value of capacitance measurement as it connects adjacent electrodes from layer 1—those electrodes had the biggest active surface. In this case, S1 seemingly offered much higher value of capacitance measurement than S2 for the whole set of testing phantoms. A similar tendency existed when configurations S1-1-9 and S2-1-9 were taken into consideration. It was a consequence of high electric field intensity peak value between two adjacent electrodes in S1, which were not separated by vertical radial screens. Vertical radial screens in S2 significantly reduced the electric field intensity value and kept a signal at a lower level. The same trend could be observed for configurations S1-1-9 and S2-1-9, where two adjacent electrodes from two adjacent sensor layers were connected. When we analyzed more distant electrode pairs, a real advantage of sensor S2 over S1 could now be observed.

Figures 8–13 presented selected 1st electrode measurement cycle extracted from the full capacitance data set for sensor S1—blue graph and S2—red graph. These normalized capacitance characteristics have been computed for all phantom Test$A$-Test$C$ configurations using the 3D ECT calibration procedure. The most representative Test$B$ phantom was investigated here to check the ability of both sensors to detect low-contrast permittivity distribution with an approximate equal distance from electrodes and from sensor center along axis $z$. The mean value of the normalized capacitance records dataset for S1-Test$B_{out}$ was 1.196, and the standard deviation was 0.181. The mean value of the normalized capacitance dataset for S2-Test$B_{out}$ was 1.098 and the standard deviation was 0.115. Therefore, this study proved that S2 outperformed S1.

The S2 capacitance-measurement signal fitted better than the calibration limits and was more stable than S1. The Figure 10 presented 1st electrode measurement cycle graph extracted from full capacitance data set for sensor S1—blue graph and S2—red graph, obtained for phantom Test$B_{in}$. This phantom demonstrated that both sensors are capable of detecting permittivity distribution in the center of the sensing area. For this test, the mean value of normalized capacitance dataset for S1 was 0.686 and the standard deviation was 0.396. The mean value of the normalized capacitance dataset for S2 was 0.721 and the standard deviation was 0.394. A capacitance-measurement data from S2 better fit the calibration limits and suffered from rapid changes of capacitances from specific electrode pairs. This test confirmed again an advantage of S2 over S1. The same trend could be observed during the remaining experimental tests for other low-contrast phantoms Test$A$ and Test$C$.

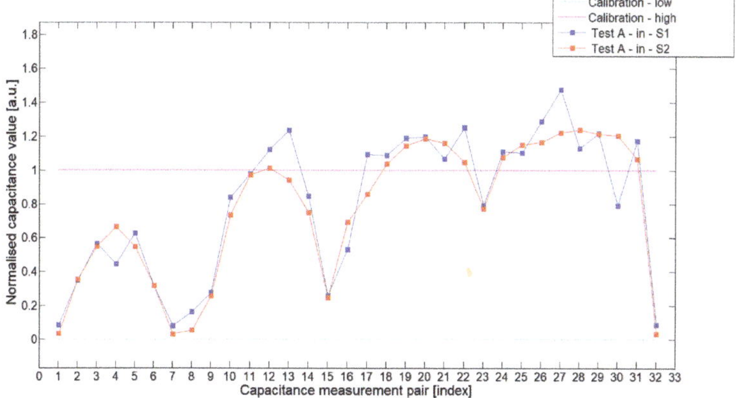


**Figure 8.** The 1st electrode measurement cycle (S1: 1->32 and S2: 1->32)for Test$A_{in}$ and S1—blue line, S2—red line. Cyan and magenta lines indicate calibration limits (0;1).

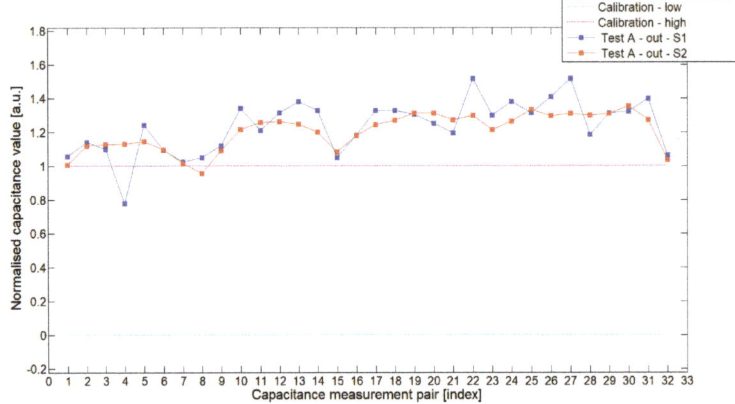

**Figure 9.** The 1st electrode measurement cycle (S1: 1->32 and S2: 1->32) for Test$A_{out}$ and S1—blue line, S2—red line. Cyan and magenta lines indicate calibration limits (0;1).

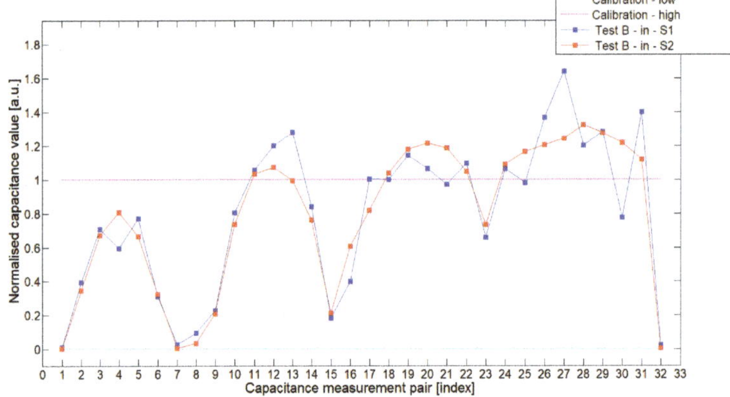

**Figure 10.** The 1st electrode measurement cycle (S1: 1->32 and S2: 1->32) for Test$B_{in}$ and S1—blue line, S2—red line. Cyan and magenta lines indicate calibration limits (0;1).

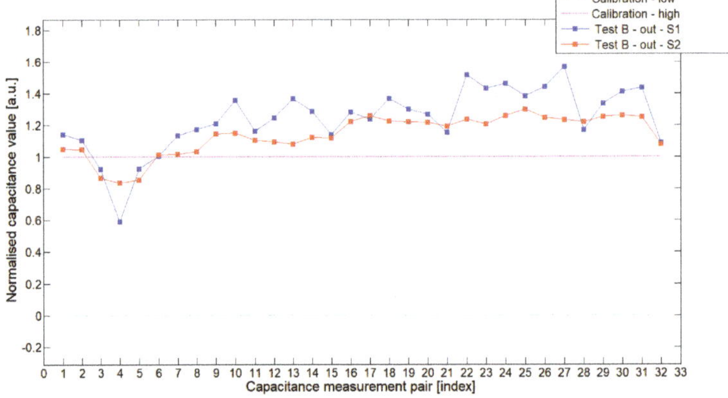

**Figure 11.** The 1st electrode measurement cycle (S1: 1->32 and S2: 1->32) for Test$B_{out}$ and S1—blue line, S2—red line. Cyan and magenta lines indicate calibration limits (0;1).

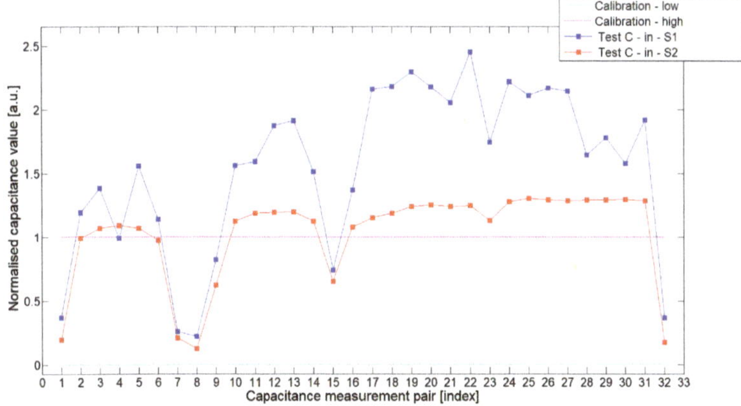

**Figure 12.** The 1st electrode measurement cycle (S1: 1->32 and S2: 1->32) for Test$C_{in}$ and S1—blue line, S2—red line. Cyan and magenta lines indicate calibration limits (0;1).

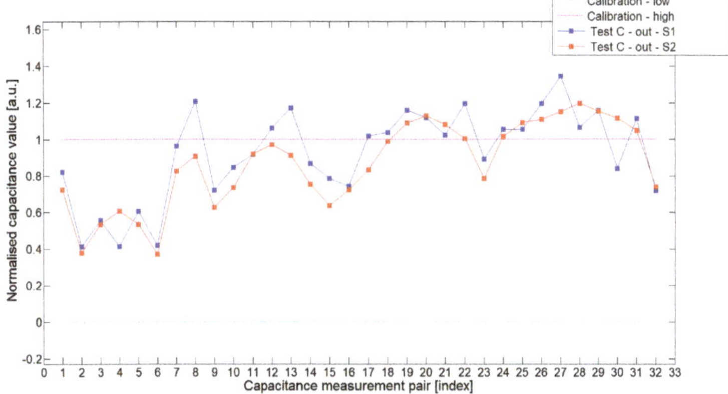

**Figure 13.** The 1st electrode measurement cycle (S1: 1->32 and S2: 1->32) for Test$C_{out}$ and S1—blue line, S2—red line. Cyan and magenta lines indicate calibration limits (0;1).

*5.2. High-Contrast Objects Investigation*

High-contrast (high permittivity distribution) gas–liquid object tests were focused on axial and spatial resolution abilities of both investigated 3D ECT sensors. The sizes of phantom diameters were selected to observe a point of object detectability, especially in the central region of devices where an electric field power is degraded significantly. Due to the static nature of tested objects, a temporal resolution was not taken into consideration in this study. As with the previous low-contrast evaluation, a standard deviation value of normalized capacitance characteristics was considered to be a "signal stability" indicator for evaluating the performance of S1 and S2 sensor. Additionally, a mean value of normalized capacitance characteristics was computed to determine a relative level of calibrated signal response to a background permittivity distribution in the presence of high-permittivity objects. The higher mean value would lead to a better detectability of smaller objects in various 3D ECT sensor areas. Tables 3 and 4 present the results of a standard deviation and mean values obtained for the 1st electrode measurement cycle (in turn with 2..32 electrodes) for both sensors S1 and S2 and all sets of test objects. To compare values, a percentage difference was calculated.

**Table 3.** The standard deviation indicators (std) calculated for the 1st electrode measurement cycle, where electrodes: 2->32 were grounded. Where 10, 15, 20, 40 stands for object diameter and P1, P2, P3 determine object positions along sensor radius, P1 = 70 mm, P2 = 35 mm, P3 = 10 mm respectively.

| std | $2 \times 40$ | $20_{P1}$ | $20_{P2}$ | $20_{P3}$ | $15_{P1}$ | $15_{P2}$ | $15_{P3}$ | $10_{P1}$ | $10_{P2}$ | $10_{P3}$ |
|---|---|---|---|---|---|---|---|---|---|---|
| S1 | 0.0883 | 0.0132 | 0.0310 | 0.1403 | 0.0083 | 0.0201 | 0.0728 | 0.0038 | 0.0143 | 0.0349 |
| S2 | 0.1296 | 0.0081 | 0.0223 | 0.1272 | 0.0057 | 0.0110 | 0.0717 | 0.0050 | 0.0097 | 0.0284 |
| % | −32% | 63% | 39% | 10% | 46% | 83% | 1.5% | −24% | 47% | 23% |

The standard deviation indicators (std) proved that sensor S2 is more stable (1.5–83%) and accurate for the majority of conducted tests except $2 \times 40$ and $10_{P1}$ tests where standard values were higher for S2 and the highest at all. These two situations only occurred for extreme cases: the largest high-contrast inclusion and the smallest one located in the less-sensitive region of the sensor. It may happen when the ET3 system range limits are achieved for a given setup of its channel amplifier gains. The optimization of channel gains for signal borderlines could help to stabilize the signal for both sensors and improve S2 3D ECT sensor overall performance.

**Table 4.** The mean value indicators (mean) calculated for the 1st electrode measurement cycle (with 2..32). Where 10, 15, 20, 40 stands for object diameter and P1, P2, P3 determine object positions along sensor radius, P1 = 70 mm, P2 = 35 mm, P3 = 10 mm respectively.

| mean | $2 \times 40$ | $20_{P1}$ | $20_{P2}$ | $20_{P3}$ | $15_{P1}$ | $15_{P2}$ | $15_{P3}$ | $10_{P1}$ | $10_{P2}$ | $10_{P3}$ |
|---|---|---|---|---|---|---|---|---|---|---|
| S1 | 0.0330 | 0.0020 | 0.0099 | 0.0332 | 0.0011 | 0.0083 | 0.0299 | −0.0024 | 0.0031 | 0.0097 |
| S2 | 0.0632 | 0.0028 | 0.0091 | 0.0683 | 0.0027 | 0.0046 | 0.0363 | 0.0011 | 0.0067 | 0.0143 |
| % | 48% | 29% | −8% | 51% | 45% | −80% | −18% | >100% | >100% | 68% |

While analyzing the general mean value of the acquired normalized signals for the 1st electrode measurement cycle, it was difficult to indicate the clear winner of this comparison except in very important tests $10_{P1}$–$10_{P3}$. All of them demonstrated better high-contrast object detection abilities of S2 sensor. These particular objects represent the smallest high-contrast inclusion. Test $10_{P1}$ was only passed by sensor S2. S2 generated output signals that could even be usable for image reconstruction algorithms. For this specific test, S1 signal response was too low and too noisy to be distinguished from lower-calibration limit (background). The remaining mean values for a given senor depended on the specific test. For 20 mm objects, there was an advantage of S2 printed sensor over S1 while this situation changed for all 15 mm objects. However, the percentage dominance for 15 mm object is on the side of the S1 sensor, and the absolute mean value for 10 mm and 15 mm objects is very low and may contain some amount of measurement noise. To look precisely at how each of the 3D ECT senor signal responses looks like a set of normalized capacitance characteristics for all test objects was computed, see Figures 14–23.

Most of these characteristics pointed out the advantage of S2 over S1. However, the signal profile for position P1 was not easy to compare due to measurement noise (Figures 15, 18 and 21) the better signal response of S2 was visible for object position P2—(Figures 16, 19 and 22) and clearly visible for object position P3 (close to the sensor internal wall) as well—(Figures 17, 20 and 23).

The conducted experiments showed that new 3D printed ECT sensor noticeably outperforms the traditionally fabricated sensor when low- and high-contrast small-sized objects (10–15 mm) located in the central region of tomographic inspection were under test. For large-sized high-contrast objects (20 mm and $2 \times 20$ mm) the performance improvement depended on its radial and axial position. However, both sensors offered similar performance for all the tested objects in the neighborhood of the electrodes where the electric field and sensitivity were strong enough to compensate wall thickness differences.

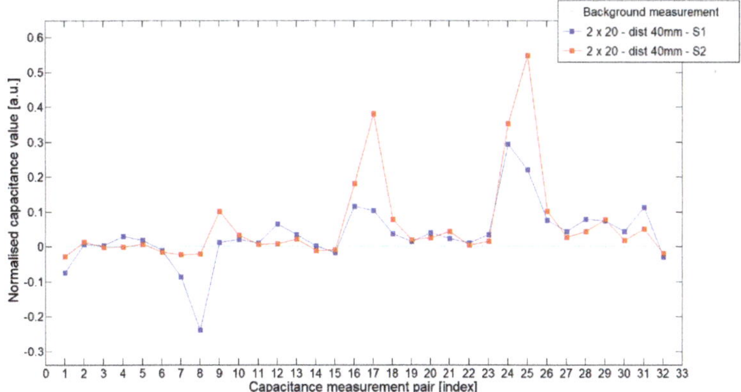

**Figure 14.** The 1st electrode measurement cycle (S1: 1->32 and S2: 1->32) for Test2$x$40 and S1—blue line, S2—red line. Cyan line indicates lower calibration limit.

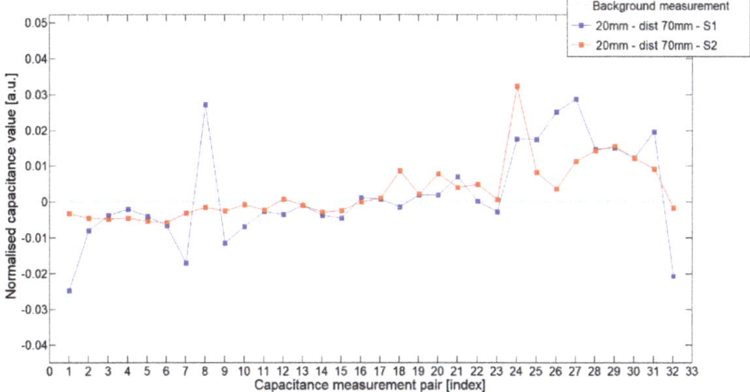

**Figure 15.** The 1st electrode measurement cycle (S1: 1->32 and S2: 1->32) for Test20$_{P1}$ and S1—blue line, S2—red line. Cyan line indicates lower calibration limit.

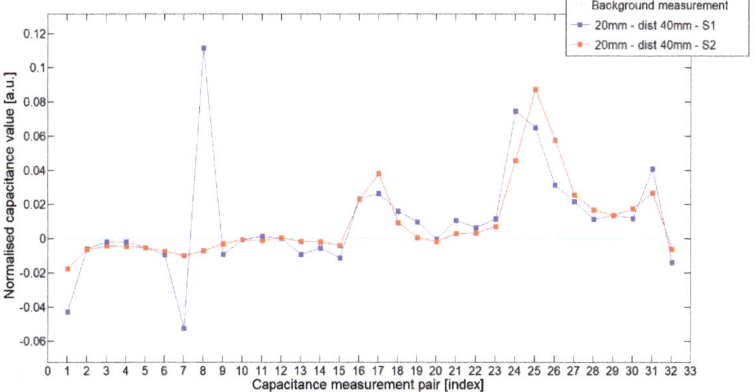

**Figure 16.** The 1st electrode measurement cycle (S1: 1->32 and S2: 1->32) for Test20$_{P2}$ and S1—blue line, S2—red line. Cyan line indicates lower calibration limit.

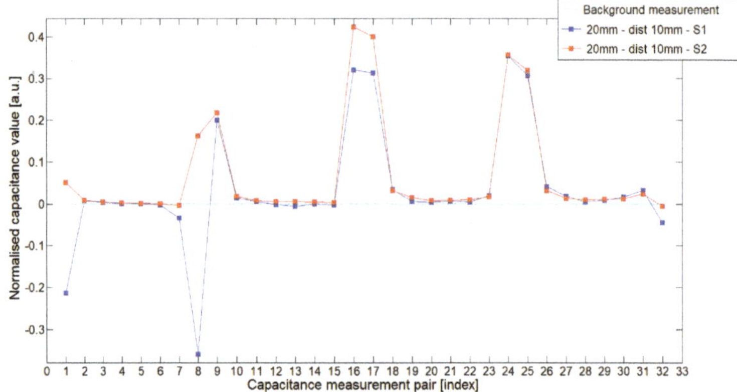

**Figure 17.** The 1st electrode measurement cycle (S1: 1->32 and S2: 1->32) for Test20$_{P3}$ and S1—blue line, S2—red line. Cyan line indicates lower calibration limit.

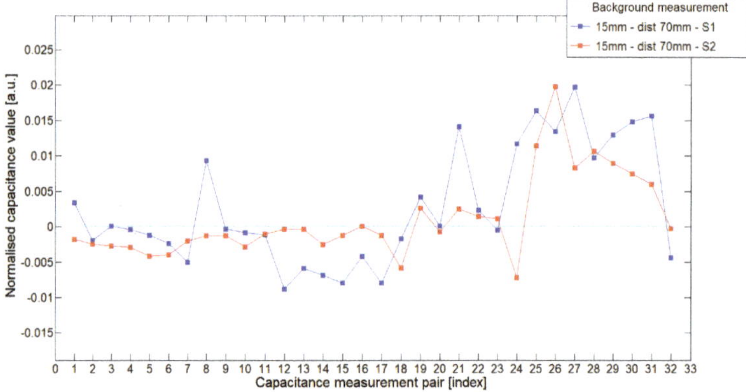

**Figure 18.** The 1st electrode measurement cycle (S1: 1->32 and S2: 1->32) for Test15$_{P1}$ and S1—blue line, S2—red line. Cyan line indicates lower calibration limit.

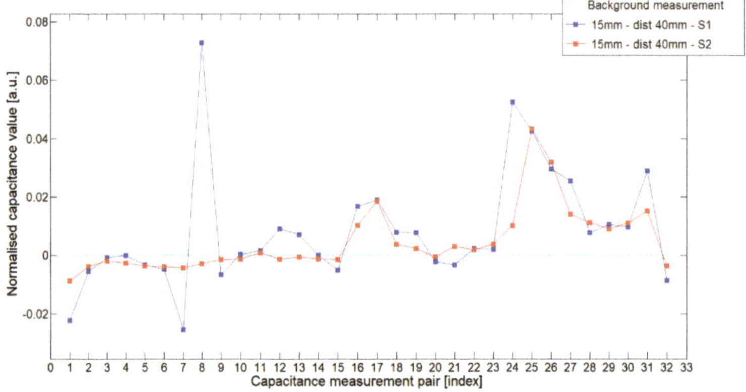

**Figure 19.** The 1st electrode measurement cycle (S1: 1->32 and S2: 1->32) for Test15$_{P2}$ and S1—blue line, S2—red line. Cyan line indicates lower calibration limit.

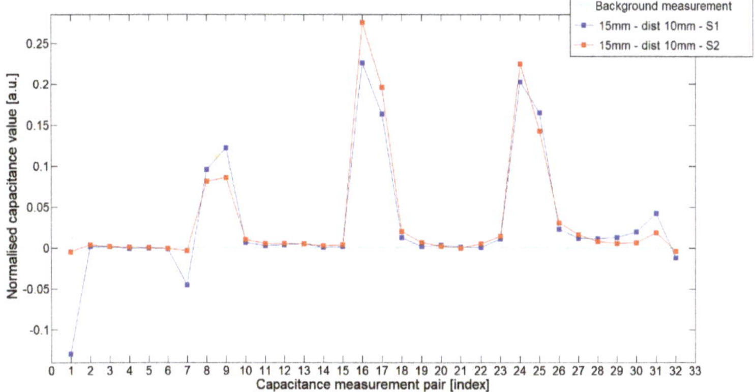

**Figure 20.** The 1st electrode measurement cycle (S1: 1->32 and S2: 1->32) for Test15$_{P3}$ and S1—blue line, S2—red line. Cyan line indicates lower calibration limit.

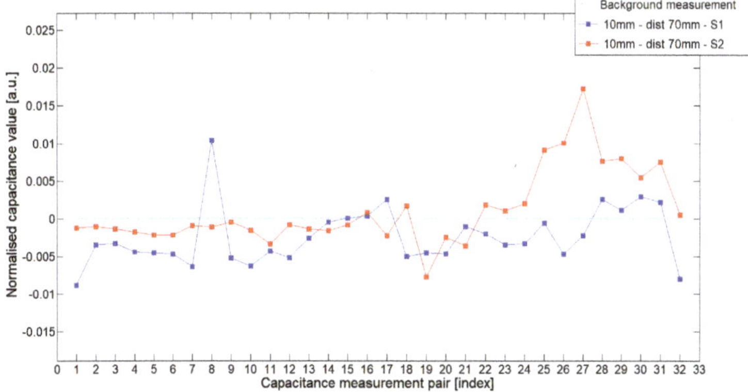

**Figure 21.** The 1st electrode measurement cycle (S1: 1->32 and S2: 1->32) for Test10$_{P1}$ and S1—blue line, S2—red line. Cyan line indicates lower calibration limit.

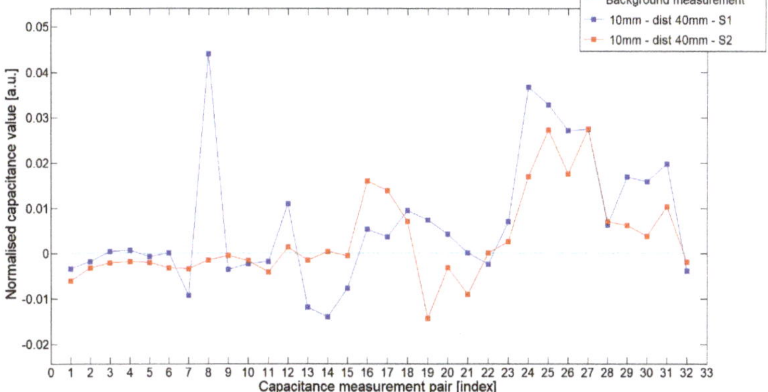

**Figure 22.** The 1st electrode measurement cycle (S1: 1->32 and S2: 1->32) for Test10$_{P2}$ and S1—blue line, S2—red line. Cyan line indicates lower calibration limit.

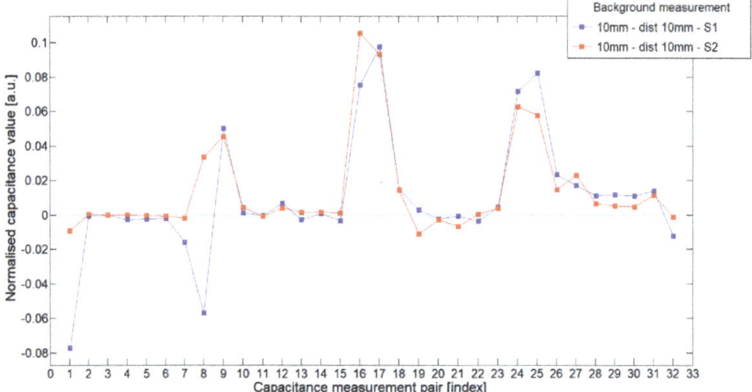


**Figure 23.** The 1st electrode measurement cycle (S1: 1->32 and S2: 1->32) for Test10$_{p3}$ and S1—blue line, S2—red line. Cyan line indicates lower calibration limit.

## 6. Conclusions and Directions for Further Work

In this paper, we demonstrated and validated a new approach for fabrication of ECT sensors. This work comprises the complete workflow for design, modeling and development of the ECT sensors dedicated to true 3D measurement. The experiments showed high precision and good performance of the proposed 3D ECT sensor concept. The proposed method employs modern 3D printing technique that offers efficient ways for the manufacture of capacitance sensors. The developed 3D printed ECT sensor has taken the most valuable features and advantages from traditionally hand-made 3D ECT constructions as well as features possible to gain from a 3D-printing technique. Output sensing structures have thin-wall levels between the electrode arrays and testing volume as well as advanced structure of internal vertical and horizontal screening system. The achieved high accuracy of the design and structure results from high-resolution 3D-printing technology. Using the 3D-printing principle, a majority of 3D ECT sensor structures can be easily designed and 3D-printed in one working day. In this study, the novel 3D-printed 3D ECT sensor model has been compared to the representative solution used widely in the past. For low-contrast and high-contrast objects, we achieved at least similar or better 3D-printed sensor overall performance. Conducted experiments showed a noticeable potential of 3D printing technology applied for the fabrication of capacitance sensors. The research presented in this study is a step forward towards high-precision and high-resolution volumetric ECT systems. The directions for future development derived from this work comprise testing sensors with a larger number of electrodes, and developing novel shapes of electrode arrays coupled with novel shielding systems. The interesting implications of the 3D-printing technique is the possibility to tailor specific designs suited to specific applications such as batch crystallization or gas–liquid separation. The ultimate challenge is to incorporate the proposed approach to measurement data problems that are still under intensive research with no easy automatic way of processing such as crowd-sourcing processing of imagery data as shown in [30] or for modern contextual data processing frameworks as shown in [17,31].

**Author Contributions:** Conceptualization, A.K. and R.B.; Methodology, R.B.; Software, A.K. and R.B.; Validation, D.S., R.B. and A.R.; Formal Analysis, R.B. and A.R.; Investigation, A.K.; Writing—Original Draft Preparation, A.K. and R.B.; Writing—Review and Editing, A.K., R.B. and A.R.; Visualization, A.R.; Supervision, R.B. and D.S.; Project Administration, D.S.; Funding Acquisition, D.S.

**Funding:** This research was partially funded by the Smart Growth Operational Programme 2014–2020, project no POIR.04.01.02-00-0089/17-00. https://rio.jrc.ec.europa.eu/en/library/operational-programme-smart-growth-2014-2020

**Acknowledgments:** This research was supported by the Smart Growth Operational Programme 2014–2020, project no POIR.04.01.02-00-0089/17-00. The project was conducted in the Institute of Applied Computer Science at the Lodz University of Technology.

**Conflicts of Interest:** The authors declare no conflict of interest.

## References


1. Rybak, G.; Chaniecki, Z.; Grudzien, K.; Romanowski, A.; Sankowski, D. Non-invasive methods of industrial processes control. *IAPGOS Inform. Control. Meas. Econ. Environ. Prot.* **2004**, *4*, 41–45. [CrossRef]
2. Che, H.Q.; Ye, J.M.; Yang, W.Q.; Wang, H.G. Measurement of the gas-solid flow in a wurster tube using 3D electrical capacitance tomography sensor. *Lect. Notes Electr. Eng.* **2019**, *506*, 367–383. [CrossRef]
3. Grudzien, K.; Romanowski, A.; Sankowski, D.; Willams, R.A. Gravitational Granular Flow Dynamics Study Based on Tomographic Data Processing. *Part. Sci. Technol.* **2007**, *26*, 67–82. [CrossRef]
4. Wajman, R.; Banasiak, R.; Mazurkiewicz, L.; Dyakowski, T.; Sankowski, D. Spatial imaging with 3D capacitance measurements. *Meas. Sci. Technol.* **2006**, *17*, 2113–2118. [CrossRef]
5. Yang, W.Q.; Peng, L. Image reconstruction algorithms for electrical capacitance tomography. *Meas. Sci. Technol.* **2003**, *14*, R1–R13. [CrossRef]
6. Marashdeh, Q.; Wang, F.; Fan, L.S.; Warsito, W. Velocity measurement of multi-phase flows based on electrical capacitance volume tomography. *Proc. IEEE Sens.* **2007**, 1017–1019. [CrossRef]
7. Plaskowski, A.; Beck, M.S.; Thorn, R.; Dyakowski, T. *Imaging Industrial Flows—Applications of Electrical Process Tomography*; CRC Press: Boca Raton, FL, USA, 1995.
8. Rymarczyk, T. Using electrical impedance tomography to monitoring flood banks. *Int. J. Appl. Electromagn. Mech.* **2014**, *45*, 489–494. [CrossRef]
9. Zeeshan, Z.; Zuccarelli, C.E.; Acero, D.O.; Marashdeh, Q.M.; Teixeira, F.L. Enhancing Resolution of Electrical Capacitive Sensors for Multiphase Flows by Fine-Stepped Electronic Scanning of Synthetic Electrodes. *IEEE Trans. Instrum. Meas.* **2019**, *68*, 462–473. [CrossRef]
10. Banasiak, R.; Wajman, R.; Sankowski, D.; Soleimani, M. Three-dimensional nonlinear inversion of electrical capacitance tomography data using a complete sensor model. *Insight Prog. Electromagn. Res. PIER* **2010**, *100*, 219–234. [CrossRef]
11. Wang, Y.; Ren, S.; Dong, F. Focusing Sensor Design for Open Electrical Impedance Tomography Based on Shape Conformal Transformation. *Sensors* **2019**, *19*, 2060. [CrossRef]
12. Wajman, R.; Fiderek, P.; Fidos, H.; Jaworski, T.; Nowakowski, J.; Sankowski, D.; Banasiak, R. Metrological evaluation of a 3D electrical capacitance tomography measurement system for two-phase flow fraction determination. *Meas. Sci. Technol.* **2013**, *24*. [CrossRef]
13. Xie, W.Q.; Huang, S.M.; Lenn, C.P.; Stoll, A.L.; Beck, M.S. Experimental evaluation of capacitance tomographic flow imaging systems using physical models. *IEE Proc. Circuits Devices Syst.* **1994**, *141*, 357–368.:19941152. [CrossRef]
14. Romanowski, A.; Grudzien, K.; Aykroyd, R.G.; Williams, R.A. Advanced Statistical Analysis as a Novel Tool to Pneumatic Conveying Monitoring and Control Strategy Development. *Part. Part. Syst. Charact.* **2006**, *23*, 289–296. [CrossRef]
15. Grudzien, K.; Romanowski, A.; Williams, R.A. Application of a Bayesian Approach to the Tomographic Analysis of Hopper Flow. *Part. Part. Syst. Charact.* **2017**, *22*, 246–253. [CrossRef]
16. Grudzien, K. Visualization System for Large-Scale Silo Flow Monitoring Based on ECT Technique. *IEEE Sens. J.* **2017**, *17*, 8242–8250. [CrossRef]
17. Romanowski, A. Big Data-Driven Contextual Processing Methods for Electrical Capacitance Tomography. *IEEE Trans. Ind. Inform.* **2019**, *15*, 1609–1618. [CrossRef]
18. Rymarczyk, T.; Sikora, J. Applying industrial tomography to control and optimization flow systems. *Open Phys.* **2018**, *16*, 332–345. [CrossRef]
19. Banasiak, R.; Wajman, R.; Soleimani, M. An efficient nodal Jacobian method for 3D electrical capacitance tomography image reconstruction. *Insight Non-Destr. Test. Cond. Monit.* **2009**, *51*, 36–38. [CrossRef]
20. Kryszyn, J.; Smolik, W.T.; Radzik, B.; Olszewski, T.; Szabatin, R. Switchless charge-discharge circuit for electrical capacitance tomography. *Meas. Sci. Technol.* **2014**, *25*. [CrossRef]

21. Mazurkiewicz, L.; Banasiak, R.; Wajman, R.; Dyakowski, T.; Sankowski, D. Towards 3D Capacitance Tomography. In Proceedings of the 4th World Congress Industrial Process Tomography, Aizu, Japan, 5–8 September 2005; pp. 546–551.
22. Wajman, R.; Banasiak, R. Tunnel-based method of sensitivity matrix calculation for 3D-ECT imaging. *Sens. Rev.* **2014**, *34*, 273–283. [CrossRef]
23. Ye, J.; Wang, H.; Yang, W. Characterization of a multi-plane electrical capacitance tomography sensor with different numbers of electrodes. *Meas. Sci. Technol.* **2016**, *27*. [CrossRef]
24. Li, Y.; Holland, D.J. Optimizing the geometry of three-dimensional electrical capacitance tomography sensors. *IEEE Sens. J.* **2015**, *15*, 1567–1574. [CrossRef]
25. Huang, A.; Cao, Z.; Sun, S.; Lu, F.; Xu, L. An agile electrical capacitance tomography system with improved frame rates. *IEEE Sens. J.* **2019**, *19*, 1416–1425. [CrossRef]
26. Hu, X.; Yang, W. Planar capacitive sensors—Designs and applications. *Sens. Rev.* **2010**, *30*, 24–39. [CrossRef]
27. Wu, M.; Ye, J.; Wang, H.; Yang, W. Evaluation of Excitation Strategy for a Large-Scale ECT Sensor with Internal-External Electrodes. *IEEE Sens. J.* **2017**, *17*, 8091–8098. [CrossRef]
28. Xu, H.; Yang, H.; Wang, S. Effect of Axial Guard Electrodes on Sensing Field of Capacitance Tomographic Sensor. In Proceedings of the 1st World Congress on Industrial Process Tomography, Buxton, UK, 14–17 April 1999; pp. 348–352.
29. Dichtl, C.; Sippel, P.; Krohns, S. Dielectric Properties of 3D Printed Polylactic Acid. *Adv. Mater. Sci. Eng. Vol.* **2017**, *2017*. [CrossRef]
30. Chen, C.; Woźniak, P.W.; Romanowski, A.; Obaid, M.; Jaworski, T.; Kucharski, J.; Grudzien, K.; Zhao, S.; Fjeld, M. Using Crowdsourcing for Scientific Analysis of Industrial Tomographic Images. *Part. Part. Syst. Charact.* **2016**, *7*, 52:1–52:25. [CrossRef]
31. Romanowski, A. Contextual processing of electrical capacitance tomography measurement data for temporal modeling of pneumatic conveying process. In Proceedings of the 2018 Federated Conference on Computer Science and Information Systems (FedCSIS), Annals of Computer Science and Information Systems, Poznan, Poland, 9–12 September 2018; IEEE: Piscataway, NJ, USA, 2018; Volume 15, pp. 283–286. [CrossRef]

Article

# X-ray Imaging Analysis of Silo Flow Parameters Based on Trace Particles Using Targeted Crowdsourcing [†]

**Andrzej Romanowski *, Piotr Łuczak and Krzysztof Grudzień**

Institute of Applied Computer Science, Lodz University of Technology,
90924 Lodz, Stefanowskiego 18/22 str., Poland
* Correspondence: androm@kis.p.lodz.pl; Tel.: +48-426-312-750
† This paper is an extended version of our paper published in Ninth World Congress on Industrial Process Tomography, 2–6 September Bath, U.K., 2018. Investigating X-ray Images for Studying Gravitational Flow in Silos Using Crowdsourcing Annotations and Analysis.

Received: 3 June 2019; Accepted: 26 July 2019; Published: 28 July 2019

**Abstract:** This paper presents a novel method for tomographic measurement and data analysis based on crowdsourcing. X-ray radiography imaging was initially applied to determine silo flow parameters. We used traced particles immersed in the bulk to investigate gravitational silo flow. The reconstructed images were not perfect, due to inhomogeneous silo filling and nonlinear attenuation of the X-rays on the way to the detector. Automatic processing of such data is not feasible. Therefore, we used crowdsourcing for human-driven annotation of the trace particles. As we aimed to extract meaningful flow parameters, we developed a modified crowdsourcing annotation method, focusing on selected important areas of the silo pictures only. We call this method "targeted crowdsourcing", and it enables more efficient crowd work, as it is focused on the most important areas of the image that allow determination of the flow parameters. The results show that it is possible to analyze volumetric material structure movement based on 2D radiography data showing the location and movement of tiny metal trace particles. A quantitative description of the flow obtained from the horizontal and vertical velocity components was derived for different parts of the model silo volume. Targeting the attention of crowd workers towards either a specific zone or a particular particle speeds up the pre-processing stage while preserving the same quality of the output, quantified by important flow parameters.

**Keywords:** measurement data analysis; targeted crowdsourcing; flow investigation tool; X-ray process tomography; radiography imaging

---

## 1. Introduction

Crowdsourcing is an emerging method for processing large amounts of data using geographically-distributed heterogeneous workers. This method has proven suitable for resolving computational problems that are difficult to solve automatically, as it is capable of coupling the data processing capabilities of automated systems with human intelligence. Another advantage of crowdsourcing is the reduced cost (in terms of money, resources, or time). Most of the tasks submitted to widely-available crowdsourcing servers (www.mturk.com, www.crowdflower.com, www.crowdmed.com) could conceivably be processed by computer systems. However, due to their complexity or uniqueness, the difficulty of achieving sufficiently high accuracy, or simply economic factors, different methods of processing such datasets are preferred, based on human orientation. Examples of such tasks, including projects pertaining to image processing, have been reported in the literature [1–3]. One interesting example is the distributed diagnosis carried out by the medical community around the world, based on medical records submitted by patients who wish to obtain a second opinion. Another is the simultaneous translation or group editing of texts on-the-go, in which human resources can compete with automatic systems [4].

Crowdsourcing facilitates the delegation of work, which usually requires the use of Internet-connected computers [5]. An open call is made for people to contribute to online tasks that do not require any special knowledge, competences, or specific training in the field. Here, the crowd refers to an undefined, but large group of participants. It turns out that most mundane tasks involved in analyzing large datasets, especially imaging data, can be broken down into chunks that are able to be processed or annotated by an average person, without expertise in a particular domain or knowledge of the experiment.

Generally, there are dedicated platforms for managing the work distributed among individual workers. Entire jobs are divided into either smaller or sub-tasks in order to facilitate the process and achieve greater efficiency. Classical categorization into macro- and micro- (or sub-) tasks is a common part of crowdsourcing methods and is possible using most popular platforms [6,7]. However, most studies focus on dividing a sequence of consecutive data elements (usually image frames) into a smaller series of adjacent frames [8]. Each entire frame is then processed by an individual worker [9]. To the best knowledge of the authors, there have been no papers describing methods that target either only selected, limited areas of individual frames or single elements within these frames (or single elements visible on particular images); although, this is technically possible with the aid of existing platforms [7].

Crowdsourcing systems coordinate the efforts of distributed workers and synthesize the results to solve problems that a single individual could not achieve on the same scale, within the same budget or time-frame. This approach makes it possible to complete problems at a level of complexity beyond the capabilities of the research personnel available. Usually, a few minutes of online training in the form of a video or illustrated manual is sufficient for a new worker to begin the task. This is important from the point of view of this research, since neither machine learning algorithms nor standard image processing methods can currently perform efficiently in the application presented here, as reported in [10,11]. On the other hand, there is a number of limitations to the crowdsourcing method, as reported especially in business contexts, but affecting research projects as well [12]. One of the most important is the so-called sloppiness of the output [13]. This may result from a range of causes, one of which is the mundane nature of the work [14].

Crowdsourcing tasks can be efficiently tailored to a range of problems relating to process tomography sensing, especially for reconstructed images. There have been previous reports of the successful application of the general method to annotate X-ray radiography data using non-expert workers [11]. This paper takes a further step, using the output results for further analysis and interpretation, in this case to obtain the vertical and horizontal components of flow velocity in silos.

The main contribution of this paper is to evaluate the approach on the basis of industrial tomographic measurement data, which was impossible to process automatically using currently-available approaches, in order to provide a meaningful quantitative description of the flow in terms of parameters. An additional contribution of this paper is the proposed method of a targeted crowdsourcing, in which each contributor is given a narrowed down task in comparison to the usual full-frame work assignment. This involved identifying the position of a single marker or a set of markers in a small, selected area of the images in a sequence of assigned frames. Such targeting of the workers' attention aims to reduce the effort required from any single participant, whilst providing data of comparable or superior accuracy to datasets processed using classical crowdsourcing systems.

The rest of the paper is organized as follows. Sections 2–4 cover the background and details of the interdisciplinary research. Section 2 discusses important related work in the field of gravitational flow of bulk solids' process measurements. Section 3 gives details of the experimental procedure for X-ray measurements and data-processing issues. Section 4 presents the basic crowdsourcing method and data processing workflow employed in this work. The results and directions for future research are discussed in Sections 5 and 6. A summary and conclusions are provided in Section 7.

## 2. Related Work

This section introduces the two key elements of the research background, namely industrial tomography for process monitoring and the basics of the gravitational flow of bulk solids.

### 2.1. Industrial and Process Tomography for Understanding Process Behavior

In the context of industrial processes, tomography is closely associated with imaging, since images provide rich information concerning the monitored process [15–17]. Optical systems are the most common for the visualization of liquid flows. They provide an image of the process using CCD/CMOS cameras. Electrical process tomography [18–22], gamma tomography [23,24], magnetic resonance tomography [25], X-ray tomography [26], or ultrasonography [27] can also be applied for this purpose. Since different types of imaging system are characterized by different modality properties (depending on the kind of process to be monitored, the type of installation, and the laboratory or industrial environment), different measurement setups will be applied. Electrical Capacitance Tomography (ECT) is best suited for the visualization of highly-dynamic processes, in which direct contact with the flowing medium is not possible, where the installation is opaque, or when the cross-section of the installation is substantial in size. However, although sufficient for control purposes, ECT imaging gives low spatial resolution. To investigate processes in the laboratory environment, using small-scale installations, other tomographic systems may provide more valuable information. For instance, X-ray tomography is a standard choice in the field of material science for investigating the structure of materials. This kind of tomography enables the analysis of phenomena that occur at the nano- or micro-scales. There are also reports in the literature on the development of X-ray systems for the visualization of processes at much larger scales, based on the low absorption of X-ray photons by a medium [28].

X-ray tomography provides much higher spatial resolution than electrical process tomography, making it more convenient and reliable as a tool for investigating processes. However, detailed imaging in 3D requires a substantial number of projections around the scanned body. Guillard et al. [29] presents a different type of X-ray imaging, namely the use of stereo-radiography for velocity and fabric fields' determination for vertically plane-symmetric flow of granular media. In contrast, the study presented in this paper shows a single X-ray radiography system in application to granular media. We demonstrate that based on 2D images (radiographs), the behavior of the material during flow can be followed by tracing the positions of the tracking particles in symmetric bulk flow.

### 2.2. Gravitational Flow of Solids

The gravitational flow of solid particles is relevant to a number of industrial processes, in which bulk solids are stored in silos. Such granular material may be of natural provenance (e.g., sands and gravels) or be generated in the process of extraction (e.g., stone breakage), as well as being deliberately processed (e.g., plastic granules). The storage of bulk material, whilst seemingly very simple, is actually a complex process requiring sophisticated analysis. The quality of material storage, as well as the ease, efficiency, and safety of unloading strongly depend on the method by which the silo is filled [30]. The initial packing density places pressure on the walls, while the grain size and diameter, as well as the direction of deformation in the granulate particle systems affect changes in the concentration of bulk material in various areas of the tank during unloading. An additional difficulty with analyzing flow is the impact of many changeable external factors, including humidity and temperature. These factors can affect the behavior of the material during the loading and unloading of silos.

Systems for measuring the process of handling granular materials and measuring the levels of silos are well established. Since the early 1960s, studies focused mainly on predicting the type of flow (i.e., mainly funnel or mass flow) and its changes [31], analyzing internal shear zones [32,33], determining fields and velocity of flow, as well as predicting the intensity of emptying silos [34]. Variable parameters include the container geometry (shape, structure), the properties of the material (particle size and shape, packing density, coefficient of friction), and external conditions (temperature, humidity). Combinations of these parameters result in different flow behaviors, especially causing mass flow (i.e., the homogeneous downward flow of bulk with relatively similar velocity) or funnel flow (i.e., forming a core funnel, with bulk flowing at the center and the rest of the bulk forming a stagnant zone) [30,35]. Measuring changes in the concentrations and levels of materials in silos, which affect both the accuracy with which the velocity and mass flow can be determined and hence the safety of silo operation, is still a matter for continuing research.

*2.3. Trace Particle Tracking Method for Flow Investigation*

In recent years, Particle Tracking Velocimetry (PTV) and Particle Image Velocimetry (PIV) have been used extensively to measure the translational velocity field of granular assemblies [36]. Various reports on applications such as in vibrated beds [37], rotating drums [38], avalanche or debris flows [39], or hoppers/silos [40] are available. There are recent interesting reports on using X-ray PTV methods using stereography for fluidized beds with low-absorption medium as well [41]. In general, PTV and PIV enable volumetric measurements within a fluid flow or visible surface planar measurements for particle assemblies. For fluid flows, it is possible to capture the movement of tiny (micro-)suspended particles. The actual calculation is conducted with the aid of image post-processing techniques and algorithms responsible for image segmentation, particle identification, spatial mapping, and temporal tracking throughout the image sequences. While PIV gives the movement of particle assemblies, PTV traces individual particles, but gives better spatial resolution with the same or lower computing effort. All of the above-mentioned studies concerned tracking spherical particles, except for one, which applied non-spherical tracer particles [42].

## 3. Sensing Equipment and X-ray Imaging

In order to investigate the two types of silo flow using an industrial X-ray tomography setup, a dedicated model of a silo was constructed. The rectangular silo model had 5 mm-thick walls made of perspex sheets, 340 mm in height, with an inner size of $150 \times 70$ mm (width $\times$ depth, respectively). The entire model consisted of an upper bin section and a lower hopper outlet section (photo on the left in Figure 1). The angle of the hopper from vertical was 60 deg. The width of the rectangular outlet along the silo depth was 5 mm, and the length was 70 mm (as it was fixed along the entire depth of the model). Silo flow measurements were conducted at INSA-Lyon, MATEISCNRSlaboratory. We used a flat silicon panel detector tomograph produced by the Phoenix X-ray company. It was equipped with a Varian 2520V (PaxScanTM) detector made of a flat silicon panel, initially developed for medical applications. It was composed of 1920 rows and 1500 lines of sensitive pixels, each of which was $127 \times 127$ $\mu m^2$. The detector can be used in a $1 \times 1$ or in a $2 \times 2$ binning mode. During the measurements, the detector worked in the $2 \times 2$ binning mode (the pixel size was $254 \times 254$ $\mu m^2$). The X-ray source (a cone beam) was an open transmission nanofocus X-ray tube with radiation emission parameters of 145 kV (voltage) and 180 $\mu A$ (current intensity). The source-detector distance was 0.577 m; the source-object distance and object-detector distance were 0.384 m and 0.193 m, respectively; and the magnification was 1.50. The geometric blurring magnitude was 2.01 $\mu m$. The exposure time was 100 ms, and the frame rate was equal to 10 fps. The beam hardening effect was counteracted using a thin copper plate (0.3 mm in thickness) mounted on an X-ray tube to filter the low energy X-rays. The resulting radiography size was $960 \times 768$ pixels. The captured sets of X-radiographs (for each experiment) were pre-processed, so as to extract and present information on changes in the X-ray radiation absorption coefficient. The X-radiographic image provided information on the X-ray attenuation within the object. Transmission radiography with a 2D flat panel detector was then used to generate a 2D map of $\mu$, the linear attenuation coefficient. Figure 2 shows the two types of radiography images obtained. As we investigated two types of silo flow, the silo was filled on two separate occasions, which resulted in different measurement records and therefore significantly different reconstructed pictures of the flat panel detector output.

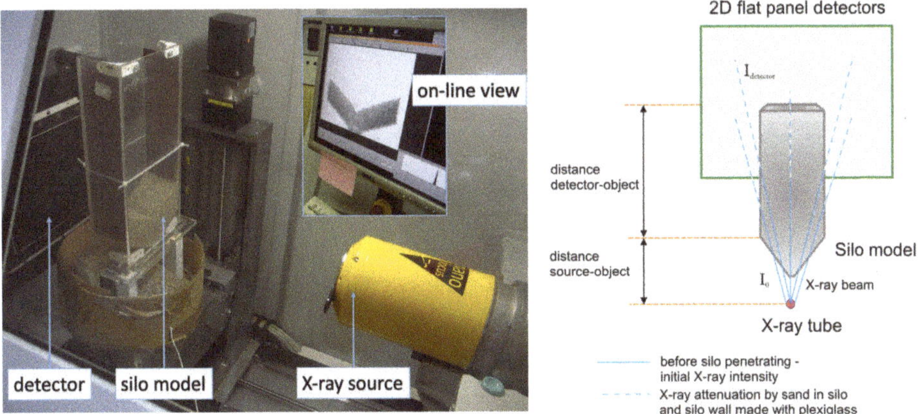


**Figure 1.** X-ray tomography imaging system: (left) photo of the silo model taken during X-ray scanning; panel detector on the far left, silo model on the left, X-ray tube on the lower right, and the corresponding online visualization included as an inset on the upper right; (right) schematic overview of the X-ray absorption model.

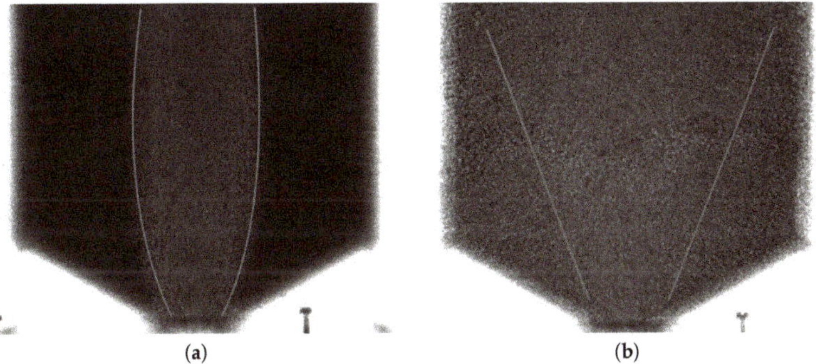


**Figure 2.** Radiography images of funnel flow in the silo: (**a**) initially dense packing; (**b**) initially loose packing. Contour lines indicate approximate flow funnel shape.

Trace particles (spherical metallic particles 2 mm in diameter) with relatively neutral buoyancy were mixed with a granular material (sand) before the experiments [43]. Initially, the particles were distributed randomly throughout the volume of the silo model. The intention was to observe their changing positions during flow, in order to derive information on the overall flow conditions in the container. By analyzing 2D radiographic images, it was possible to estimate the paths of particles moving in 2D space. To investigate the 3D nature of the flow, it would have been necessary to analyze a 3D tomography image. This paper focuses on 2D analysis only. The main reason for choosing 2D radiography is its much higher temporal resolution, which enables the measurement of dynamic processes, such as gravitational bulk flow. X-ray imaging in 3D would require many more measurements per time unit, and therefore is not suitable in this case.

One notable drawback of the applied imaging technique, however, lies in the flatness of the generated image. Some features of the observed process may have been concealed by objects closer to the emitter. Due to this "shadowing" effect, some of the objects may be partially visible (i.e., their shape is distorted and the area of the object is smaller than in reality) or totally invisible in certain random individual or consecutive frames. Such behavior must be taken into consideration in the subsequent analysis, as it causes issues for automatic processing of trace particle images using both classical image processing algorithms and modern approaches such as neural network-based methods [10].

### 3.1. X-ray Measurement Data Processing

Radiography images were taken of funnel flow in the cases of dense and loose initial packing densities. The flow area in each case was different, since for initially dense packing, the funnel should be narrower than in the case of loose packing. In addition, the contrast between the funnel zone and the remainder of the material in the radiography images was much higher in the case of dense flow. Figure 2 shows two radiography images for initially dense and loose packing densities. As may be imagined, analyzing a series of such images could result in different rates of eyestrain and hence different anticipated demands and cognitive workload. More details on the visualization of funnel flow can be found in [44].

Figure 3 illustrates the difficulty of extracting the trace particles from the background noise. Due to limitations in both the spatial and temporal resolutions of the collected flow data, the markers were often distorted, smudged, or blurred to the point where they became undetectable by any morphological or shape-metric approach. Therefore, detecting the particles by automatic means is very problematic, and even for humans is extremely difficult.

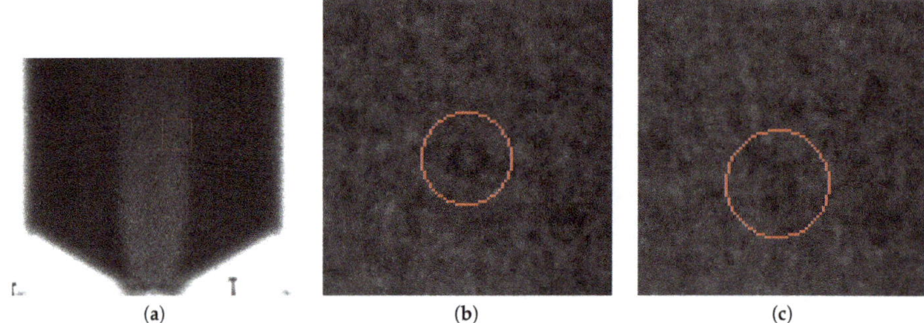

**Figure 3.** Difficulty of distinguishing markers in radiography images: (**a**) frame with the investigated area marked, (**b**) zoomed-in marker, and (**c**) the same marker on the next frame.

The combination of inherent graininess at such scales and the grainy nature of the observed material itself probably makes extraction of the relevant features in the image an insurmountable task for classical image processing methods. However, as previously shown [11] and observed again in the course of this research, human operators can be taught to distinguish such markers in a relatively short time. Furthermore, they can attempt to estimate the approximate position of the marker given the previous or next frame. A human operator can also anticipate the existence of an object based on a wider temporal sequence and locate it even if it is significantly occluded by the surrounding material in one or more of the recorded frames. In this work, we investigated different approaches for using crowdsourcing to annotate images. We developed a simple application with a front-end graphical interface implemented in Python and Pyside2. With the aid of this tool, it was possible to rapidly distribute the raw datasets and assign tasks to separate workers.

### 3.2. Experimental Methodology

Downward velocity was calculated based on the positions of particles provided by the crowdsourcing system (details on the crowd workflow are given later in this paper). In order to compare the velocity for different heights above the silo outlet, the area of the silo was divided into nine sections. This division was established based on expert knowledge of the gravitational flow process, with each zone containing part of the funnel flow. Figure 4 shows an example zone grid, predefined to determine the velocity components. The funnel flow area contained three central zones (B, E, H) and six side segments (A, C, D, F, G, I). Analysis of the speed of the flow of loose material was carried out based on the established trajectories of trace particles found in the zones. The analysis was completed for both loose and dense initial packing densities.

Three crowdsourcing data processing methods were applied. The first method was an established classical procedure (the "classical" method), according to which each worker processed all the particles visible in the picture on each image in the assigned sequence. The second was a zone-particle tracking method, whereby each worker focused on and processed a limited number of trace particles located only within a single zone of each image in the assigned sequence. For this method (the "zone-targeted" method), we used the same zone grid as for the comparison of flow velocity, as shown on the left in Figure 4. Finally, a single-particle method (the "single-particle-targeted" method) was applied, in which each worker processed only one, selected trace particle, tracking it through consecutive images of the assigned sequence, ignoring all other particles. The picture on the right-hand side in Figure 4 illustrates the process of tracing a single particle in the single-particle targeted method.

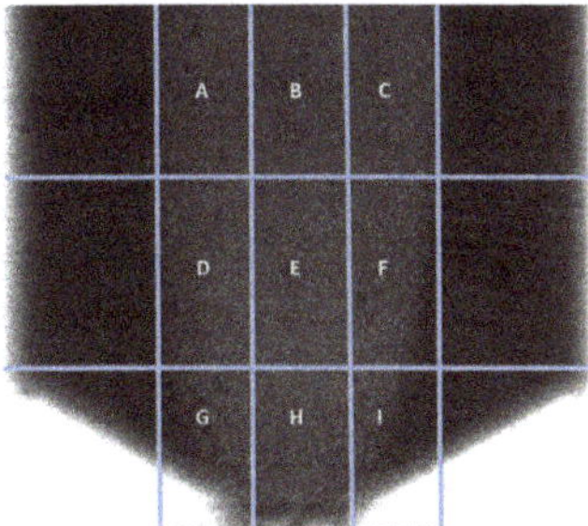
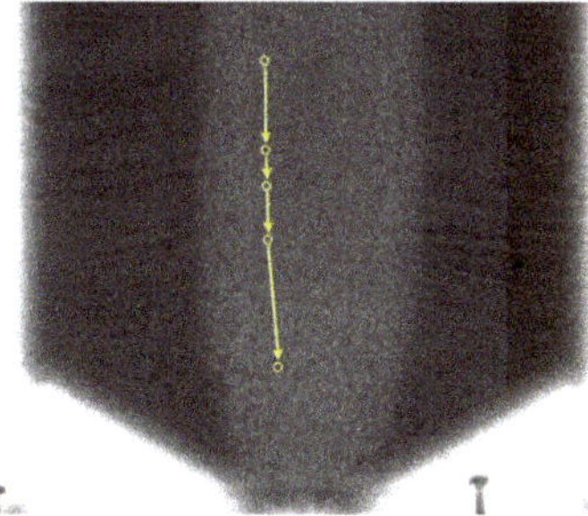

**Figure 4.** Comparison of the zone-targeted mode and single-particle-targeted method of crowdsourcing: (left) grid showing velocity calculation zones used as target zones in the zone-targeted tracking method; (right) illustration of single-particle-targeted method for tracking an arbitrary trace particle.

The first, classical method was an implementation of the well-known crowdsourcing paradigm, while the other two (i.e., targeted zone and single-particle methods) were novel and are evaluated within this study. Both require an additional initialization step, which involved preparing (marking) the first frame of the assigned image sequence in order to indicate the zone of interest (for the targeted zone method) or either selecting the starting particle or giving instructions on how to select the particle of interest (for the single-particle-targeted method).

## 4. Crowdsourcing as an Effective Image Processing Engine

### 4.1. Crowdsourcing System Workflow

The general workflow for processing the X-ray trace particle images is presented in Figure 5. The solution employed neither a dedicated system, nor a high capacity big data framework, yet the modular design supported easy extension [45,46]. First, radiography images captured by the X-ray imaging device were transferred to a common database, as shown on the top row of the figure. A measurement database (as represented at the center of the middle row, Figure 5) provided the source of the data that could be used by the operator (either a human operator or automatic system) to make decisions regarding further processing. The next possible steps were (#1) to employ an expert to analyze the data visually (analysis module on the left-hand side of the middle row, Figure 5) or (#2) to send them to the crowdsourcing system (right-hand side in the middle row, Figure 5) for distribution and processing (bottom row, Figure 5).

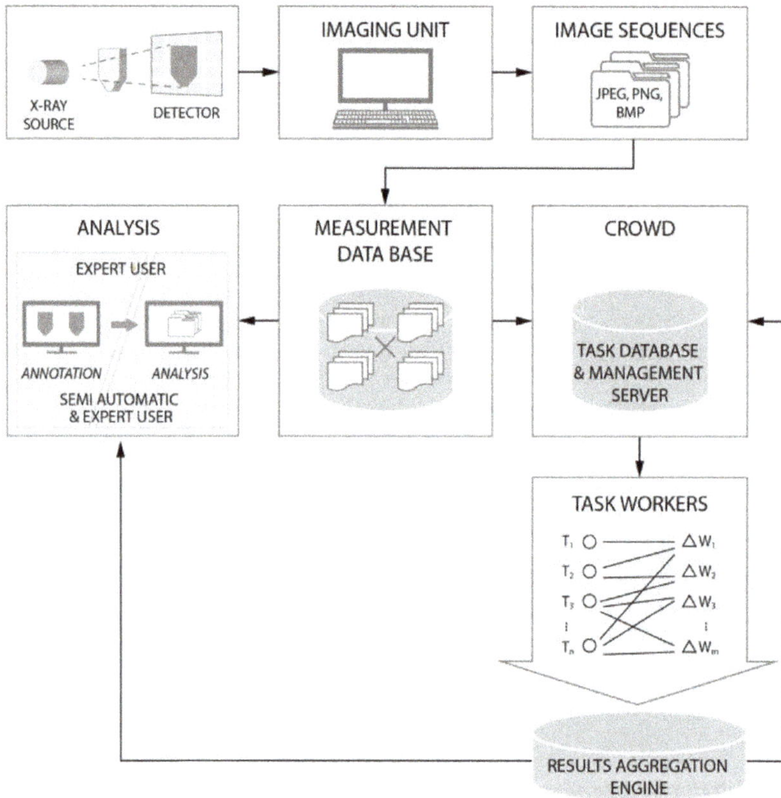



**Figure 5.** Complete workflow diagram of the data processing and analysis system for tracking particles during bulk solid flow monitored by X-ray imaging.

Option (#1) was a typically mundane task. The volumetric parameters of the flow were the required output. Suppose there were 1000+ images in sequence, showing up with 50 tracking particles. At least those in the center (usually 20+ particles) needed to be analyzed in each picture. The analysis consisted of tracing the movement of the particles in consecutive frames. Heterogeneous damping of radiation resulted in non-uniform representation of the tracing particles on adjacent images, which created difficulties for automatic image processing algorithms to detect them efficiently, as described previously in Section 3.1. Due to the fact that this was a highly repetitive procedure, a gradual drop in the quality of the obtained parameters was to be expected [47].

Option (#2) involved uploading selected datasets to a crowdsourcing system for distribution among workers. Different strategies for dividing, managing, and verifying the results are used in commercial crowdsourcing systems. However, a simple approach was used in this work [12]. The task of each worker was to process a sequence of frames (images), always showing a fixed, predefined fragment of the full experimental sequence (for example, workers may process sequences of several tens of frames, up to 100 frames). Since each frame needed to be annotated in such a way as to mark all the trace particles that can be distinguished by a human on a single image, each complete annotation may consist of multiple sub-annotations, carried out by different volunteers on partially- or fully-overlapping areas of the frame. The positions of the markings (the central coordinates of the sphere) were recorded for each particle and each frame. Usually, many workers process the same frame frame sequence fragments. Hence, these fragments may overlap. However, these overlapping fragments did not have to be the same frame fragments, nor did the frame sequence lengths processed by different workers have to be equal, as the start and end points of the sequence fragments may vary (as illustrated in Figure 6). The proposed targeted crowd sourcing approach introduced the additional condition of processing only a confined, limited part of each frame. Yet, this did not interfere with the described procedure, and different areas of focus within the frames may be assigned to different workers. We applied both methods (full-frame and partial processing) in this work, and a comparison of the results is presented later in the text.

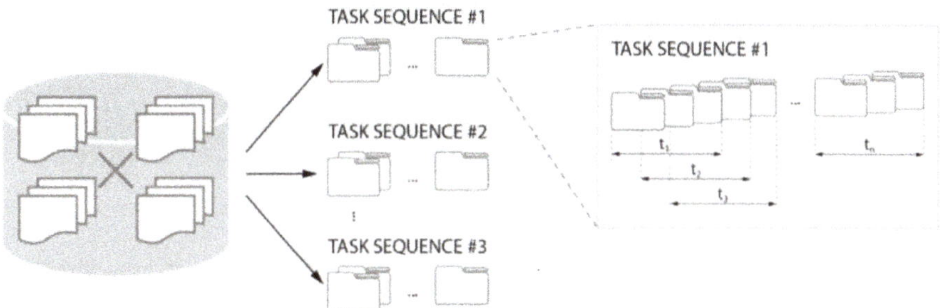


**Figure 6.** Schematic algorithm for task replication and dispatching in the crowdsourcing system.

When the worker finished with a single frame, the marked particles were saved and the markings transferred to the next picture in the sequence. In the next picture, the worker could adjust the positions of the markings indicating the particles, since generally there were only slight shifts between the particles on consecutive images as the bulk flow proceeded. The markings from the previous frames were reported to be helpful, as the trace particles may not be equally exposed on consecutive images [48]. Workers could remove the markings that no longer belong to any trace particle on the current picture in the sequence. Once the fragments were processed, the results (the coordinates of the annotated particles) were aggregated, as illustrated in the bottom row of Figure 5. The average coordinates were calculated, based on superposition of the trace particles marked by different workers. Some of these results (especially for the first frames) were transferred back to the management database until all the scheduled tasks had been completed. In the final step, as illustrated in Figure 5, the output of the crowdsourcing was fed into the interpretation stage. This was where it was analyzed by the domain expert or, ideally, treated as an input for an automatic support system used by the expert. In the latter case, it could be further processed based on the numerical data received from the crowdsourcing system. Otherwise, it was possible for the expert to analyze the images previously annotated by the crowd workers directly.

We used three different approaches (full-frame, zone targeted, and single-particle targeted), as described in Section 3.2. However, none of those different methods affected the overall crowdsourcing data workflow. Only minor changes to the initial training given to the workers before starting the task were necessary in order to instruct them on how to proceed in each particular mode.

*4.2. Crowdsourcing System Output*

The output of the crowdsourcing system was two-fold. Firstly, it exported images with the tracked particles marked (annotated) on each frame. The aggregated trajectories of the particles may also be marked. Secondly, numerical data conveying the exact positions of each particle in each frame were exported from the system. Figure 7 presents the results of the crowdsourced annotation of a sample frame. It is worth noting that whilst each user was generally capable of placing the label on the trace particle, there was some discrepancy between the exact locations of the markers. Hence, these labels needed to be aggregated in order to obtain their final positions. This aggregation can be carried out either using statistical methods or, should the required end result be only visual, through simple morphological operations.

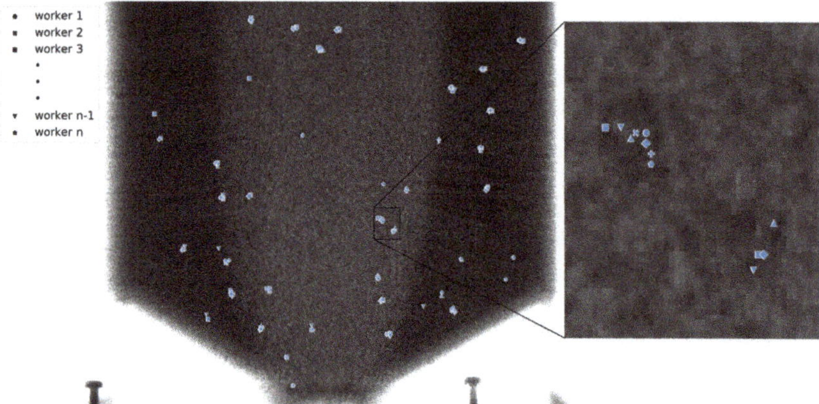

**Figure 7.** Example of the annotated X-ray image with noticeable discrepancies between the labels positioned by different workers. (Left) Full image view. (Right) Zoom-in showing slight shifts between the centers of markers set by different workers.

There are different approaches to the process of annotating the images. The usual way is to mark the region of interest with a circle of a specific, highly-contrasting color, or to place a marker at the approximate center of the region [11]. Some other designs of the user interface utilize a range of colors that convey additional information, such as the marking already processed in the current or previous frames, shifts from previous positions, etc. [48].

## 5. Results and Discussion

### 5.1. Flow Velocity Determination

The first goal of this work was to test crowdsourcing as a valid method for analyzing experimental radiography data, supporting the determination of meaningful process parameters. Therefore, a quantitative scenario for calculating the velocity of each marked particle based on the position extracted from the system output was designed. A pilot study was conducted with two domain experts and up to $n = 22$ distributed participants processing the assigned fragments as crowd workers. The workers were presented with a series of chunks taken from experimental datasets showing silo flows. The lengths of the chunks ranged from 10–140 frames, but typically contained at least 40 frames. For calculations, we used the superimposed positions of between seven and 19 individual trace particles, overlapping on separate frames. In order to compare flow for different conditions (dense, loose) and for different heights, particles located at the center of the silo were chosen. This approach reduced the displacement of particles from the main path of the flowing particle. The following four tables present calculated velocity results for each silo zone for initially dense (Tables 1 and 2), as well as loose (Tables 3 and 4) packing density conditions. Tables 1 and 3 give an overview of the obtained numbers in pixels per frame, while Tables 2 and 4 give the actual calculated velocity in mm per s. Intermediate state results are shown in pixels in order to give the reader a deeper understanding of the consecutive steps in the algorithmic procedure of using crowdsourcing for velocity determination and to demonstrate its consistency.

Table 1, which presents initially dense packing conditions, shows the four components of velocity: Vx: horizontal Velocity (upper left corner), Vy: vertical Velocity (upper right corner), Vt: total Velocity (bottom left corner), Vp: pointer index Velocity (bottom right corner); while Table 2 gives a simplified view of Vx and Vy only, but given in mm/s. Tables 3 and 4 give similar results for initially loose packing conditions, accordingly.

**Table 1.** Velocity components for dense initial packing density presented for nine zones, from A–I. Each quadrant presents Vx (V, velocity), Vy, Vt (t, total), or Vp (p, pointer), clockwise from the upper-left corner. Vx, Vy, and Vt are given in pxpf (pixels per frame).

| 0.8213 | 4.9124 | 0.1602 | 6.8481 | −0.7188 | 4.7917 |
|---|---|---|---|---|---|
| A | | B | | C | |
| 4.9806 | 16.49% | 6.8500 | 2.34% | 4.8453 | −14.83% |
| 0.8131 | 5.0736 | −0.1438 | 8.3339 | −0.6240 | 6.7173 |
| D | | E | | F | |
| 5.1383 | 15.82% | 8.3352 | −1.73% | 6.7462 | −9.25% |
| 1.1250 | 4.7500 | −0.3187 | 12.5440 | −1.09651 | 5.2293 |
| G | | H | | I | |
| 4.8814 | 23.05% | 12.5480 | −2.54% | 5.3430 | −20.52% |

**Table 2.** Velocity components for dense initial packing density presented for nine zones, from A–I. Each quadrant presents horizontal (Vx on the left) and vertical (Vy on the right) components only. Vx and Vy values are given in mm/s.

| 1.3901 | 8.3142 | 0.2711 | 11.5905 | −1.2165 | 8.1099 |
|---|---|---|---|---|---|
| A | | B | | C | |
| 1.3762 | 8.5871 | −0.2434 | 14.1052 | −1.0562 | 11.3690 |
| D | | E | | F | |
| 1.9041 | 8.0394 | −0.5394 | 21.2306 | −1.8558 | 8.8506 |
| G | | H | | I | |

**Table 3.** Velocity components for loose initial packing density presented for seven zones, from A-F and H (Zones G and I show no results for the loose condition since they were outside the funnel). Each quadrant presents Vx, Vy, Vt, or Vp, clockwise from the upper-left corner. Vx, Vy, and Vt are given in pxpf (pixels per frame).

| 0.5263 | 2.3509 | 0.1062 | 6.1258 | −0.4750 | 1.5750 |
|---|---|---|---|---|---|
| A | | B | | C | |
| 2.4091 | 21.85% | 6.1267 | 1.73% | 1.6451 | −28.87% |
| 0.5415 | 3.4333 | 0.2735 | 7.6415 | −0.5631 | 2.2733 |
| D | | E | | F | |
| 3.4758 | 15.58% | 7.6464 | 3.58% | 2.3420 | −24.04% |
| | | 1.6844 | 18.5511 | | |
| G | | H | | I | |
| | | 18.6274 | 9.04% | | |

**Table 4.** Velocity components for loose initial packing density presented for nine zones from A–I (zeros for G and I). Each quadrant presents horizontal (Vx on the left) and vertical (Vy on the right) components only. Vx and Vy values are given in mm/s.

| 0.9808 | 3.9789 | 0.1797 | 10.3679 | −0.8039 | 2.6657 |
|---|---|---|---|---|---|
| A | | B | | C | |
| 0.9165 | 5.8109 | 0.4629 | 12.9332 | −0.9530 | 3.8475 |
| D | | E | | F | |
| 0.0000 | 0.0000 | 2.8509 | 31.3978 | 0.0000 | 0.0000 |
| G | | H | | I | |

As anticipated, the higher overall velocity was observed in the center zones of each row (B, E, H) than in the side zones (A, D, G, C, F, I). The highest velocity was observed at the bottom of the hopper component of the silo: Zone H [34,49]. The comparison of dense and initially loose packing density flow, based on velocity component analysis (Tables 2 and 4), produced a significant difference in Zone H, up to 50% in the overall velocity values. While the central zones for both conditions were similar (Zones E, B), the side zones (Zones D, F, A, C) differed (by more than 65% for Zones C and F), since the character and shape of the funnel varied. The increasing differences between the initially dense and initially loose packing conditions may be explained primarily by the varying size of the flow area (funnel area; see Figure 2). The wider the funnel, especially in the upper part of the silo (Zones A, B, C), the greater the difference (assuming the silo outlet is the same size). These differences were also visible in the values of horizontal component velocity, where in the case of initially loose packing density, the absolute values inside of side zones were generally higher than for initially dense packing density (Zones A, C, D, F). Figure 8 shows a velocity distribution map derived on the basis of the results obtained. Groups of similar velocity vectors were arranged in circles. It can be noted that similar results were obtained for the three main vertical Zones, A, D, and G (left of the funnel), C, F, and I (right of the funnel), and finally, B and E (upper center of the funnel). These results were consistent regardless of the condition, i.e., they were similarly situated on the velocity map for both loose and dense initial packing densities. The only exception was the velocity for H, which was the lowest funnel zone, just above the outlet where the particles gained the highest velocity. Therefore, given that the results for $H_d$ (dense) and $H_l$ (loose) were still within a moderate range of values conforming to theoretical expectations, the efficacy of the method was proven.

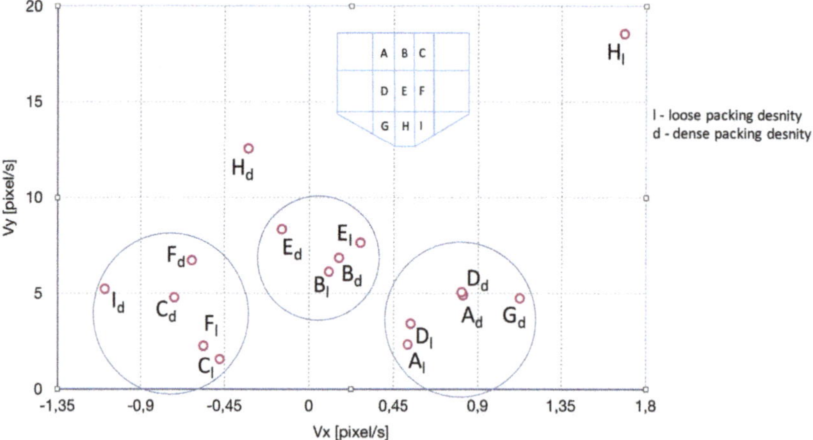


**Figure 8.** Trace particle-based silo flow velocity distribution map. Small circles show velocity values for distinct silo zones (from A–H), as indicated on a schema drawn in the upper part of the picture. Large circles show groups of similar results. The lower index d indicates the initially dense packing condition, while index l indicates the initially loose packing condition.

The results presented here were for the single-particle method, but the accuracy and precision calculated for all three methods were satisfactory and did not differ significantly. First of all, we considered accuracy compared to the ground truth baseline prepared by two experts. All the methods achieved comparable results, with variations of no more than 2%. Precision taken as the repeatability of results also reached 98%. Next, we examined the duration, i.e., the length of time required to complete the task by workers using different methods.

Table 5 shows aggregated times taken by workers to complete the tasks. The columns show the results for classical, zone, and single-particle methods, respectively. The rows show average results for processing a single frame, 10 consecutive frames, or 100 frames (from top to bottom, respectively). SD indicates the Standard Deviation. The single-particle targeted method was the fastest, as was to some extent expected. However, it should be noted that it consumed approximately 10-times less time than the classical method and almost four-times less time than the zone particle method. These factors were greater than anticipated, since there were no cases in which there were 10 trace particles; the average maximum number oscillated around four or five for most of the populated zones. The zone particle method was more than twice as fast as the classical method in all cases. Given the ability of crowdsourcing to parallelize jobs, it may be possible to speed up the entire process significantly. It is also worth noting that the SD was significant, since some frames or frame sequences were much more difficult to process (or simply required more time to process).

**Table 5.** Comparison of annotation time for different crowd work strategies. Time values are averaged and rounded to two decimal places and given in s.

| | Classical | | Zone Targeted | | Single-Particle Targeted | |
|---|---|---|---|---|---|---|
| | Time (avg) | SD | Time | SD | Time | SD |
| 1 frame | 58.47 | 36.26 | 18.57 | 15.13 | 5.94 | 4.20 |
| 10 frames | 584.69 | 262.51 | 185.70 | 121.65 | 59.41 | 12.30 |
| 100 frames | 5800.53 | 1952.39 | 1845.58 | 978.37 | 591.57 | 68.43 |

Table 6 shows average processing times for different zones. The most important zones were the central funnel flow zones, i.e., B, E, and H. However, no significant differences were visible, since both the processing time and SD remained close to the average values.

**Table 6.** Comparison of annotation timed for different zones. Average times rounded to two decimal places and given in s.

| Zone | Time | SD |
|---|---|---|
| A | 9.60 | 2.70 |
| B | 13.33 | 3.67 |
| C | 47.07 | 18.24 |
| D | 9.11 | 3.62 |
| E | 19.04 | 6.38 |
| F | 8.56 | 5.18 |
| G | 12.80 | 4.62 |
| H | 15.66 | 4.92 |
| I | 31.98 | 15.78 |

By quantitative analysis of radiographic images with the aid of the crowdsourcing system, it is possible to obtain a profile of the granular material velocity during the silo discharging process. The results provided additional knowledge about granular flows, making detailed comparative analysis of flow dynamics possible. Such analysis can be conducted on the basis of the calculated velocity profile derived from X-ray imaging data. The results obtained in our study are in agreement with previously-reported data [34,49].

The proposed crowdsourcing system enabled the distribution of the imaging data (image sequences) for different flow fragments (see Figure 6 for the task allocation algorithm). The results showed better quality particle detection for frames pre-marked based on previous images in the sequence. More details on the development of the crowdsourcing method and system were given in [11,48]. In contrast, workers reported decreasing efficiency due to rising fatigue related to physical and cognitive workload demands over time when they worked with longer fragments of image sequences. Therefore, in the future, it would be interesting to investigate whether it would be beneficial to work with the system at random times chosen by the workers, of limited durations, possibly adjusted to the specific needs of the workers. Further development of the system itself, as well as of the crowdsourcing methodology for tomographic imaging analysis will be continued in the next stages of this research.

## 5.2. Qualitative Assessment: NASA TLX

In order to assess the workload of the participants, we performed NASA Task Load Index (TLX) tests. The participants completed a self-assessment rubric, in which they evaluated six main factors related to the given tasks, namely mental, physical, and temporal demand, how they perceived their performance in terms of quality and effort, and finally, the level of frustration induced by the task. These factors approximated to some extent the measurement of task complexity, the user experience, and the usability of the proposed approach, all in relation to the background of individual workers.

Diagrams of the TLX results are presented in Figure 9, separately for the two conditions for the processed X-ray measurement results, i.e., dense silo filling (on the left-hand side) and loose silo filling (on the right-hand side). The colored bars for each TLX category represent results obtained using the three methods. The blue bar (always on the left in each group) shows the performance of the baseline crowdsourcing method, i.e., results achieved by crowd workers annotating all the trace particles in each frame. The orange bar (always in the middle) illustrates the performance of the zone tracking method, i.e., results achieved by crowd workers annotating trace particles bounded by a single area, as depicted in Figure 4. The green bar (always on the right) illustrates the performance of the single-particle tracking method, i.e., results achieved by crowd workers annotating only a single particle of their choice, taken from a particular, indicated area.

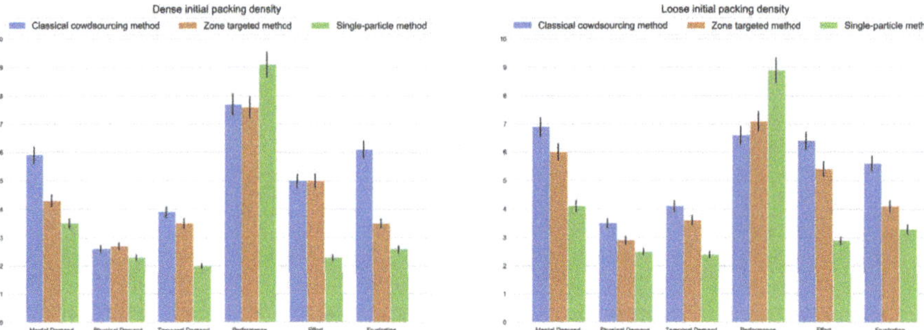


**Figure 9.** NASA Task Load Index (TLX) plot for the classical method (blue bar, left-hand side in each triple post), zone-targeted crowdsourcing method (orange bar, middle position in each triple post) and the single-particle focused crowdsourcing method (green bar, right-hand side in each triple post) for two initial silo filling conditions, i.e., dense (left) and loose (right) initial packing density.

The NASA TLX index tests showed a significant decrease in mental and temporal demand, as well as a drop in job frustration for both conditions (dense and loose silo filling), for both proposed methods compared to the classical crowdsourcing method. The results were better (a larger drop) for the single-particle tracking method than for the zone-tracking targeted method. However, the decrease in physical demand was slightly more prominent for the single-particle method and was significant for both methods only in the case of the loose filling condition. A different effect can be observed in the category of effort. Effort was reported to decrease significantly, mainly for the single-particle method (a drop of more than 50%) compared to the classical baseline method. Performance went up significantly, by more than 10%, but only for the single-particle method. Interestingly, performance increased by more than 10% only for the single-particle method. Performance was not perceived to be significantly different for the zone-tracking method, no matter the condition, yet it was perceived to be worse than the classical method in the case of dense filling.

## 5.3. Discussion Summary

The methodology presented here provides a practical way to analyze reconstructed images, when automatic methods for feature extraction using classical computer vision algorithms are not efficient. The proposed method based on crowdsourcing was verified using real measurement data, and the results were in agreement with those obtained by other research methods. Specifically, the velocity vectors calculated using the crowdsource-annotated data were verified. The positions of trace particles were annotated in sequences of radiography data over time and were used to determine velocity vectors in three different parts of a silo container during bulk solid flow. In this paper, we showed a method for how to determine the flow velocity based on horizontal and vertical velocity components of trace particles.

In our comparison of three approaches to crowdsourcing the processing of image sequences, the proposed zone-targeted and single-particle targeted methods performed well, giving task completion time benefits. Further investigation is needed to prove the usefulness of the methodology for other applications.

## 6. Future Work

Our results showed definite potential for further applications of the targeted crowdsourcing methods. Methods that decrease overall workload are needed to reduce common limitations of crowdsourcing, such as sloppiness and the low percentage of quality results [13,14]. More research is required on how to apply these targeted methods to different tomography sensing problems, which are difficult to process and interpret automatically. Future work could also attempt to couple these methods with novel Virtual Reality (VR), Augmented Reality (AR), and Mixed Reality (MR) visualization technologies [50], in order to design novel interfaces for future-of-work and Industry 4.0 mash-ups of human operators working in AI-driven automated industrial process installations [51,52]. It is worth emphasizing that the output of the crowdsourcing methods is a labeled dataset that could later be used as input data for the further training of machine learning algorithms. Given a large enough dataset, it is anticipated that the process could be automated, at least to some extent, for classes of similar images.

## 7. Conclusions

In this work, we presented a method for extracting process parameters using crowdsourcing. The proposed data processing workflow was applied to study gravitational silo flow, measured by X-ray radiography, and proved to be a reliable and useful way to process tomography data. Crowdsourcing proved to be efficient for pre-processing raw images captured by an industrial standard 2D X-ray radiography sensor. The aggregated average output from the crowdsourcing system can be taken as the input for further, automatic or semi-automatic calculations, as demonstrated here in the case of calculations of the axial velocity of trace particles moving along a silo during unloading. An additional benefit of the proposed targeted crowdsourcing method is that it minimizes the cognitive workload, enabling similar research tasks to be completed more efficiently.

**Author Contributions:** Conceptualization, A.R. and K.G.; Data curation, P.Ł. and K.G.; Formal analysis, A.R. and K.G.; Investigation, A.R. and K.G.; Methodology, A.R. and K.G.; Software, P.Ł., K.G.; Validation, A.R. and K.G.; Visualization, A.R.; Writing—original draft, A.R.; Writing—review & editing, A.R., K.G. and P.Ł.

**Funding:** This work was financed by the Lodz University of Technology, Faculty of Electrical, Electronic, Computer and Control Engineering, Institute of Applied Computer Science as a part of statutory activity within TomoKIS group and Ubicomp.pl research initiative.

**Acknowledgments:** Authors would like to thank Dr. John Speller for his valuable help with English proofreading.

**Conflicts of Interest:** The authors declare no conflict of interest.

## References


1.  Irshad, H.; Montaser-Kouhsari, L.; Waltz, G.; Bucur, O.; Nowak, J.; Dong, F.; Knoblauch, N.W.; Beck, A.H. Crowdsourcing image annotation for nucleus detection and segmentation in computational pathology: Evaluating experts, automated methods, and the crowd. In *Pacific Symposium on Biocomputing Co-Chairs*; World Scientific: Singapore, 2014; pp. 294–305.
2.  Ghadiyaram, D.; Bovik, A.C. Crowdsourced study of subjective image quality. In Proceedings of the 2014 48th Asilomar Conference on Signals, Systems and Computers, Pacific Grove, CA, USA, 2–5 November 2014; pp. 84–88.
3.  Cabezas, F.; Carlier, A.; Charvillat, V.; Salvador, A.; Giro-i Nieto, X. Quality control in crowdsourced object segmentation. In Proceedings of the 2015 IEEE International Conference on Image Processing (ICIP), Quebec City, QC, Canada, 27–30 September 2015; pp. 4243–4247.
4.  Bernstein, M.S.; Little, G.; Miller, R.C.; Hartmann, B.; Ackerman, M.S.; Karger, D.R.; Crowell, D.; Panovich, K. Soylent: A Word Processor with a Crowd Inside. *Commun. ACM* **2015**, *58*, 85–94. [CrossRef]
5.  Brabham, D.C. Crowdsourcing as a model for problem solving: An introduction and cases. *Convergence* **2008**, *14*, 75–90. [CrossRef]
6.  Cheng, J.; Teevan, J.; Iqbal, S.T.; Bernstein, M.S. Break It Down: A Comparison of Macro- and Microtasks. In *Proceedings of the 33rd Annual ACM Conference on Human Factors in Computing Systems*; ACM: New York, NY, USA, 2015; pp. 4061–4064. [CrossRef]
7.  Luz, N.; Silva, N.; Novais, P. A survey of task-oriented crowdsourcing. *Artif. Intell. Rev.* **2015**, *44*, 187–213. [CrossRef]
8.  Kittur, A.; Smus, B.; Khamkar, S.; Kraut, R.E. CrowdForge: Crowdsourcing Complex Work. In *Proceedings of the 24th Annual ACM Symposium on User Interface Software and Technology*; ACM: New York, NY, USA, 2011; pp. 43–52. [CrossRef]
9.  Zheng, H.; Li, D.; Hou, W. Task Design, Motivation, and Participation in Crowdsourcing Contests. *Int. J. Electron. Commer.* **2011**, *15*, 57–88. [CrossRef]
10. Grudzień, K.; Gonzalez, M.H.D.L.T. Detection of tracer particles in tomography images for analysis of gravitational flow in silo. *Image Process. Commun.* **2013**, *18*, 11–22. [CrossRef]
11. Chen, C.; Woźniak, P.W.; Romanowski, A.; Obaid, M.; Jaworski, T.; Kucharski, J.; Grudzień, K.; Zhao, S.; Fjeld, M. Using crowdsourcing for scientific analysis of industrial tomographic images. *ACM Trans. Intell. Syst. Technol. (TIST)* **2016**, *7*, 52. [CrossRef]
12. Law, E.; Gajos, K.Z.; Wiggins, A.; Gray, M.L.; Williams, A. Crowdsourcing As a Tool for Research: Implications of Uncertainty. In *Proceedings of the 2017 ACM Conference on Computer Supported Cooperative Work and Social Computing*; ACM: New York, NY, USA, 2017; pp. 1544–1561. [CrossRef]
13. Kazai, G.; Kamps, J.; Milic-Frayling, N. Worker Types and Personality Traits in Crowdsourcing Relevance Labels. In *Proceedings of the 20th ACM International Conference on Information and Knowledge Management*; ACM: New York, NY, USA, 2011; pp. 1941–1944. [CrossRef]
14. Lyu, L.; Kantardzic, M.; Sethi, T.S. Sloppiness mitigation in crowdsourcing: Detecting and correcting bias for crowd scoring tasks. *Int. J. Data Sci. Anal.* **2019**, *7*, 179–199. [CrossRef]
15. Beck, M.; Williams, R. Process tomography: A European innovation and its applications. *Meas. Sci. Technol.* **1996**, *7*, 215. [CrossRef]
16. Sideman, S.; Hijikata, K. *Imaging in Transport Processes*; Begell House: Danbury, CT, USA, 1993; Volume 621.
17. Romanowski, A.; Grudzien, K.; Chaniecki, Z.; Woźniak, P. Contextual processing of ECT measurement information towards detection of process emergency states. In Proceedings of the 2013 13th International Conference on Hybrid Intelligent Systems (HIS), Gammarth, Tunisia, 4–6 December 2013; pp. 291–297.
18. Banasiak, R.; Wajman, R.; Jaworski, T.; Fiderek, P.; Fidos, H.; Nowakowski, J.; Sankowski, D. Study on two-phase flow regime visualization and identification using 3D electrical capacitance tomography and fuzzy-logic classification. *Int. J. Multiph. Flow* **2014**, *58*, 1–14. [CrossRef]
19. Wajman, R.; Fiderek, P.; Fidos, H.; Jaworski, T.; Nowakowski, J.; Sankowski, D.; Banasiak, R. Metrological evaluation of a 3D electrical capacitance tomography measurement system for two-phase flow fraction determination. *Meas. Sci. Technol.* **2013**, *24*, 065302. [CrossRef]
20. Rymarczyk, T. New methods to determine moisture areas by electrical impedance tomography. *Int. J. Appl. Electromagn. Mech.* **2016**, *52*, 79–87. [CrossRef]

21. Rymarczyk, T. Using electrical impedance tomography to monitoring flood banks. *Int. J. Appl. Electromagn. Mech.* **2014**, *45*, 489–494. [CrossRef]
22. Rymarczyk, T.; Adamkiewicz, P.; Duda, K.; Szumowski, J.; Sikora, J. New electrical tomographic method to determine dampness in historical buildings. *Arch. Electr. Eng.* **2016**, *65*, 273–283. [CrossRef]
23. Tjugum, S.; Hjertaker, B.; Johansen, G. Multiphase flow regime identification by multibeam gamma-ray densitometry. *Meas. Sci. Technol.* **2002**, *13*, 1319. [CrossRef]
24. Hampel, U.; Bieberle, A.; Hoppe, D.; Kronenberg, J.; Schleicher, E.; Suhnel, T.; Zimmermann, F.; Zippe, C. High resolution gamma ray tomography scanner for flow measurement and non-destructive testing applications. *Rev. Sci. Instrum.* **2007**, *78*, 103704. [CrossRef]
25. Fukushima, E. Nuclear magnetic resonance as a tool to study flow. *Annu. Rev. Fluid Mech.* **1999**, *31*, 95–123. [CrossRef]
26. Bieberle, M.; Fischer, F.; Schleicher, E.; Koch, D.; Menz, H.J.; Mayer, H.G.; Hampel, U. Experimental two-phase flow measurement using ultra fast limited-angle-type electron beam X-ray computed tomography. *Exp. Fluids* **2009**, *47*, 369. [CrossRef]
27. Figueiredo, M.; Goncalves, J.; Nakashima, A.; Fileti, A.; Carvalho, R. The use of an ultrasonic technique and neural networks for identification of the flow pattern and measurement of the gas volume fraction in multiphase flows. *Exp. Therm. Fluid Sci.* **2016**, *70*, 29–50. [CrossRef]
28. Fischer, F.; Hampel, U. Ultra fast electron beam X-ray computed tomography for two-phase flow measurement. *Nucl. Eng. Des.* **2010**, *240*, 2254–2259. [CrossRef]
29. Guillard, F.; Marks, B.; Einav, I. Dynamic X-ray radiography reveals particle size and shape orientation fields during granular flow. *Sci. Rep.* **2017**, *7*, 8155. [CrossRef]
30. Schulze, D. *Powders and Bulk Solids. Behavior, Characterization, Storage and Flow*; Springer: Berlin/Heidelberg, Germany, 2008; pp. 35–74.
31. Michalowski, R. Flow of granular material through a plane hopper. *Powder Technol.* **1984**, *39*, 29–40. [CrossRef]
32. Tejchman, J. *Confined Granular Flow in Silos: Experimental and Numerical Investigations*; Springer Science & Business Media: Berlin/Heidelberg, Germany, 2013.
33. Wójcik, M.; Tejchman, J. Modeling of shear localization during confined granular flow in silos within non-local hypoplasticity. *Powder Technol.* **2009**, *192*, 298–310. [CrossRef]
34. Choi, J.; Kudrolli, A.; Bazant, M.Z. Velocity profile of granular flows inside silos and hoppers. *J. Phys. Condens. Matter* **2005**, *17*, S2533. [CrossRef]
35. Tüzün, U.; Nedderman, R. An investigation of the flow boundary during steady-state discharge from a funnel-flow bunker. *Powder Technol.* **1982**, *31*, 27–43. [CrossRef]
36. Fu, S.; Biwolé, P.H.; Mathis, C. A Comparative Study of Particle Image Velocimetry (PIV) and Particle Tracking Velocimetry (PTV) for Airflow Measurement. *Int. J. Mech. Mechatron. Eng.* **2015**, *9*, 40–45.
37. Tai, C.; Hsiau, S. Dynamic behaviors of powders in a vibrating bed. *Powder Technol.* **2004**, *139*, 221–232. [CrossRef]
38. Khakhar, D.V. Rheology and Mixing of Granular Materials. *Macromol. Mater. Eng.* **2011**, *296*, 278–289. [CrossRef]
39. Pudasaini, S.P.; Hsiau, S.S.; Wang, Y.; Hutter, K. Velocity measurements in dry granular avalanches using particle image velocimetry technique and comparison with theoretical predictions. *Phys. Fluids* **2005**, *17*, 093301. [CrossRef]
40. Slominski, C.; Niedostatkiewicz, M.; Tejchman, J. Application of particle image velocimetry (PIV) for deformation measurement during granular silo flow. *Powder Technol.* **2007**, *173*, 1–18. [CrossRef]
41. Chen, X.; Zhong, W.; Heindel, T.J. Using stereo XPTV to determine cylindrical particle distribution and velocity in a binary fluidized bed. *AIChE J.* **2019**, *65*, 520–535. [CrossRef]
42. Chung, Y.; Hsiau, S.; Liao, H.; Ooi, J. An improved PTV technique to evaluate the velocity field of non-spherical particles. *Powder Technol.* **2010**, *202*, 151–161. [CrossRef]
43. Melling, A. Tracer particles and seeding for particle image velocimetry. *Meas. Sci. Technol.* **1997**, *8*, 1406–1416. [CrossRef]
44. Babout, L.; Grudzien, K.; Maire, E.; Withers, P.J. Influence of wall roughness and packing density on stagnant zone formation during funnel flow discharge from a silo: An X-ray imaging study. *Chem. Eng. Sci.* **2013**, *97*, 210–224. [CrossRef]

45. Romanowski, A. Big Data-Driven Contextual Processing Methods for Electrical Capacitance Tomography. *IEEE Trans. Ind. Inform.* **2019**, *15*, 1609–1618. [CrossRef]

46. Rymarczyk, T.; Sikora, J. Applying industrial tomography to control and optimization flow systems. *Open Phys.* **2018**, *16*, 332–345. [CrossRef]

47. Floru, R.; Cail, F.; Elias, R. Psychophysiological changes during a VDU repetitive task. *Ergonomics* **1985**, *28*, 1455–1468. [CrossRef]

48. Jelliti, I.; Romanowski, A.; Grudzień, K. Design of Crowdsourcing System for Analysis of Gravitational Flow using X-ray Visualization. In Proceedings of the 2016 Federated Conference on Computer Science and Information Systems, Gdansk, Poland, 11–14 September 2016.

49. Staron, L.; Lagrée, P.Y.; Popinet, S. Continuum simulation of the discharge of the granular silo. *Eur. Phys. J. E* **2014**, *37*, 5. [CrossRef]

50. Knierim, P.; Kiss, F.; Schmidt, A. Look Inside: Understanding Thermal Flux Through Augmented Reality. In Proceedings of the 2018 IEEE International Symposium on Mixed and Augmented Reality Adjunct (ISMAR-Adjunct), Munich, Germany, 16–20 October 2018; pp. 170–171. [CrossRef]

51. Fraga-Lamas, P.; Fernández-Caramés, T.M.; Blanco-Novoa, Ó.; Vilar-Montesinos, M.A. A Review on Industrial Augmented Reality Systems for the Industry 4.0 Shipyard. *IEEE Access* **2018**, *6*, 13358–13375. [CrossRef]

52. Eschen, H.; Kötter, T.; Rodeck, R.; Harnisch, M.; Schüppstuhl, T. Augmented and Virtual Reality for Inspection and Maintenance Processes in the Aviation Industry. *Procedia Manuf.* **2018**, *19*, 156–163. [CrossRef]

*Article*

# Smart Water Meter Using Electrical Resistance Tomography [†]

**Chenning Wu [1,2,*], Martin Hutton [2] and Manuchehr Soleimani [1]**

[1]  Engineering Tomography Laboratory (ETL), Department of Electronic and Electrical Engineering, University of Bath, Bath BA2 7AY, UK
[2]  Ashridge Engineering Ltd, Okehampton EX20 1BQ, UK
*  Correspondence: cnw24@bath.ac.uk
†  This paper is extended from the paper entitled 'Low cost design of Electrical Impedance Tomography hardware system for industrial application', published in proceedings of the 9th World Congress on Industrial Process Tomography, Bath, UK, 2–6 September 2018.

Received: 20 May 2019; Accepted: 8 July 2019; Published: 10 July 2019

**Abstract:** Smart flow monitoring is critical for sewer system management. Obstructions and restrictions to flow in discharge pipes are common and costly. We propose the use of electrical resistance tomography modality for the task of smart wastewater metering. This paper presents the electronics hardware design and bespoke signal processing to create an embedded sensor for measuring flow rates and flow properties, such as constituent materials in sewage or grey water discharge pipes of diameters larger than 250 mm. The dedicated analogue signal conditioning module, zero-cross switching scheme, and real-time operating system enable the system to perform low-cost serial measurements while still providing the capability of real-time capturing. The system performance was evaluated via both stationary and dynamic experiments. A data acquisition speed of 14 frames per second (fps) was achieved with an overall signal to noise ratio of at least 59.54 dB. The smallest sample size reported was 0.04% of the domain size in stationary tests, illustrating good resolution. Movements have been successfully captured in dynamic tests, with a clear definition being achieved of objects in each reconstructed image, as well as a fine overall visualization of movement.

**Keywords:** electrical resistance tomography; smart water meter; wastewater management

---

## 1. Introduction

Process tomography provides measurements of concentration distribution and flow profile within a process instrument by collecting data via remote sensors. Electrical tomography is one of the most extensive tomographic modalities that can provide cross-sectional profiles of the distribution of materials or velocities in a process vessel or supply information about transient phenomena in the process. The advantages of such technology, such as high temporal resolution, low cost, no radiation hazard and non-intrusive/non-invasive, have made electrical tomography a promising technology in monitoring and analyzing various industrial flows [1]. Electrical tomography can be further divided into Electrical Capacitance Tomography (ECT) and Electrical Impedance Tomography (EIT), based upon the dielectric properties of the continuous phase in the domain. EIT measurements provide information on the real and imaginary parts of the impedance, as well as the phase angle. However, in the process industries, applications commonly rely on the conductivity difference between two phases to obtain the concentration profiles. Therefore, Electrical Resistance Tomography (ERT) are dominantly employed in imaging conductive flow measurements [2].

Urban wastewater is defined as the mixture of domestic wastewater from kitchens, bathrooms and toilets; other sources of wastewater are industries discharging into sewers and rainwater run-off from roads and other impermeable surfaces such as roofs, pavements and roads draining to sewers [3]. In the UK, the underground sewer system, with a total length of 347,000 km, carries over 11 billion liters of wastewater every day [4]. However, the operational conditions of sewer systems can change over time due to blockages caused by sediment and fats in the untreated wastewater, and structural changes associated with ageing; hence, it is vital to gather sufficient information on the current condition of a sewer system to proactively prevent the service failures. Water companies are now looking for new ways of reducing blockages, and thus the chances of flooding, through means such as modelling to identify hot-spots, monitoring for progressive operational deterioration, and intervening proactively [5]. It has been concluded through statistical analysis that the majority of blockages occur in pipes of diameter 225 mm or less and mostly upstream of a junction [6]. Moreover, the flow velocities can affect the risk of blockages, vice versa. Therefore, flow monitoring at the locations with high likelihood of blockages is essential for continuously inspecting the sewer and identifying the intermittent sewer blockages. Currently, conventional non-tomographic technologies that are only capable of providing flowrate measurements, such as ultrasonic Doppler velocity profilers, electromagnetic meters, Coriolis mass meters, and Venturi meters, are commercially available. However, the wastewater flow behaviour comprises more than one phase due to the complexity of its compositions, which cannot be differentiated by traditional flow meters. Process tomographic scanners, as a result, can be good alternatives by providing real-time cross-sectional images on the phase distribution.

The Environment Agency Monitoring Certification Scheme (MCERTS) regulates the minimum rated operating conditions for fluid conductivity shall be 50 $\mu$S cm$^{-1}$ to 1200 $\mu$S cm$^{-1}$ [7], which makes ERT a good candidate for playing an important role in monitoring wastewater flow. One of the major concerns in sewers is blockages caused by built-up grease, sanitary products, food scraps, etc. ERT can distinguish between different concentrations by their conductivities; further identification of transitions between homogeneous and heterogeneous flow regime can be implemented by direct/indirect interpretation of measurements. Since obstacles in pipeline systems can disturb the dynamic flow behavior [8], blockages can change the flow within the sensing area. Therefore, by properly characterizing wastewater flow regimes, early detection of sewer blockages can be achieved.

ERT has been heavily applied in multiphase flow studies by the means of estimating the phase fraction through the mixture's dielectric properties of fluid [9,10] without distorting the flow field, including identifying the flow regimes [11–13], visualizing multiphase flow [14,15], mass flow rate measurement [16]. Additionally, ERT sensors can estimate the velocity profiles by cross-correlation between two identical sensors [17,18] The operation principle of ERT is injecting current or applying voltage to electrodes mounted on the boundary of the domain and measuring the voltage or current from the rest of the electrodes. Various ERT hardware systems have been developed as research tools for process engineering studies [10,14,15,19–22]; a few electrical tomographic systems have also been commercially developed, for instance, MPFM4R&D ECT system from Tech4Imaging [23], SONARtrac ECT flow meter from CiDRA [24], Rocsole ECT pipe sensor from Rocsole [25] and the FLOW-ITOMETER ECT/ERT sensors from itoms [26].

In this paper, an ERT hardware system will be developed, dedicated to wastewater applications. A top-down view of the proposed design will be presented and explained, and each module included in the design will be described in detail. Lab-scaled stationary and dynamic experiments were conducted to verify the accuracy of the measurements. Image reconstructions, including a quantitative image quality analysis, as well as numerical analysis, namely Signal to Noise Ratio (SNR) and cross-correlation, will be presented to evaluate the system performances.

## 2. System Overview

An overview of a 16-channel ERT hardware system is presented in Figure 1. The system incorporates: (1) A sensor array equally spaced around the domain periphery; (2) An effective data acquisition system (DAQ) device; (3) A personal computer that interacts with the DAQ, exacting and processing information.

The proposed DAQ system is a 16-channel serial ERT device, as shown in Figure 2, operating at 50 kHz using the adjacent driving and measuring protocol. It features the user configurable current injection at 50 kHz over the range of 6 mA to 18 mA, as well as the load-adaptive capability. The DAQ system is designed in a modular manner, and may be broken down into four modules:

- An excitation source;
- An electrode switching module;
- A reception module;
- A central control unit functioned by an STM32 microcontroller

These four modules are prototyped on three separate boards, i.e., excitation and reception board (top layer), switching board (middle layer) and microcontroller board (bottom layer). The advantages of this architecture are easy assembly, high mobility, and compactness.

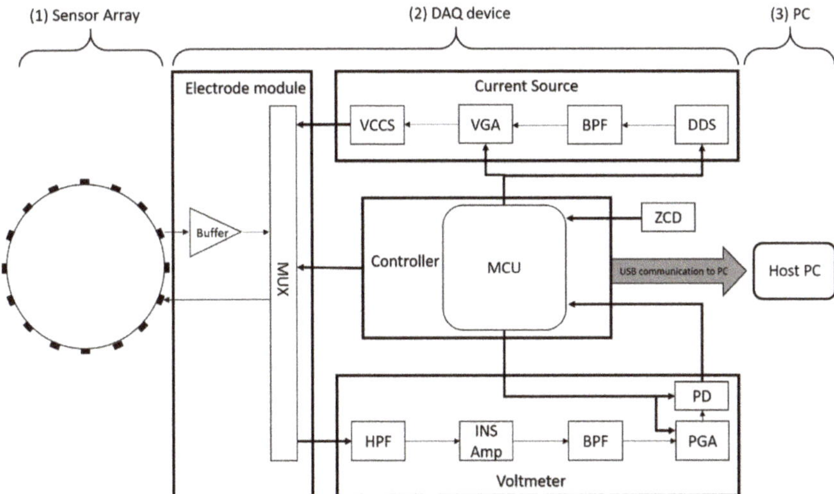


**Figure 1.** System overview.

**Figure 2.** 16 channel ERT device.

## 3. Design and Methods

### 3.1. Excitation Source

Current, as opposed to voltage, is chosen as an excitation signal to the periphery of the conducting medium on the premise that a current source has a high output impedance, so that the effects of electrode contact impedance will be negligible. The current source presented in this paper is composed of a signal generator, a band-pass filter, a variable gain amplifier and a voltage controlled current converter.

Excitation signals in alternative waveforms have great advantages in reducing the effect of electrode polarization and isolating the direct current offset potential. Sinusoidal waveforms further provide access to full electrical impedance demodulation, and hence, the capability of spectroscopic measurements. Low-cost monolithic devices which can generate accurate waveforms via direct digital synthesis have recently become commercially available. Hence, in the reported design, a Direct Digital Synthesizer (DDS) is selected to produce a sinusoidal signal. A DDS can benefit the system with its fine resolutions at low frequencies (up to 200 kHz), inherent stability, and compactness. However, an unfiltered DDS output signal is biased and rich in spurious content introduced by a Digital to Analogue Converter (DAC). Thus, a fast settling Sixth-order Butterworth Bessel band-pass filter with a passband of 20 kHz centred at 50 kHz follows the DDS.

A variable gain amplifier (VGA) is used to allow for a flexible current excitation within the range of 6 mA to 18 mA. The VGA is essentially a non-inverting amplifier incorporating with a programmable digital potentiometer (DigiPot) and fixed value resistors. The value of DigiPot determines the gain of the VGA and is set via user commands at the beginning of collections referring to a look-up table.

One of the difficulties in designing an ERT system for process applications is the wide range of conductivities, varying from several $\mu S\,cm^{-1}$ to a few hundred $mS\,cm^{-1}$. Hence, a current source that is able to provide a constant current supplying over a wide range of load impedances is a necessity. A dual op-amp voltage controlled current source developed by [27], which is capable of preserving high output impedance over a wide range of operating frequencies, is built around commercially available current feedback amplifiers in our system. The output current is a result of input voltage, $V_{in}$ and the sensing resistor $R_{sense}$, following the equation:

$$I_{out} = V_{in} / R_{sense} \tag{1}$$

*3.2. Electrode Switching Module*

In this system, a serial collection protocol is employed; furthermore, it is critical to have a fast switching mechanism to achieve an overall data acquisition speed. Due to the non-idealities of multiplexers, including the non-zero on-resistance, charge injection, various settling time and retained charge, an appropriately designed switching circuit is essential to not only fulfil fast signal transients but also give the least chance of distorting receiving signals.

A multiplexer with the lowest possible on-resistance is desirable to maintain good linearity. However, the low on-resistance also means a large charge injection, which will result in a voltage glitch occurring at the multiplexer input when switching channels. To compensate for this, the source impedance of a multiplexer should be kept as low as possible. Therefore, a buffer amplifier with a high input impedance and an extremely low output impedance should be placed at each multiplexer input to settle a full-scale step.

A total of four 16:1 multiplexers are exploited, of which two are for the selection source and sink ports and the remaining two for selecting differential voltage ports. Four quadruple low-cost precision Junction gate Field-Effect Transistor (JFET) input operational amplifiers are introduced for buffering the received signal from the electrodes.

*3.3. Reception Module*

3.3.1. Signal Conditioning

Common-mode errors have been considered as being among the most significant sources of measurement errors. As described previously, the proposed design uses a single-end current source topology, the impedance between the grounded side of the current source and the amplifiers in the reception module is the main factor that introduces common-mode signal [28]. To eliminate common-mode errors, two techniques are implemented: (1) an instrumentation (INS) amplifier with a high common-mode rejection ratio (CMRR) at the desired frequency (i.e., 50 kHz) is selected to determine the differential voltage between two voltage signals; (2) fast high pass filters at the inputs of the instrumentation amplifier are used to remove the Direct Current (DC) component of the incoming voltage signals without compromising the acquisition speed.

An Eighth-order Butterworth bandpass filter (BPF) follows the INS amplifier to filter out any distortion that is introduced by the INS amplifier. The BPF is centered at 50 kHz with a 40 kHz passband (−3 dB) and 200 kHz stopband (−40 dB), which results in an approx. 100 µs step response time.

The voltage responses to the current injection can vary over a wide dynamic range from a few millivolts to several volts, depending on the size of the vessel as well as the medium/inclusion in the vessel. To accommodate for diverse applications of the system, another VGA of the same topology as the one employed in the excitation source is incorporated after the BPF, so that the measurements can fall within the acceptable range of the Analog to Digital Converter. This is achieved by a calibration process which happens both at the startup of the system operation and along the whole acquisition process, so that the system can adapt to any unpredictable changes.

### 3.3.2. Peak Detection

In an ERT system, only the in-phase component is required for reconstructing the conductivity distribution. An analogue precision peak detection circuit is implemented to obtain the amplitude of the differential voltage, as displayed in Figure 3. The maximum value of the incoming signal is captured by charging up a capacitor, $C_1$. Specifically, when the input signal is rising, $C_1$ is charged to a new peak level; whereas $D_1$ and $D_2$ prevent $C_1$ to be discharged when the signal is falling. It is also vital to have a reset switch, $Q_1$, coupling the capacitor, which is performed by an N-channel enhanced Metal Oxide Semiconductor Field-Effect Transistor, so that the capacitor can be discharged and ready for the next series of signals. The active-high reset signal comes from the microcontroller and will be further discussed in Section 3.4.

The proposed peak detection technique is capable of generating high precision measurements with a high signal to noise ratio. The simplicity, and hence, the cost efficiency of the system, are achieved by replacing the complex phase shift demodulation technique with such peak detection circuit when compared with the conventional ERT device.

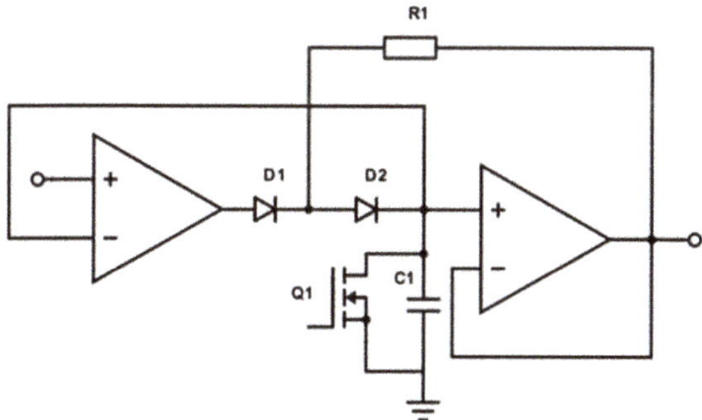

**Figure 3.** Peak detection circuit.

### 3.3.3. Analogue to Digital Converter (ADC)

The general dynamic range of collected signals is from 1:10 to 1:40 [29]; hence, with a minimum 1% accuracy, a 12-bit analogue to digital conversion is necessary. The built-in 12-bit ADC is used to continuously read voltage measurements using Direct Memory Access (DMA) mechanism, which allows for an automatic data transfer from ADC to memory without the usage of the Central Processing Unit (CPU). A total of 35 ADC readings are taken and 5 extrema are eliminated so that each voltage measurement is a result of an average over 30 ADC readings. The averaging technique benefits the system with a better signal to noise ratio by the factor of $\sqrt{N}$ (N is the number of times of averaging).

### 3.4. Central Control Unit

An STM32F4 microcontroller handles the commands to programmable integrated circuits (ICs), data exchange between a host PC and the device, and data processing. The sequence of operation follows the flowchart in Figure 4a below. After the system initialization, based on the choice of injection current level from the user command, a VGA gain on the injection side ($VGA_i$) is set for current. At the same time, an initial gain is also set for the reception side VGA ($VGA_r$), which is normally the smallest value in the look-up table. A calibration process is then performed after a whole frame of data is collected and a corresponding gain for $VGA_r$ is calculated. Following that, a startup operation that allows for 5 frames (each frame contains 208 switches) of switching is carried out before the data transmission happens. This startup operation enables the system to settle; hence, a better quality of data will be obtained in the later collection procedure. The authentic measurements then start on the command received at the end of the startup stage. A re-calibration check takes place each time a whole frame of readings is fed into the ADC. If any readings in this frame have reached 4095, then the calibration process will happen again to set a new $VGA_r$ gain accordingly.

The sequence of one measurement follows that shown in Figure 4b. The timing is essentially governed by a series of pulses of the stimulation frequency, i.e., 50 kHz, which is generated by the zero crossing detection of the input signal. The multiplexer switches at every 17 cycles of pulses and leaves the system operating at 14 fps. The peak detection reset signal occurs after 9 cycles from the switching instance to give signals sufficient settling time. Subsequently, the ADC conversion happens at cycle 11; the reading and averaging of measurements finishes in 4 cycles.

Communication between the host PC and the ERT device is accomplished via Universal Serial Bus (USB) micro-AB at full speed (12 Mbps). On the host terminal, data could be received and stored either by REALTERM terminal emulator software or MATLAB. It's worth noting that only after a whole frame of 208 measurements all taken, this one frame of data is transmitted to the host PC at once, as indicated in Figure 4c.

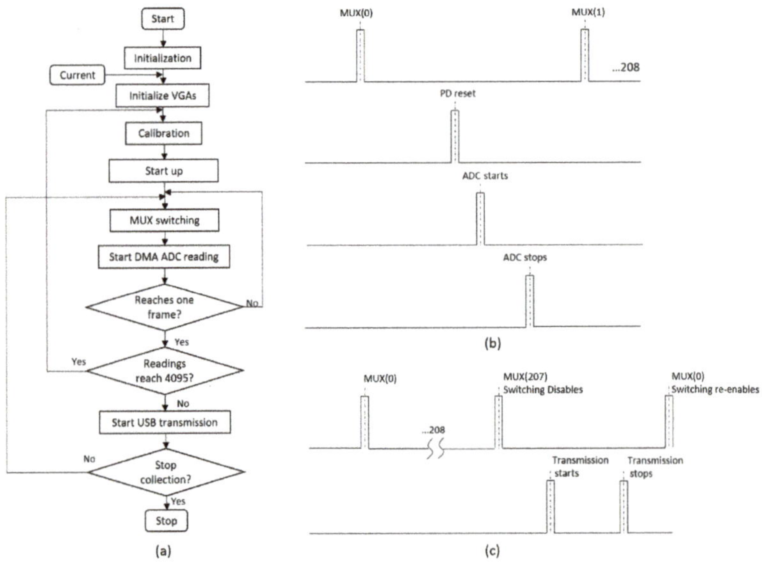


**Figure 4.** (a) Operation procedure of the microcontroller command (b) Timing simulation between two switching instances (c) Timing simulation between two frames of collection.

## 4. Experiments

To evaluate the performance of the proposed system, multiple experiments were carried out in a cylindrical phantom of 25 cm diameter with 16 electrodes attached evenly on the periphery, as indicated in Figure 5. In all tests, saline is used as a background medium, and is prepared with the solution of NaCl and tap water.

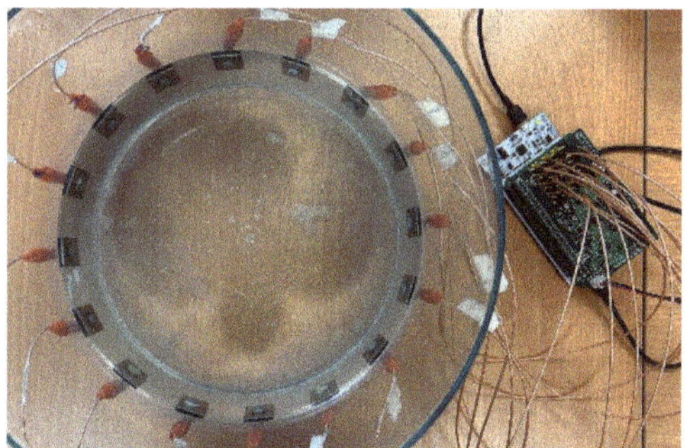

**Figure 5.** The EIT system for experiments.

Image reconstructions aiming at recovering conductivity distributions from boundary measurements were performed by field electrical modelling for forward model and inversion algorithm. The electrical conductive field model is governed by the derivation of Maxwell's equation:

$$\nabla \cdot (\sigma \nabla \mathbf{u}) = 0, \tag{2}$$

where $\sigma$ is the conductivity, $\mathbf{u}$ is the electric potential distribution, and $\nabla$ is the divergence operator. For the appropriate formulation of the system, the complete electrode model (CEM) is used to constrain the boundary conditions described by:

$$\mathbf{U}_k = \mathbf{u} + Z_k \sigma \frac{\partial \mathbf{u}}{\partial \hat{n}} \text{ on } e_k; \ k = 1, 2, \ldots, K, \tag{3}$$

where $\mathbf{u}$ is the potential field, $Z_k$ is the contact impedance of the kth electrode, $\hat{n}$ is the outward unit normal vector and $e_k$ denotes the part of boundary that corresponds to the kth electrode. The finite element method (FEM) is a numerical discretizing method commonly used in EIT, and it discretizes the domain of interest into small elements to turn a continuous problem into a discrete problem, and hence solves the forward model. For stable image reconstruction, a linear inverse problem is solved using the Jacobian Matrix J and a Laplacian regularization function leading to matrix R used with regularization parameter $\gamma^2$ which is empirically selected. An image of change in electrical conductivity $\Delta\sigma$ can be obtained from differential voltage measurements $\Delta \mathbf{u} = \mathbf{u}_i - \mathbf{u}_b$, where $\mathbf{u}_i$ and $\mathbf{u}_b$ are the measurements with and without inclusions respectively, following Equation (4):

$$\Delta\sigma = (\mathbf{J}^T\mathbf{J} + \gamma^2\mathbf{R})^{-1}\mathbf{J}^T\Delta\mathbf{u} \tag{4}$$

### 4.1. Background Test

To help understand the quality of measurements collected from the proposed 16 channel ERT system, a uniform background test is firstly conducted under the lowest possible current injection, 6 mA, and the voltage measurements $\mathbf{u}_b$ are captured in Figure 6a. Signal-to-Noise Ratio (SNR) evaluates the precision of measurements by indicating the ability to produce the same results under the unchanged conditions [30]; it can be calculated by:

$$SNR = 10 \log \frac{[\bar{v}]_i}{SD[v]_i} \tag{5}$$

where $SD[v]_i$ is the standard deviation of measurements and $[\bar{v}]_i$ is the mean value of total measurements. For one projection from the first excitation using electrodes 1 and 2, the SNR of the corresponding 13 measurements is displayed in the bar chart in Figure 6b. As expected, a higher SNR occurs at adjacent electrode measurements that have a larger trans-impedance, while a lower SNR occurs at opposite-electrode measurements that have a smaller trans-impedance. The overall average SNR over 208 measurements is 59.54 dB, demonstrating the good reliability of the device.

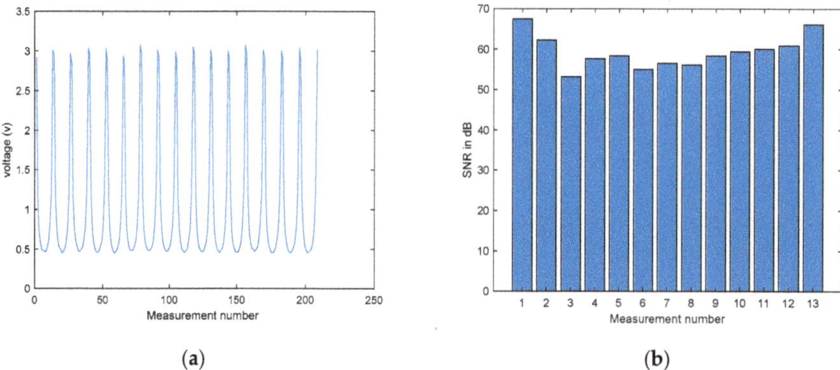

(a)                                               (b)

**Figure 6.** (a) Background dataset plot (b) SNR plot of 13 measurements in background test.

*4.2. Stationary Tests*

### 4.2.1. Single Sample Tests

Four sets of experimental tests using plastic (Teflon) rods in the sizes of 2 cm (small), 3 cm (medium) and 4 cm (large) diameter, and a plastic ballpoint pen in the size of 5 mm (Xsmall) are reported in this section. In the large and medium sample cases, 6 mA current was injected; whereas 16 mA current was used to stimulate the system in the small and Xsmall sample tests. This is because the signal changes induced by small inclusions are not as significant as those by larger inclusions and are more vulnerable to the background noise. Each object was placed at various locations within the tank, and results are displayed and compared with the real distributions in Tables 1–3.

As indicated from Tables 1–4, the reconstructed images are well defined within the view region, and can reflect the location variations when compared with the real distributions. However, notable image distortion can be observed as objects move further away from the boundary of the domain. This is caused by the inherent ill-posed nature of ERT that it has a high sensitivity to the changes occurring near the boundary. This poses a more significant effect when the targets are rather small and the changes in impedance by objects could be more severely contaminated by noises occurring at the boundary. Hence, even with a stronger excitation, images still struggle to preserve the shape and size of the real images in the small sample test (Table 3) and the Xsmall sample test (Table 4). It is worth noting that even the extra small object, whose size is only 0.04% of the area of the tank, is still detectable by the system. This is benefited from the high SNR of the ERT system and gives the system the confidence of capturing small obstacles and hence potential blockages at an early stage in the field applications.

**Table 1.** Large sample tests.

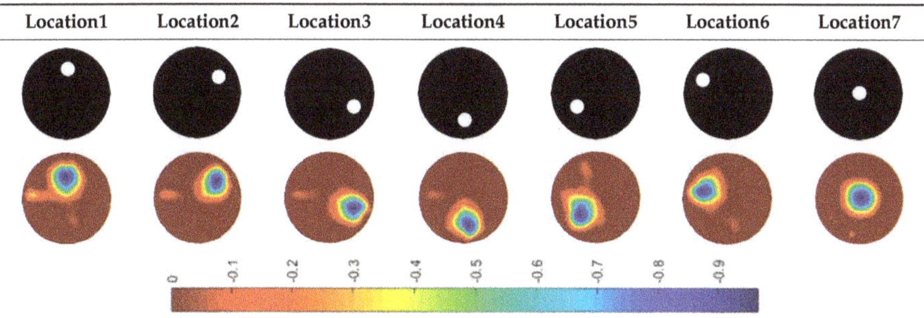

**Table 2.** Medium sample tests.

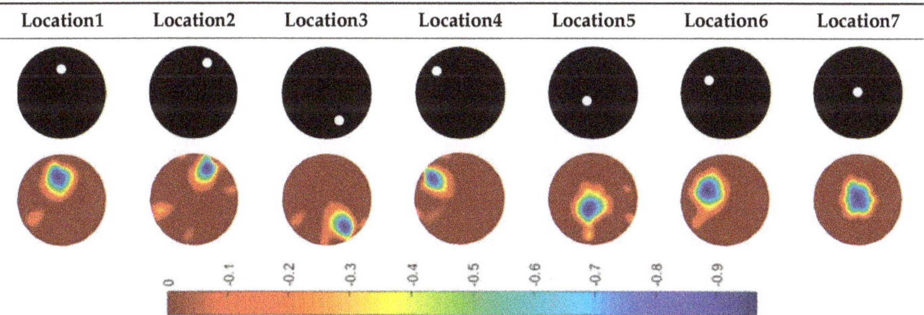

**Table 3.** Small sample tests.

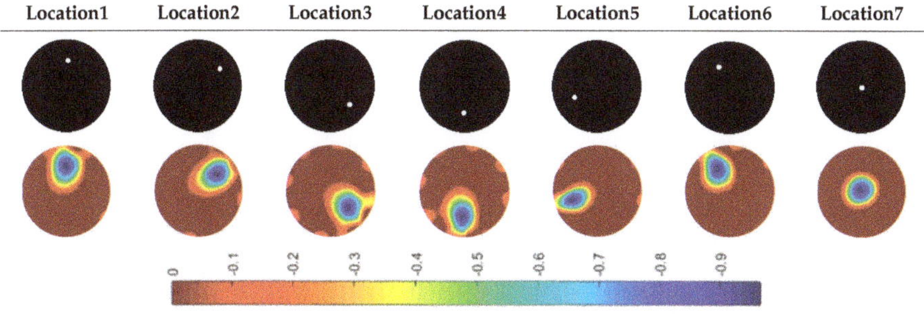

**Table 4.** Extra small (Xsmall) sample tests.

Quantitative image quality analysis is reported to further compare the reconstructed images with the real distribution. Two evaluation parameters are employed here, i.e., Position Error (PE) and Shape Deformation (SD) from [30], and are plotted against locations. Each image is made up of $200 \times 200$ pixels and can be represented by a column vector $\hat{x}$; a threshold of one-fourth of the maximum amplitude is applied for the detections of the most visually significant effects:

$$\left[\hat{x}_q\right]_i = \begin{cases} 1, & \left[\hat{x}_q\right]_i \geq \frac{1}{4}\max(\hat{x}) \\ 0, & \text{otherwise} \end{cases} . \tag{6}$$

1.   Position Error

Position error describes the mismatch between the centre of mass of the real distribution $P_o$ and the reconstructed image $P_q$:

$$\text{PE} = \left| P_o - P_q \right| \tag{7}$$

Hence, PE is preferably as small as possible so that it could provide reliable information on the location of blockages in the smart-metering applications. The PE of the four objects are plotted in Figure 7a. Position errors in all tests are managed to be kept below 1 cm, which indicates good tracking of potential blockages.

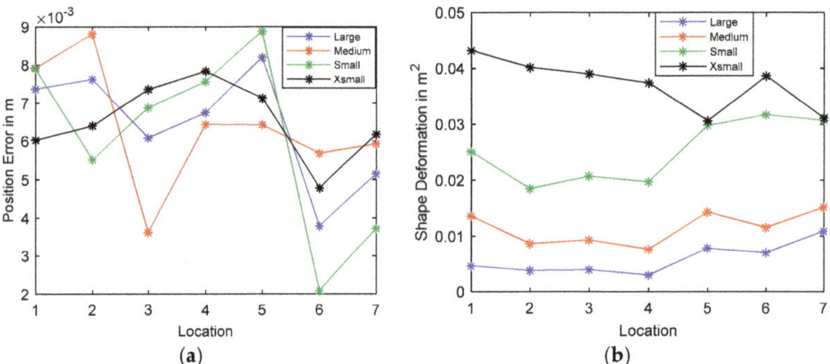


**Figure 7.** Reconstruction accuracy plots (**a**) Position Error; (**b**) Shape Deformation.

2.   Shape Deformation

Shape deformation measures the fraction of the reconstructed one-fourth amplitude that fails to fit within a circle of the area equivalent to the real image:

$$\text{SD} = \sum_{k \notin C} \left[\hat{x}_q\right]_k \Big/ \sum_{k} \left[\hat{x}_q\right]_k \tag{8}$$

where C is a circle centered at $P_q$ with an area equal to $A_q$. SD should also be low and uniform as large SD may result in incorrect interpretation of inclusions. The SD plots of four samples at various locations are presented in Figure 7b. The ability to preserve the shape and size of objects declines as the objects get smaller, as expected. Moreover, the variation of SD becomes larger as the size of objects increases. The SD performance could be further improved by applying an advanced reconstruction algorithm in offline post-process or applying a size orientated threshold technique.

4.2.2. Multiple Sample Tests

Another set of tests with more than two samples in the tank are also performed in Table 5. The results give a good insight of the distinguishability of the system. When the objects are placed equally close to the boundary, all of them can reveal themselves in the recovered images; however, smaller objects tend to have lower amplitude responses. Difficulties arise when any of the objects were placed in the center of the view region, and objects are effectively merged into one in the reconstructed images (i.e., Location3 and Location6 in Table 5).

**Table 5.** Multiple sample tests.

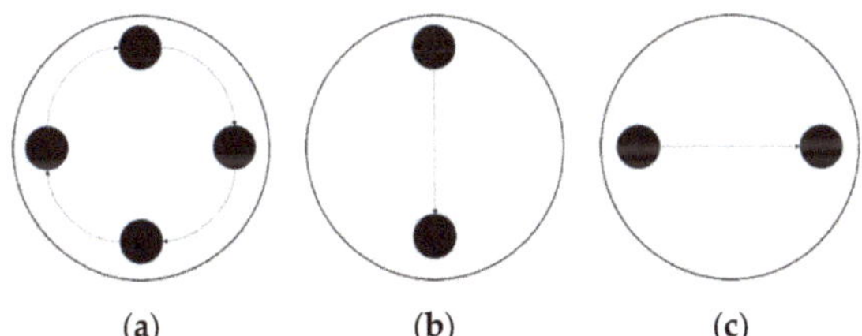

| Location1 | Location2 | Location3 | Location4 | Location5 | Location6 | Location7 |
| --- | --- | --- | --- | --- | --- | --- |

*4.3. Dynamic Tests*

In order to study the feasibility of flow monitoring with the framerate offered by the proposed ERT hardware system, dynamic tests were carried out by continuously moving an object in this section. For the purpose of simplification, 2-dimensional flow simulation, which is accomplished using measurements obtained from one electrode plane, is considered.

Experiments were conducted using the same phantom setup as single stationary tests in 0 employing the large sample (4 cm diameter plastic rod). Two types of movement are studied: the circular movement where the sample moves along the inner wall of the tank clockwise (Figure 8a); the cross movement where the sample moves across the region along the diameter from one side to another (Figure 8b,c).

**Figure 8.** Movement indications of (**a**) circular (**b**) vertical cross (**c**) horizontal cross.

4.3.1. Circular Movement

The circular movement provides a good insight into employing ERT into monitoring dynamics within process equipment such as impeller-based mixers, hydro-cyclones and centrifugal separators [31]. Several reconstructed images along the movement path are presented in Table 6. The sample is well defined, even under the continuous movement, and this illustrates that the data capture speed is sufficiently high compared to the rate of movements.

To demonstrate the sample movement over the given time margin, a 3D volume model is introduced in Figure 9. In Figure 9a, the top view of the sample motion describes the circular movement within the phantom and a helical 3D volume model in Figure 9b provides the time taken for such movement to complete. The 3D modelling helps not only visualize the movement of targets but obtain volume information given the angular speed derived from the system framerate (will be further discussed in Section 5.1).

**Table 6.** Large sample circular movement test reconstructed images.

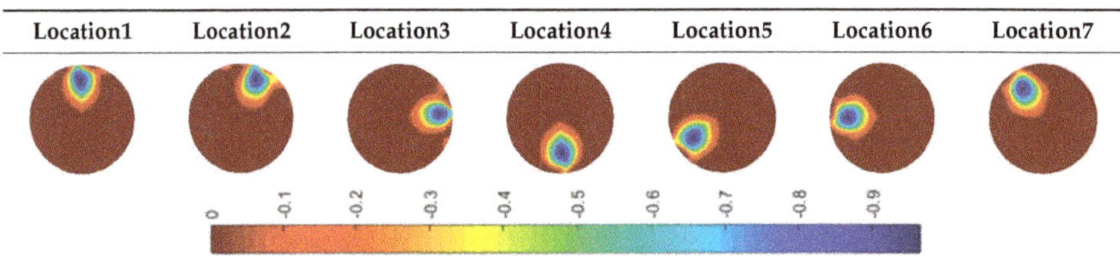

| Location1 | Location2 | Location3 | Location4 | Location5 | Location6 | Location7 |

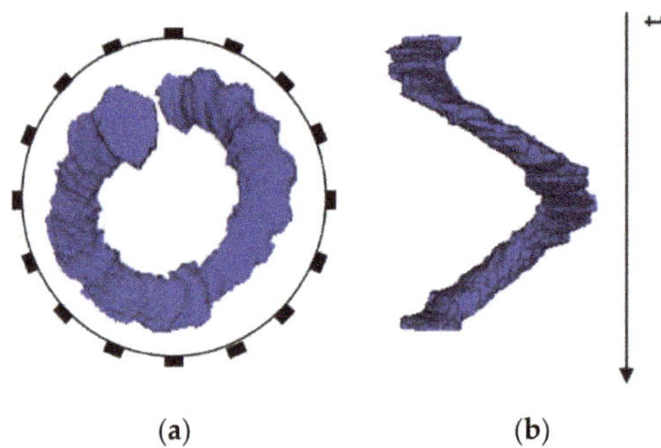

(a)     (b)

**Figure 9.** Circular motion 3D volume model (**a**) top view (**b**) helical 3D view with time axis.

### 4.3.2. Along Diameter Movements

The 2D cross movements can be extended into 3D processes where two electrode planes can be implemented; flow patterns in horizontal slurry transport pipelines [32] can then be investigated. Again, seven critical slices of images of each type of cross movements are displayed in Tables 7 and 8. The results prove that the movements are successfully monitored.

**Table 7.** Large sample vertical cross movement test reconstructed images.

| Location1 | Location2 | Location3 | Location4 | Location5 | Location6 | Location7 |

**Table 8.** Large sample horizontal cross movement test reconstructed images.

| Location1 | Location2 | Location3 | Location4 | Location5 | Location6 | Location7 |

## 5. Cross-Correlation

The resulting ERT images displayed in Section 4.3 provide good insights into the phase distribution as well as transients in two phase flow. Subsequently, a velocity field can be inferred by computing correlations between images of different time instants based on a sequence of reconstructed tomographic images as illustrated in Figure 10a. In a system with one ring of electrodes, the cross-sectional plane of interest is divided into finite elements, which is defined as the pixel correlation method [32], as shown in Figure 10b. Each reconstructed image is composed of these pixels, with each unit having a value indicative of the resistivity of the region it occupies. Then, the profile of individual pixels (i.e., the characteristic vector) can be obtained from the variations of its resistivity measurements over time. By discretely cross-correlating between two targeting pixels' profiles (Equation (8)), the resulting cross-correlation $R_{AB}(\tau)$ peaks at the time representing the delay between signals $V_A$ and $V_B$, i.e., $\tau$ [31].

$$R_{AB}(\tau) = \frac{1}{N} \sum_{n=1}^{N} V_A(n \Delta t) V_B(n \Delta t + \tau), \tag{9}$$

where signal $V_A$ and $V_B$ are the characteristic vectors of pixels A and B that the cross-correlation is applied to; N is the length of vector, and $\Delta t$ is the time-step.

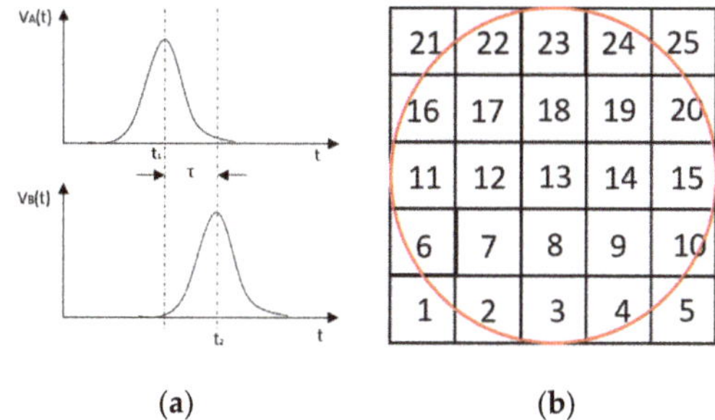


**Figure 10.** (**a**) examples of signals from pixel A and B; (**b**) indication of pixel units division within the view region.

### 5.1. Circular Movement

The circular movement starts from pixel 23 and completes one lap in 366 frames. Therefore, pixel 23 is set as a reference whose characteristic vector against frame number is plotted in Figure 11. The pixel profile plots provide information of the time instances when the sample enters and leaves the corresponding pixels. In the example of pixel 23, as described in Figure 11, the sample was placed at pixel 23 before the beginning of capturing and started moving out of the pixel after frame 10 until frame 57 when the object was completely out of this pixel; the object appeared back in pixel 23 from frame 324. The circular movement finishes one lap within 366 frames, which can be calculated as 26.14 s given that the data capture speed is 14 fps. This is comparable to the actual time spent on completing one lap, which is measured by a timer, i.e., 26.38 s.

Six pixels along the circular path are chosen to correlate with the reference pixel and results are displayed in the time-shifting manner in Table 9. Transit time can then be referred to the frame number of the peak in each cross-correlation plot. The time-step of adjacent frames, which is the reverse of the framerate of DAQ device is given, and the speed calculation is hence attainable.

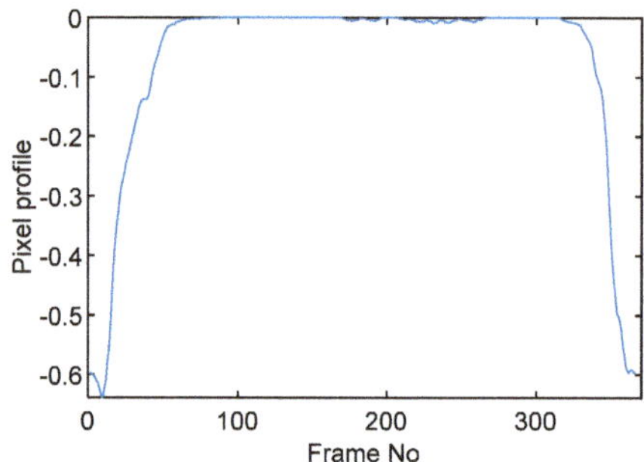

**Figure 11.** Characteristic value vector plot of pixel 23.

**Table 9.** Pixel characteristic vector plots and cross-correlation plots of circular movement.

| | Pixel 20 | Pixel 15 | Pixel 9 |
|---|---|---|---|
| **Pixel profile plot** | 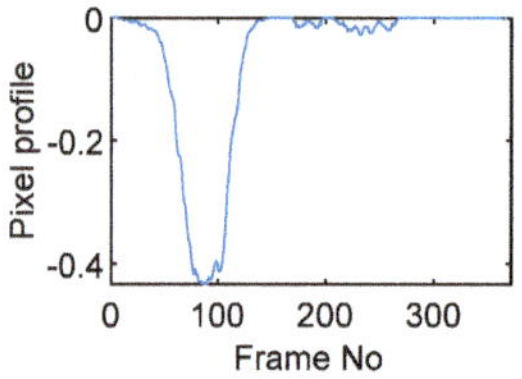 | 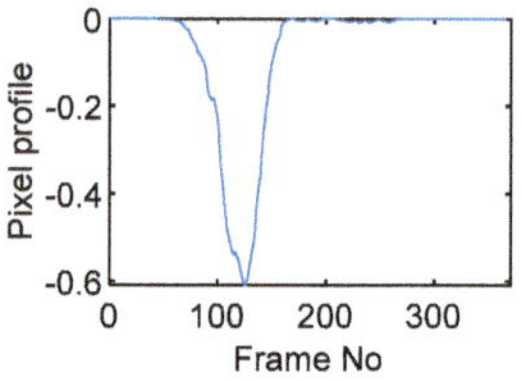 | 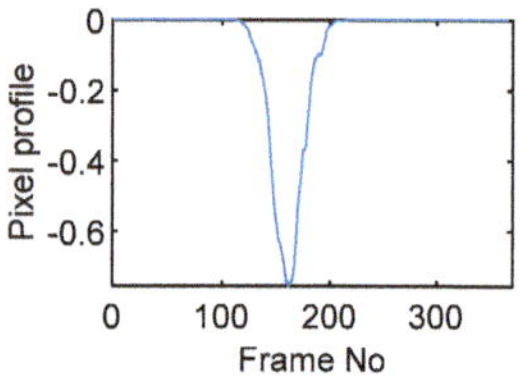 |
| **Cross-correlation** | 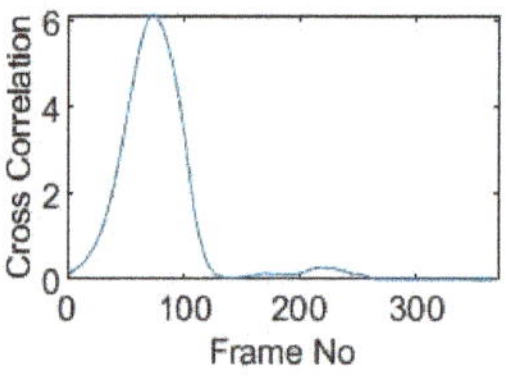 | 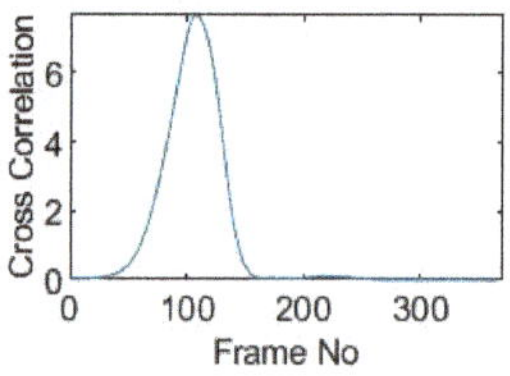 | 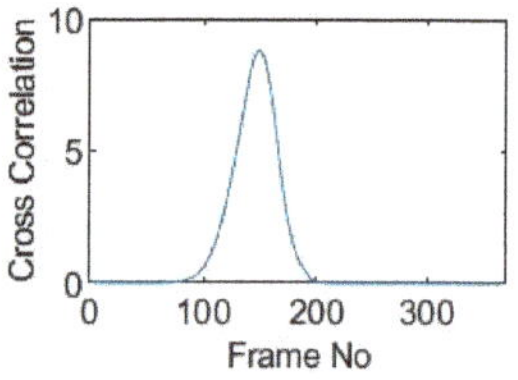 |

| | Pixel 3 | Pixel 7 | Pixel 11 |
|---|---|---|---|
| **Pixel profile plot** | 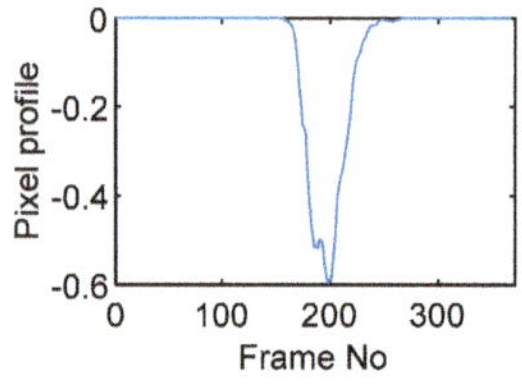 | 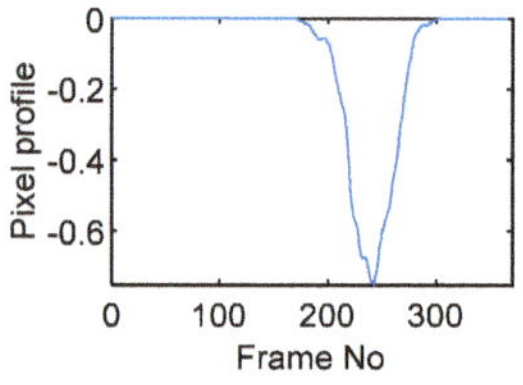 | 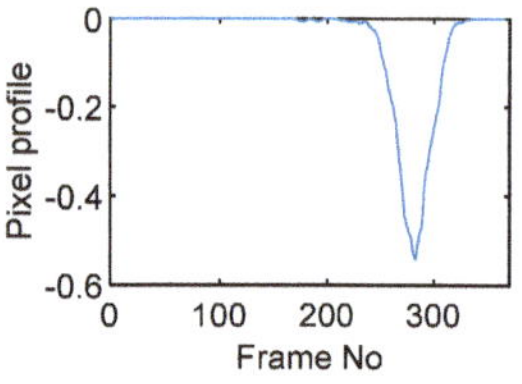 |
| **Cross-correlation** | 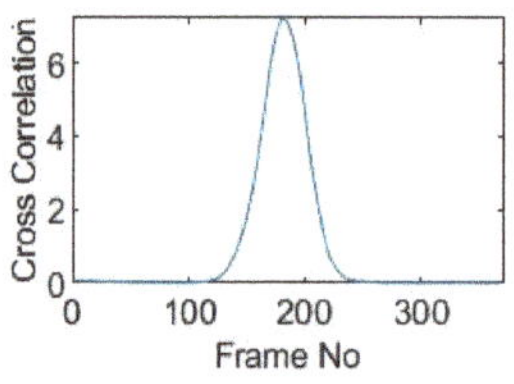 | 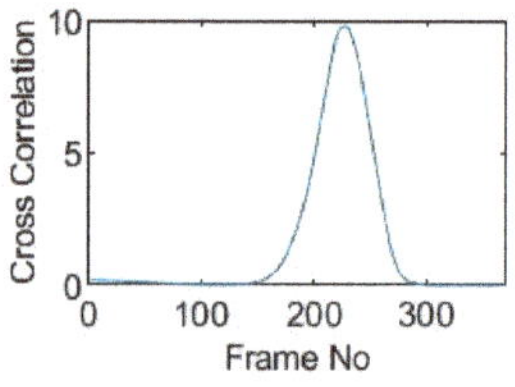 | 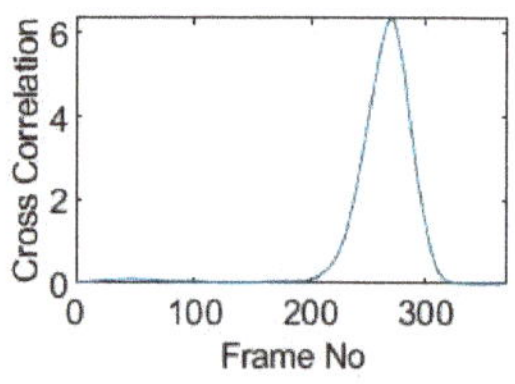 |

The frame numbers of the peaks in the cross-correlation plots in Table 9 are 72, 107, 147, 181, 225 and 269 in the displayed order. The increase in the peak frame numbers in the cross-correlation plots indicates the time shifting.

The angular speed can be performed by:

$$\omega = \frac{\theta}{\tau \times v_{DAQ}},\tag{10}$$

where $\theta$ is the angular distance between two pixels in radius, $\tau$ is the transit time in frames and $v_{DAQ}$ is the data acquisition speed which is 0.072 s per frame. For simplification, a quarter of one lap, which is from pixel 23 to pixel 15, is considered. The cross-correlated transit time, $\tau$, is 107 frames; hence, the angular speed can be calculated as 0.207 rad/s. The accuracy of the speed calculation can be analyzed in comparison with the measured angular speed. The measured time taken for the object to travel over the angular distance of $\pi/2$ is 7.81 s, which yields an average angular speed of 0.201 rad/s. Therefore, the relative error of using the cross-correlation method is 2.98% when compared with the measured speed; this is acceptable and proves that the strategy of using cross-correlation to obtain the speed measurements is feasible. The error is partially contributed by the imaging algorithm as the cross-correlation is essentially accomplished from a sequence of reconstructed images and the image quality can play an important role in the success of cross-correlation results. Also, the random error of timing the instance when the object arrives at 90-degree location can result in the difference between the measured time and the cross-correlated time, thereby giving the corresponding speeds.

## 5.2. Along Diameter Movements

Tests were also conducted by moving the object across the domain in vertical and horizontal directions, as mentioned previously. In these tests, it started at pixel 23 in the vertical direction and arrived at the opposite side of the tank, i.e., pixel 3, within 121 frames; started at pixel 11 in the horizontal direction and arrived at pixel 15 in 121 frames of time. Thus, pixel 23 and 11 are set as reference and their characteristic vectors are plotted in Figure 12a,b respectively.

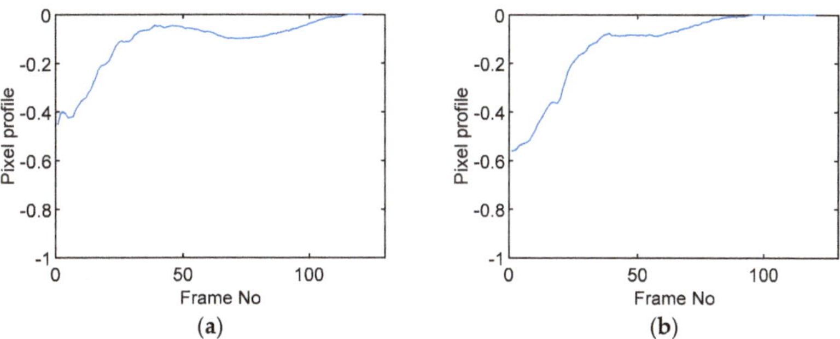

**Figure 12.** Characteristic value vector plot of (**a**) pixel 23; (**b**) pixel 11.

Three pixels along the moving path in each case are chosen to cross-correlate against the reference pixels. Similar to the circular movement, the cross-correlation plots peak at the frame numbers that represent the travel time spent on moving from the reference pixels to the corresponding pixels. Results are shown in Tables 10 and 11.

The frame numbers of the peaks of pixel 18, 13 and 8 cross-correlating to the reference pixel are 9, 51 and 82 respectively in the vertical movements; the frame numbers of the peaks in the cross-correlation plots of pixel 12, 13 and pixel 14 against pixel 11 are 41, 73 and 97.

The speed at which the sample travels along the path can be derived by the same principal as Equation (9) but with the metric distance between two targeting pixels instead. To illustrate, the cross-correlated average speed is calculated at which the sample travels from the reference pixel to the centre (pixel 13) in both the vertical and horizontal directions. Again, the accuracy analysis is shown in Table 12.

**Table 10.** Pixel characteristic vector plots and cross-correlation plots of vertical cross movement.

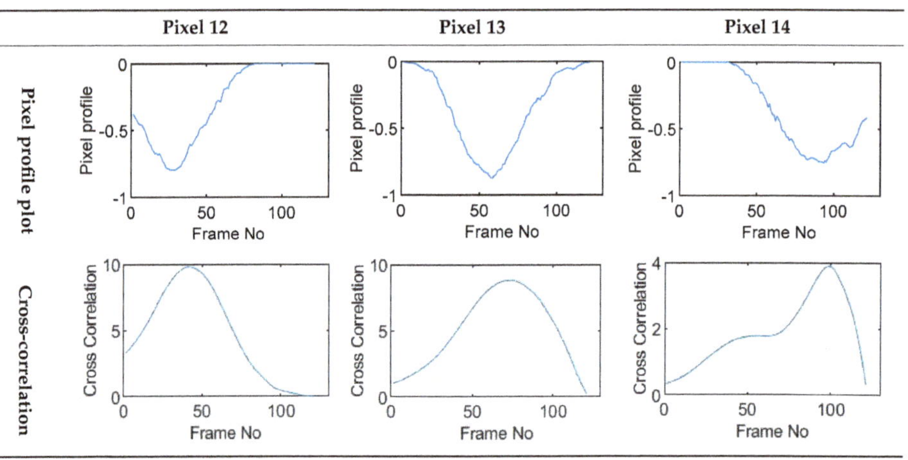

**Table 11.** Pixel characteristic vector plots and cross-correlation plots of horizontal cross movement.

**Table 12.** Speed results of cross movements.

|  | Measured Speed (m/s) | Cross-Correlated Speed (m/s) | Relative Error (%) |
|---|---|---|---|
| Vertical | 0.0265 | 0.0274 | 3.39 |
| Horizontal | 0.0186 | 0.0192 | 3.22 |

The relative errors of along diameter movements are slightly higher than those of circular movements, yet can still provide the confidence of employing cross-correlation to measure speeds. One of the reasons for to the larger relative errors is the image quality. As discussed, the image quality is one of the most vital factors that determine the accuracy of the cross-correlated speeds. Circular movements occur near the periphery of the tank whereas the cross movements involve the object travelling across the whole domain. The inherent difficulties of recovering the objects placed away from the center lower the accuracies.

## 6. Conclusions

Monitoring water supplies is economically and environmentally important, however, it is also costly. Blockages of pipes or flow restrictions are common; whether it is wet wipes in sewage networks, or gravel and silt clogging in pipes, it is important to know where these blockages occur (or are likely to occur), to facilitate maintenance and minimize disruptions to the networks. This work reports an ERT system demonstrating the use of sensors to image pipes with different flow constituents including solids and liquids. The impact of this work is an embedded sensor monitoring the quality and rate of flow in water and sewage networks, feeding into water quality and availability improvements.

The device incorporates a serial measurement structure with a common mode error cancellation method to allow for a cost-effective but fast data collection design. Furthermore, it features the flexibility of injecting current from 6 mA to 18 mA whilst being able to respond to the wide dynamic load range. A data collection rate of 14 fps, as well as an overall mean SNR of more than 59.54 dB, has been achieved. The system performances in both the static and dynamic range were studied in lab-scale experiments. With the smallest sample size reported being 0.04% of the phantom, the system managed to locate inclusions with a position error of less than 1 cm. Also, successful motion tracking was achieved by clear visualization of the movements, as well as numerical speed calculation with less than 4% relative errors.

**Author Contributions:** Conceptualization, M.S.; Methodology, C.W.; Software, C.W.; Supervision, M.H. and M.S.; Writing—original draft, C.W.; Writing—review & editing, M.S.

**Funding:** This research is part of a Knowledge Transfer Partnership project between the University of Bath and Ashridge Engineering Ltd, funded by Innovate UK grant #509928.

**Conflicts of Interest:** The authors declare no conflict of interest.

## References

1. Aw, S.R.; Rahim, R.A.; Rahiman, M.H.F.; Yunus, F.R.M.; Goh, C.L. Electrical resistance tomography: A review of the application of conducting vessel walls. *Powder Technol.* **2014**, *254*, 256–264. [CrossRef]
2. Zhao, Y.L.; Wang, M.; Yao, J. Electrical Impedance Tomography Spectroscopy Method for Characterising Particles in Solid-Liquid Phase. In Proceedings of the 8th international symposium on measurement techniques for multiphase flows, Guangzhou, China, 13–15 December 2013.
3. *Waste Water Treatment in the United Kingdom-2012: Implementation of the European Union Urban Waste Water Treatment Directive-91/271/EEC*; Department for Environment, Food and Rural Affairs: London, UK, 2012.
4. *Sewage Treatment in the UK*; Department for Environment, Food and Rural Affairs: London, UK, 2002.
5. Bin Ali, M.T.; Horoshenkov, K.V.; Tait, S.J. Rapid detection of sewer defects and blockages using acoustiv-based instrumentation. *Water Sci. Technol.* **2011**, *64*, 1700–1707. [CrossRef]
6. Arthur, S.; Crow, H.; Pedezert, L. Understanding blockage formation in combined sewer networks. *Proc. ICE Water Manag.* **2008**, *161*, 215–221. [CrossRef]
7. *Performance Standards and Test Procedures for Continuous Water Monitoring Equipment-Part 3: Performance Standards and Test Procedures for Water Flowmeters*; Environment Agency: Bristol, UK, 7 April 2014; p. 60.
8. Kim, Y.; Simpson, A.; Lambert, M. Behavior of orifices and blockages in unsteady pipe flows. In Proceedings of the 9th Annual Symposium on Water Distribution Systems Analysis, Tampa, FL, USA, 15–19 May 2007.
9. Rasteiro, M.G.; Silva, R.C.C.; Garcia, F.A.P.; Faia, P.M. Electrical Tomography: A review of Configurations and Applications to Particulate Processes. *KONA Powder Particle J.* **2011**, *29*, 67–80. [CrossRef]
10. Jia, J.; Wang, M.; Schlaberg, H.I.; Li, H. A novel tomographic sensing system for high electrically conductive multiphase flow measurement. *Flow Meas. Instrum.* **2010**, *21*, 184–190. [CrossRef]
11. KarKi, B.; Faraj, Y.; Wang, M. Electrical conductivity based flow regime recognition of two phase flows in horizontal pipeline. In Proceedings of the 8th World Congress on Industrial Process Tomography Iguassu Falls, Paraná, Brazil, 26–29 September 2016.
12. Zhang, Y.; Chen, Y. *A Novel PCA-GRNN Flow Pattern Identification Algorithm for Electrical Resistance Tomography System*; Springer: Berlin/Heidelberg, Germany, 2012; pp. 249–254.

13. Tan, C.; Dong, F.; Wu, M. Identification of gas/liquid two-phase flow regime through ERT-based measurement and feature extraction. *Flow Meas. Instrum.* **2007**, *18*, 255–261. [CrossRef]

14. Tan, C.; Dong, F.; Xu, C.; Zhang, Z.; Ren, S. Electrical resistance tomography system for two-phase flow monitoring. *J. Southeast Univ. Nat. Sci. Ed.* **2011**, *41*, 125–129. [CrossRef]

15. Dong, F.; Xu, C.; Zhang, Z.; Ren, S. Design of Parallel Electrical Resistance Tomography System for Measuring Multiphase Flow. *Chin. J. Chem. Eng.* **2012**, *20*, 368–379. [CrossRef]

16. Oliveira, J.L.G.; Passos, J.C.; Verschaeren, R.; Geld, C.v.d. Mass flow rate measurements in gas–liquid flows by means of a venturi or orifice plate coupled to a void fraction sensor. *Exp. Therm. Fluid Sci.* **2009**, *33*, 253–260. [CrossRef]

17. Henningsson, M.; Östergren, K.; Dejmek, P. Plug flow of yoghurt in piping as determined by cross-correlated dual-plane electrical resistance tomography. *J. Food Eng.* **2006**, *76*, 163–168. [CrossRef]

18. Sharifi, M.; Young, B. Qualitative visualization and quantitative analysis of milk flow using electrical resistance tomography. *J. Food Eng.* **2012**, *112*, 227–242. [CrossRef]

19. Mi, W.; Yixin, M.; Holliday, N.; Yunfeng, D.; Williams, R.A.; Lucas, G. A high-performance EIT system. *IEEE Sens. J.* **2005**, *5*, 289–299. [CrossRef]

20. Cilliers, J.J.; Xie, W.; Neethling, S.J.; Randall, E.W.; Wilkinson, A.J. Electrical resistance tomography using a bi-directional current pulse technique. *Meas. Sci. Technol.* **2001**, *12*, 997–1001. [CrossRef]

21. Dickin, F.; Wang, M. Electrical resistance tomography for process applications. *Meas. Sci. Technol.* **1996**, *7*, 247–260. [CrossRef]

22. Zhang, Z.; Dong, F.; Xu, C. Data acquisition system based on CompactPCI bus and FPGA for electrical resistance tomography. In Proceedings of the 2011 Chinese Control and Decision Conference (CCDC), Mian Yang, China, 23–25 May 2011; pp. 3538–3543.

23. Tech4imaging; Multi-Phase Flow Meters. Available online: https://www.tech4imaging.com/multiphase-flow/multiphase-flow-meter/ (accessed on 22 February 2019).

24. CiDRA. SONARtrac Flow & Entrained Air Measurement. Available online: https://www.cidra.com/cidra-products/sonartrac-flow-measurement (accessed on 22 February 2019).

25. Rocsole. Rocsole Flow Water. Available online: https://www.rocsole.com/applications/flow-watch (accessed on 22 February 2019).

26. Itoms. Flow-ITOMETER for visualizing flows. Available online: https://www.itoms.com/products/flow-visualiser/ (accessed on 22 February 2019).

27. Wang, M. Electrical Impedance Tomography on Conducting Walled Process Vessels. Ph.D. Thesis, University of Manchester, Manchester, UK, 1994.

28. Boone, K.G.; Holder, D.S. Current approaches to analogue instrumentation design in electrical impedance tomography. *Physiol. Meas.* **1996**, *17*, 229–247. [CrossRef] [PubMed]

29. Lee, J.; Ha, U.; Yoo, H. 30-fps SNR equalized electrical impedance tomography IC with fast-settle filter and adaptive current control for lung monitoring. In Proceedings of the 2016 IEEE International Symposium on Circuits and Systems (ISCAS), Montreal, QC, Canada, 22–25 May 2016; pp. 109–112.

30. Adler, A.; Arnold, J.H.; Fau-Bayford, R.; Bayford, R.; Fau-Borsic, A.; Borsic, A. Fau-Brown, B.; Brown, B. Fau-Dixon, P.; Dixon, P.; et al. GREIT: A unified approach to 2D linear EIT reconstruction of lung images. *Physiol. Meas.* **2009**, *30*, S35. [CrossRef] [PubMed]

31. Etuke, E.O.; Bonnecaze, R.T. Measurement of angular velocities using electrical impedance tomography. *Flow Meas. Instrum.* **1998**, *9*, 159–169. [CrossRef]

32. Dong, F.; Xu, Y.; Hua, L.; Wang, H. Two Methods for Measurement of Gas-Liquid Flows in Vertical Upward Pipe Using Dual-Plane ERT System. *IEEE Trans. Instrum. Meas.* **2006**, *55*, 1576–1586. [CrossRef]

 *sensors*

*Article*

# Nanoparticle Assisted EOR during Sand-Pack Flooding: Electrical Tomography to Assess Flow Dynamics and Oil Recovery [†]

**Phillip Nwufoh *, Zhongliang Hu, Dongsheng Wen and Mi Wang ***

School of Chemical and Process Engineering, University of Leeds, Leeds LS2 9JT, UK

* Correspondence: pm11pn@leeds.ac.uk (P.N.); m.wang@leeds.ac.uk (M.W.)

† This paper is an extended version of our paper "Tomography to Visualize Nanoparticle- Assisted Multiphase Flow in Porous Media" published in Proceedings of the 9th World Congress on Industrial Process Tomography. Bath, UK, 2–6 September 2018.

Received: 29 May 2019; Accepted: 5 July 2019; Published: 10 July 2019

**Abstract:** Silica nanoparticles have been shown to exhibit many characteristics that allow for additional oil to be recovered during sand-pack flooding experiments. Additionally various imaging techniques have been employed in the past to visually compare flooding procedures including x-ray computed tomography and magnetic resonance imaging; however, these techniques require the sample to be destroyed or sliced after the flooding experiment finishes. Electrical resistance tomography (ERT) overcomes these limitations by offering a non-destructive visualization method allowing for online images to be taken during the flooding process by the determination of spatial distribution of electrical resistivity, thus making it suitable for sand-packs. During the scope of this research a new sand-pack system and methodology was created which utilized ERT as a monitoring tool. Two concentrations, 0.5 wt% and 1.0 wt%, of $SiO_2$ nanoparticles were compared with runs using only brine to compare the recovery efficiency and explore the ability of ERT to monitor the flooding process. Electrical resistance tomography was found to be an effective tool in monitoring local recovery efficiency revealing 1.0 wt% $SiO_2$ to be more effective than 0.5 wt% and brine only runs during the scope of this research. A new method involving the slope function in excel was used to compare the effects of nanofluids on resistivity trends also revealing information about the rate of recovery against time. $SiO_2$ nanofluid recovery mechanisms such interfacial tension reduction and viscosity enhancement were then considered to explain why the nanofluids resulted in greater oil recovery.

**Keywords:** imaging techniques; multiphase flow; nanoparticles; enhanced oil recovery; tomography; sand-pack flooding

---

## 1. Introduction

Both electrical conductivity and resistivity measurements have been conducted in the past to characterize porous media flow, especially during soil infiltration measurements [1–3]. The resistivity of a porous media is governed by the electrical resistivity of the constituent phases, saturation content, fabric, and porosity. These resistivity measurements have been conducted to a lesser extent in regards to flooding processes for oil recovery. Local resistivity distributions can provide information on phase behavior and recovery efficiency during sand-pack flooding experiments for nanoparticle induced, enhanced oil recovery. Typically x-ray computed tomography and magnetic resonance imaging are used to obtain images of flow patterns during flooding experiments. Although these options provide better spatial resolutions, electrical resistance tomography (ERT) also has many benefits including excellent temporal resolutions, great practicability, and its ability to perform reconstructions in real time.

Flooding experiments for enhanced oil recovery typically consists of a porous media that's largely saturated with oil, which is then flushed with brine until there is a plateau in oil recovery. Once all of the movable oil is recovered, a fraction of bypassed oil will still be present within the pores in what is termed residual oil. To reduce the residual oil saturation, different enhanced oil recovery techniques must be applied including thermal, gas, and chemical injections. The scope of this research focuses on a branch of chemical injection termed nano-EOR in which nanoparticles are introduced to increase oil recovery after the water-flooding stage is deemed to be no longer efficient. Nanoparticles have many desirable characteristics which allow for a reduction in residual oil saturation, including (i) their ability to travel through the smallest of pore throats due to their minute size (1–100 nm) which allows them to penetrate deep into reservoir rocks, (ii) their large specific surface area, resulting in an increased contact area of nanoparticles with the oil phase, and better interaction between phases [4], (iii) their high multi-functionality allowing them to be fabricated for a specific task, thus allowing for nanoparticle characteristics such as particle coatings and morphology to be altered [5–7], and lastly, (iv) their ability to change both multiphase and rock–fluid behaviors, allowing for the interfacial tension and wettability of the system to be altered [8–11].

There are many experimental challenges associated with using ERT as a monitoring tool during nano-EOR flooding experiments. Firstly, a sand-pack setup had to be constructed which incorporates ERT around the perimeter of the vessel. Then the limitations of ERT itself had to be considered such as the lack of spatial resolution (5% of vessel diameter) which meant the nanoparticles themselves could not be detected during the measurements as the resolution in this case is on the millimeter scale, and the nanoparticles on the nano-scale. Therefore, instead of considering the flow behavior of the nanoparticles themselves, the change in flow behavior or oil recovery rate resulting from the presence of nanoparticles will be considered and if large enough should be able to be detected by ERT. A major limitation of ERT is the need for the medium to be conductive so an electrical continuity exists between the sensors and the medium. Previously, researchers have considered conductive flows through porous media such as water through soil. However performing such experiments with oil is more challenging due to the electrically discontinuous nature of oil making acquiring a signal much more difficult. The advent of new sensing methods, particularly the conductive ring and resistor network arrangement, have increased the measurement range for ERT allowing for more accurate measurements in highly intermittent environments such as oil saturated porous medias.

The aims here are to construct a sand-pack setup which incorporates traditional ERT sensors around the sand-pack. Then ERT will be used to monitor oil recovery during nano-EOR flooding experiments, and to validate the oil recovery results using the local resistivity plots and tomograms. Additionally, ERT will also be used to reveal any differences in flow behavior and oil recovery brought about by the presence of nanoparticles by observing changes in local resistivity trends and conductivity profiles with time.

## 2. Materials and Methods

### 2.1. The System

A sand-pack flooding system was developed in order to utilize tomography as a visualization tool to determine the effect of nanofluid on porous flow (Figure 1). In order to achieve this a column made of perplex (5 cm diameter) was fitted with dual plane ERT sensors and filled with building sand to represent the porous media. A vibrating electric oscillator (50 Hz), capable of variable amplitude settings (max amplitude >0.15 mm) was situated underneath the column to provide a well compacted sand-pack. A gravity feed system was developed in order to introduce the fluids at a constant pressure head, and solenoid timer valves allowed for pulses of brine, with and without nanoparticles, to be introduced. The solenoid valves were opened for periods of 5 min at a time, after a period of 1 h and 15 min, during which portions of about 10 mL of fluids were released from the medical drainage bags. A collection table was situated beneath the table consisting of a number of beakers arranged on a collection tray, which was placed on top of a rotation motor. The rotation angle, time spent at each angle, and number of times that the cycle is completed, could all be set by the user making it suitable for different flooding scenarios.

The dual plane ERT sensors consisted of 16 electrodes (12.7 mm × 25.4 mm) made from stainless steel which were connected to the data acquisition system via 36 pin connectors (IEE-488) on the front of the system. The data acquisition system was the P2000 ERT system, a system designed by Industrial Tomography Systems, based in Manchester, UK. For the scope of these experiments, an injection current of 15 mA was used and a frame capture rate of 1 fps was adequate to capture the flooding process over the range of 4.5 h. An adaptor cable was used to create a more homogenous sensitivity distribution by mimicking the technology incorporated in resistor-network and conductive ring sensing apparatuses [11]. This allowed for the measurement of more complex multiphase flows with an electrical discontinuous phase such as the three phase oil-brine-nanofluid flows considered in this setup.

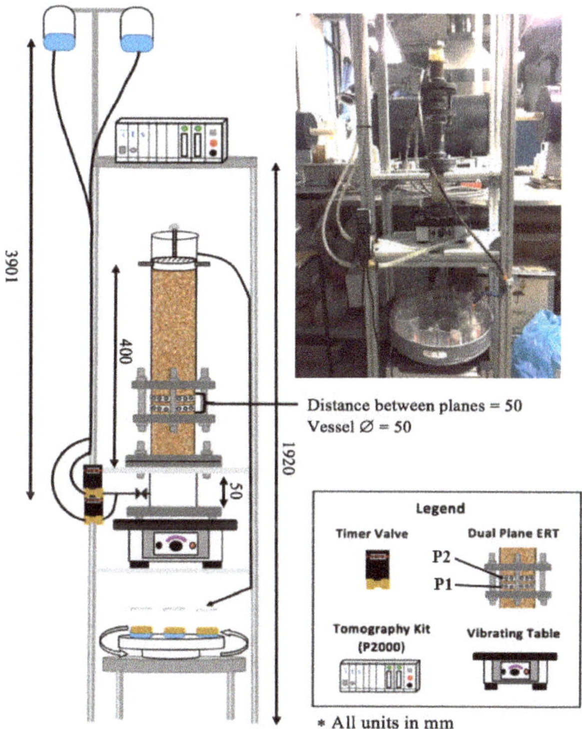


**Figure 1.** Sand-pack setup for nano-EOR experiments, including a gravity feed and effluent collection system. Where P1 and P2 represent the lower and upper sensor planes, respectively.

## 2.2. Gravity Feed System

In order for the system to work effectively, the flow rate must be near identical for both phases and kept constant as a large change in flowrate could alter the conductivity distribution within the sensing plane, thus making the results invalid. Conventionally, the use of an injection pump along with holding vessels for each fluid is used in flooding setups to introduce various fluids. However, these setups are relatively expensive and require the valves to be opened and closed manually when changing flooding fluids. The use of a gravity feed system along with timer valves allows the fluids to be introduced automatically at near constant flow rates, allowing for 5 min pulsations of each flooding fluid to be introduced without having to manually open and close the valves during every pulsation. The gravity feed was simply constructed by connecting 2000 mL medical drainage bags (Beambridge Medical Ltd., Guildford, UK) to 10 mm tubing which fed to the inlet chamber via timer valves. The pressure drop was calculated by using measuring tape to quantify the hydraulic head in feet, and converted to atmospheres using a conversion factor where 304.8 mm = 2989.1 Pa. The pressure drop for 3091 mm was found to be 38,260 Pa, which was used for all the experimental runs.

A liquid or object that is free to move usually moves spontaneously from a state of higher potential energy to a state of lower potential energy. The same applies for water in a porous medium, such that a unit volume or mass of water will tend to migrate from an area of higher potential energy, such as in Figure 2, highlighting the total heads at the inlet and outlet. The driving force equation is used to explain the water flow from the drainage bags to the outlet, and quantify the force of displacement, given by the total potential (HH) at two points ($HH_A$, $HH_B$) divided by the distance ($L_{AB}$) between the points:

$$df = -\frac{dHH}{dx} = \frac{dHH_A - dHH_B}{L_{AB}} \tag{1}$$

Amending the above equation and solving for driving force gives:

$$df = -\frac{dHH}{dx} = \frac{dTH_{TOP} - dTH_{BOTTOM}}{L_{AB}}$$
$$df = \frac{3231 - 450}{400} = 6.95 \tag{2}$$

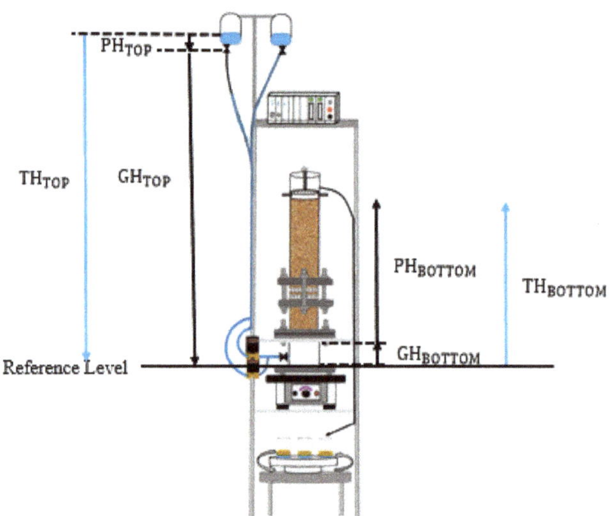


**Figure 2.** A model to explain fluid flow from gravity drainage system through the sand-pack taking into account the total heads at the top and bottom of the system. Where TH represents the total head and is the summation of GH, the gravitational head and PH, the pressure head.

### 2.3. Vibrating Oscillator

The packing method is a crucial factor for EOR-flooding affecting the nature of the flow, porosity, and permeability. Almost all reservoir rocks are composed of sedimentary rocks in the porosity range of 10–40% in sandstones and 5–25% in carbonates, whilst permeability is found to vary much more from a fraction of a millidarcy to several darcies. In order to provide a packing that is representative of reservoir values, an eclectic oscillator was incorporated into the setup so that the compaction could take place without the user having to hold the column over a vibrating mechanism, which is how a sand-pack is conventionally compacted. Also when the user is manually holding a sand-pack, there is a higher possibility of error in packing as the user will change the position of the pack, making them less uniform and prone to heterogeneities in different layers. The vibrating oscillator being incorporated into the setup overcomes these limitations as the column is always fixed in the same position during compaction and the compaction time can last for much longer as the user does not have to hold the column. Therefore, compaction can last several hours and the user can return when the packing is complete, which also ensures the packing is repeatable and uniform before each run. The vibrating oscillator itself (model JT-51B) ran on a frequency of 50 Hz and was capable of variable amplitude settings (max amplitude ≥15 mm). The maximum amplitude was used during the range of the experiments and the compaction time was kept to 1.5 hours. The repeatability of the packing method is shown in Figure 3 below which reveals the relationship between permeability and compaction time. The packing method was found to be highly repeatable with a maximum error of 6% over three runs at 15 min compaction time, which lowers as the compaction time increases to eventually less than 2% error after 90 min of compaction.

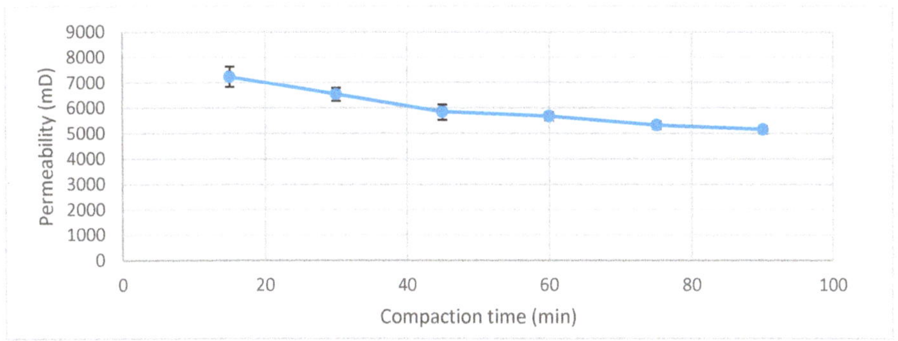


**Figure 3.** The compaction time against permeability tested in 15 min intervals for 3 runs at each compaction time, revealing the high repeatability of the packing method especially after 1 h of packing.

### 2.4. Effluent Collection System

One of the main challenges faced when constructing this particular sand-pack column was how to quantify the recovered oil. Traditional effluent collection methods include graduated pipettes, fraction collectors, mass flow meters, and digital scales [12]. A new effluent collection system and methodology was developed based on the concept of a fraction collector.

A rotating turntable, purchased from Comxim Ltd., was the base of the effluent collection table and acted as a rotating motor. A collection tray was then fitted on top of the rotating motor which housed a number of collection beakers. The beakers were arranged in such a way so that no fluid was lost when they changed position (Figure 4). The table's rotational settings could be altered so that the rotation angle, time spent at each angle, and number of times the cycle is completed could all be set before the experiment. The exact quantity of oil was measured by pouring fluids from the collection beakers into graduated cylinders after the experiment. A small amount of excess oil is left on the sides of the beakers which is also accounted by simply subtracting the mass of an empty beaker from the mass of the beaker with excess oil, and converting to volume since the density of oil is known. The details of the rotating tunable are listed in Table 1.

**Figure 4.** Effluent collection table with beakers arranged on the top capable of rotating automatically in timed increments.

**Table 1.** The parameters for the effluent collection table.

| Effluent Collection Table Parameters | | | | | | | |
|---|---|---|---|---|---|---|---|
| Product Model | Diameter (mm) | Height (mm) | Transmission Gear Material | Control Method | Speed Range (Sec/Rev) | Net Weight (Kg) | Handling Load (Kg) |
| MT370L20 | 370.8 | 8 | POM + Metal | Infrared + Bluetooth | 15–31.5 | 5.5 | 20 |

*2.5. Sand-Pack Flooding Preparations*

Electrical Resistance Tomography (ERT) works by obtaining a conductivity/resistivity distribution in the domain of interest by using electrodes that induce currents or voltages, and electrodes that measure the resultant currents or voltages. Therefore, in order to effectively visualize the effect of nanofluids on oil recovery, the conductivities of both the brine phase and nanofluid phase must be near identical so any changes in conductivity distribution are a function of oil recovery and not the conductivity contrast between the two flooding fluids. To achieve this, small quantities of NaCl solution were added to each phase to ensure their conductivities matched. The tracer dilution method was used to quantify how much NaCl solution is needed to be added to the tap water and nanofluid phases, given by:

$$\Delta t \left( C_0 \, q_v + C_1 \, q_1 \right) = \Delta t \left( C_2 (q_v + \, q_1) \right) \tag{3}$$

where $\Delta t$ is the change in time, $C_0$ is the initial conductivity of DI water, $C_1$ is the conductivity of the NaCl solution to be added to the DI water, $C_2$ is the post mixing conductivity of the brine, $q_v$ is the volumetric flow before mixing, and $q_1$ is the volumetric flow after mixing.

Since, $q_v = \frac{volume}{time}$, the equation above can be rewritten in terms of volume:

$$C_0 V_v + C_1 V_1 = C_2 \left( V_v + V_1 \right) \tag{4}$$

where $V_v$ is the total volume of DI water before mixing and $V_1$ is the volume of NaCl solution to be added to DI water.

The conductivities of both the brine and nanofluid phases were matched using Equation (4), and a Eutech CyberSca PC 6500 bench conductivity meter was used to validate the results. The findings are revealed in Table 2.

**Table 2.** The conductivities of flooding fluids both before and after the addition of tracer NaCl solution.

| Fluid Type | Volume (mL) | Pre-Mixing Conductivity (mS/cm) | Mixing Brine Volume and Conductivity | Post Mixing Conductivity (mS/cm) | |
|---|---|---|---|---|---|
| | | | | Tracer | Probe |
| DI Water | 2000 | 0.033 | 100 mL @ 0.6 mS/cm | 0.061 | 0.059 |
| 0.5 wt% $SiO_2$ | 1000 | 0.041 | 100 mL @ 0.025 mS/cm | 0.060 | 0.058 |
| 1.0 wt% $SiO_2$ | 1000 | 0.044 | 100 mL @ 0.22 mS/cm | 0.060 | 0.061 |

Next, the sand was prepared by cleaning and drying it before saturating it with known volumes of oil and water. This was achieved with prior knowledge of the average pore volume for a range of sand-packs as a function of compaction time and intensity. The porosities of the sand-packs were found to be in the range of 35–37% and the permeability between 5–6 Darcy's (Table 3). Once the average pore volume was obtained (found to be approximately 282.42 mL), the sand was mixed with 1/3 water and 2/3 oil to fill this pore volume and ensure the sand was fully saturated. If the sand is not fully saturated, the presence of air bubbles could show as areas of low conductivity on the tomograms and could be mistaken for oil. The compaction process was then initiated by switching on the electric oscillator and leaving it for 1.5 h. Lastly, before the fluids were introduced, the parameters were set for the ITS P2000 system, solenoid timer valves, and the rotating collection table.

**Table 3.** The parameters used to calculate the pressure drop and permeability.

| Sand-Pack Parameters | |
|---|---|
| Pressure in (psi) | 19.9 |
| Pressure out (psi) | 14.9 |
| Distance (cm) | 40 |
| Area ($cm^2$) | 667.59 |
| Discharge (mL/min) | 2 |
| Viscosity (Pa·s) | 0.89 |
| Permeability (mD) | 5225 |

*2.6. Sand-Pack Flooding Procedure*

The flooding procedure differs slightly from the conventional method as the sand-pack is already saturated with known volumes of oil and water prior to packing. Since the initial water and oil saturations were already known, the traditional drainage process was skipped, where oil is injected into the sand-pack, and instead flooding began at the first forced imbibition process in which brine was injected at a constant rate of 2 mL/min for 0.8 pore volumes (PVs) to displace the oil phase. Next the nanofluid pulsations were introduced at the same flow rate by setting the timer of valves to allow for 5 min intervals of nanofluid and brine for 1.3 PVs. The accumulated oil was then collected in a number of beakers housed on a collection tray which rotated automatically in timed intervals. This allowed for the cumulative oil recovery to be accounted automatically without having to observe the recovery in timed increments over the duration of 4.5 h. This coupled with the gravity feed system to introduce the fluids meant that the whole flooding process could be run automatically, with the fluids being introduced as well as the collected in timed increments without the user having to be present. When the flooding was complete, the recovered oil was then accounted for in each beaker by pouring their contents into a graduated cylinder as described in Section 2.4.

*2.7. Materials*

De-ionized water was used as the base fluid for dispersing the nanoparticles, as well as the formation liquid for all the experiments. Laboratory grade sodium chloride purchased from Sigma Aldrich was used to make the brine solutions. The density of the brine was measured at $1000 \pm 0.01$ g/cm$^3$, pH $6.46 \pm 0.3$, and the dynamic viscosity $0.87 \pm 0.01$ mPa·s at 25 °C. The oil phase was a light mineral oil from Fisher Scientific with a density of 0.84 g/cm$^3$ and a dynamic viscosity of 39.95 mPa ± 0.11 mPa·s at 25 °C. Silica nanoparticles with a diameter of 80 nm were purchased from get nanomaterials and were used for all experiments. The sand used was a fine building sand purchased from Wickes with an average grain size between 2–4 mm.

## 3. Results and Discussion

### 3.1. Flow Imaging

Relative changes of conductivity as a function of time were used to visualize the flow behavior of oil, water, and nanofluid phases as flooding progresses to provide information on local recovery efficiency. The tomograms shown in Figure 5 began with a relatively high concentration of oil phase (in blue color), which was shown to be displaced by the aqueous phase (in red color) over time. During both runs the concentration of oil was initially seen to deplete from the center of the column, leaving behind zones of bypassed oil towards the outer perimeter as time progressed. The conductivity scale was selected between 0.98 mS/cm and 1.14 mS/cm to ensure there was an adequate contrast to differentiate between the low conductivity phase (oil) and the high conductivity phases (brine and nanofluids). To the left of the tomograms are 15 min time stamps highlighting relative changes in conductivity as time progressed.

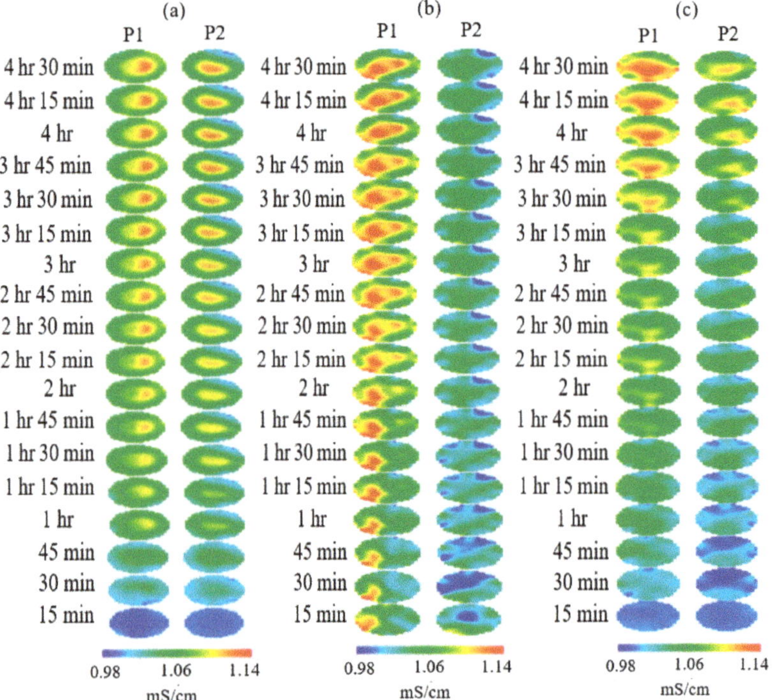


**Figure 5.** The dual plane tomograms for the flooding runs no. 1 involving (**a**) no NF (**b**) 0.5 wt% SiO$_2$ and (**c**) 1 wt% SiO$_2$ revealing the effect of 10 nanofluid pulsations after 1 h and 15 min of water-flooding.

The objective of the study was to ultize the ERT to qualitatively compare the recovery efficiency during different flooding scenarios. Figure 5 above reveals the tomograms for brine-only, 0.5 wt% SiO$_2$, and 1.0 wt% SiO$_2$. All three of the tomogram sets show zones of bypassed oil after the initial water flood, leaving a good opportunity to observe the effects of silica nanofluid injection. The run using only brine clearly shows a much lower recovery efficiency along the measuring planes, when compared with both runs using silica nanofluid. For the brine only run, after 1 h and 15 min, the local oil recovery becomes almost stagnant in both planes and does not show large variations in flow behavior and flooding efficiency. When this run is compared with the silica nanofluid runs, a different picture is painted revealing the ability of nanofluids to open new flow pathways and alter the rate of recovery. The nanofluids were introduced after 1 h and 15 min during both nanofluid runs. When considering the 1.0 wt% run, it is apparent that the nanofluid injection displaced some residual oil below the sensors which then accumulated in oil pockets at the 4 o'clock position. The tomograms for 0.5 wt% $SiO_2$ show two zones of residual oil at the 1 o'clock and 11 o'clock positions, which are both gradually stripped away, until only a very faint pocket of oil can be seen at the 1 o'clock position. The run using 1 wt% $SiO_2$ reveals the nanofluids to be more effective in removing local oil residue than both the brine only and 0.5 wt% nanofluids, as confirmed by the resistivity plots (Section 3.2).

The tomograms are validated by the resistivity and slope function plots. There is a sudden change in recovery efficiency between 3 h 15 min and 3 h 30 min, in which the flow channels begin to open and spread laterally. This rapid change is believed to be the result of nanoparticles altering the flow pattern and allowing flooding fluids to infiltrate pore channels which were previously bypassed. However, other recovery mechanisms could explain the differences in flow structure (Section 3.5). These quick changes in the tomograms are also related to the plots for resistivity and cumulative oil recovery, with the resistivity plot showing quick variations at the same time and the plot cumulative oil recovery plot showing additional recovered oil between 1.2–1.5 pore volumes (PVs) (between 180–210 min) for the 1.0 wt% $SiO_2$ and between 1.2–1.6 PVs (between 200–240 min) with regards to the 0.5 wt% $SiO_2$. This behavior can be attributed to a number of recovery mechanisms such as interfacial tension (IFT) reduction, viscosity enhancement, or the presence of nanoparticles altering the pore network and flow structure of the sand-pack. A recovery analysis was conducted later in the paper to determine the influence of some of these mechanisms. The conductivity profiles are also displayed in Figures 6–8 and reveal the changes to flow dynamics with time across the central region of the tomograms. The effects of nanofluids on oil recovery are highlighted by revealing larger increases in conductivity trends after the introduction of nanofluids which were not present during the runs without nanofluids. The trends show conductivity to increase as the flooding progresses and oil is displaced, followed by a near convergence of the trends as recovery efficiency begins to stagnate. The plots reveal the ability of ERT to add to the physical understanding of the flow dynamics and oil displacement process with time.

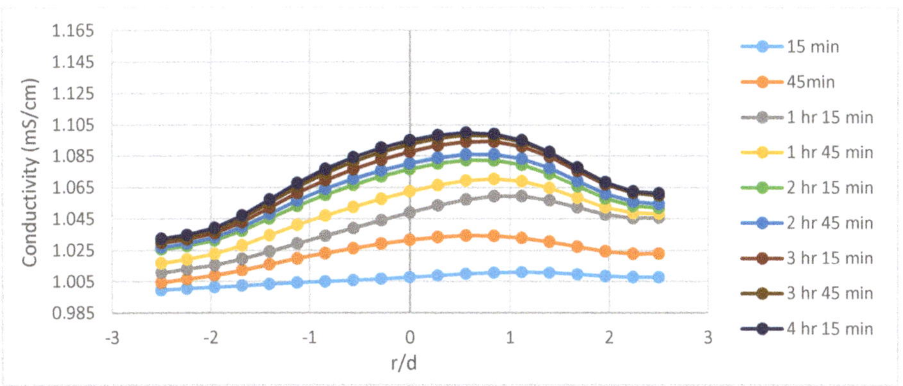


**Figure 6.** The conductivity profiles for P1 during run no. 1 with no nanoparticles, taking the central horizontal region of the tomograms into account.

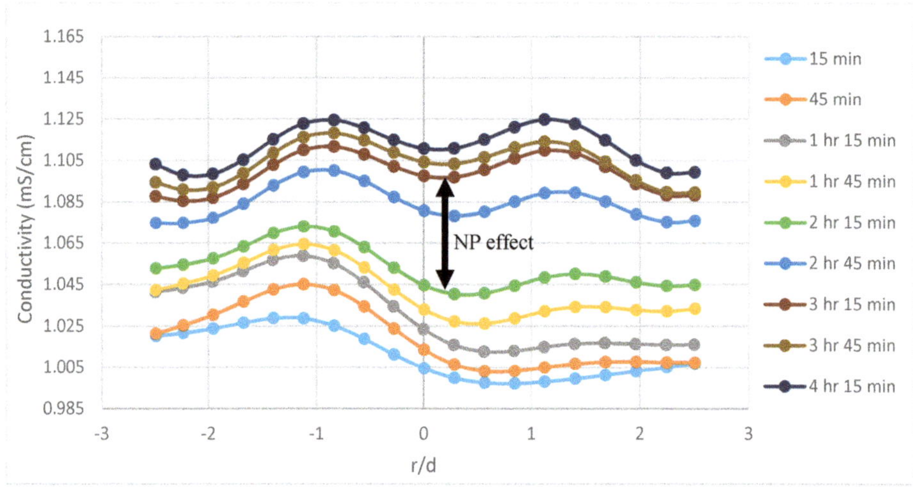



**Figure 7.** The conductivity profiles for P1 during run no. 1 with 0.5 wt.% $SiO_2$, taking the central horizontal region of the tomograms into account. The effects of nanofluid are highlighted with an arrow.

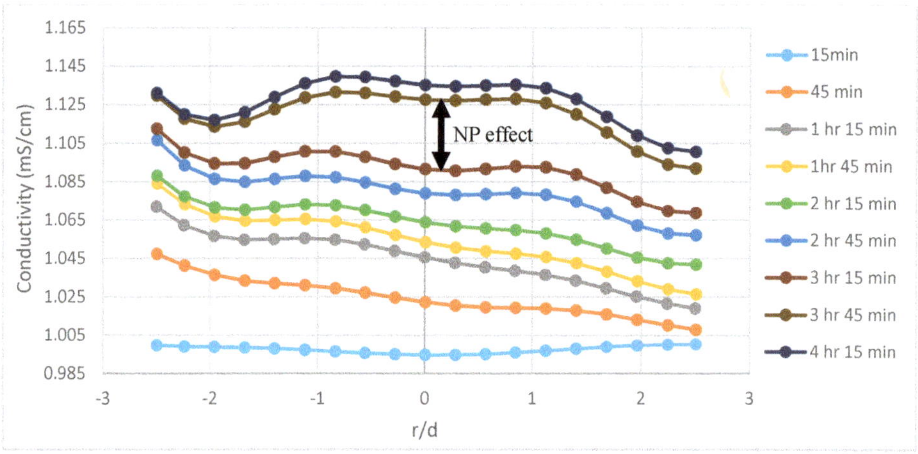



**Figure 8.** The conductivity profiles for P1 during run no. 1 with 1.0 wt.% $SiO_2$, taking the central horizontal region of the tomograms into account. The effects of nanofluid are highlighted with an arrow.

### 3.2. Resistivity vs. Time

The ability of ERT to monitor the local flooding efficiency of different fluids was assessed during the scope of the experiments by considering resistivity plots as a function of time (Figure 9). The resistivities were found to gradually decrease during all the runs as oil was displaced from the sensing planes and replaced with brine. From the data it is evident that the local recovery efficiency was highest during the 1 wt% $SiO_2$ run, followed by the 0.5 wt% $SiO_2$ run and, lastly, the brine only run. The effects of nanofluids are clearly shown by the gradients of the plots, with both the 0.5 wt% and 1.0 wt% runs revealing sudden downward trends after the introduction of nanofluids which were not present during the brine runs. The plot also reveals 1 wt% $SiO_2$ to have a quicker and slightly more profound effect on oil recovery than the 0.5 wt% $SiO_2$ run. For example, taking run no.1 into account, the sudden downward trend during the 1 wt% $SiO_2$ run occurs approximately 55 min after the introduction of nanofluids, whilst the downward gradient occurs approximately 115 min after nanofluid injection. Additionally, since one pulsation of nanofluid is approximately 10 mL, the volume of flooding fluids before a change in gradient could be approximated, which was found to be 5.5 pulsation cycles of brine/nanofluids each (where every pulsation = 10 mL) for the case of 1 wt% $SiO_2$ and 12 pulsation cycles of brine/nanofluid for the 0.5 wt% $SiO_2$ run. The quicker rate of recovery may be a combination of IFT reduction, nanofluid viscosity enhancement, or temporary log-jamming. Another important feature to note is the local resistivity curves correlate well with the oil recovery data. For the brine only runs, the resistivity decreases until no more oil is produced after approximately 170 min of flooding. Contrastingly, the runs with nanofluid are found to enhance oil recovery, consequently lowering the resistivity below the runs with only brine. An anomaly can be observed in the resistivity data between 195–205 min for the 1 wt% $SiO_2$ run, in which the resistivity is found to rise sharply and then decrease back to the original trend. This is most likely the displacement of a large oil zone, revealing resistivity to rise as the oil pocket begins to flow and then decrease as it leaves then sensing zone. This assumption is also confirmed by the tomograms (Section 3.1) which reveal a large zone of oil being displaced at the same time.

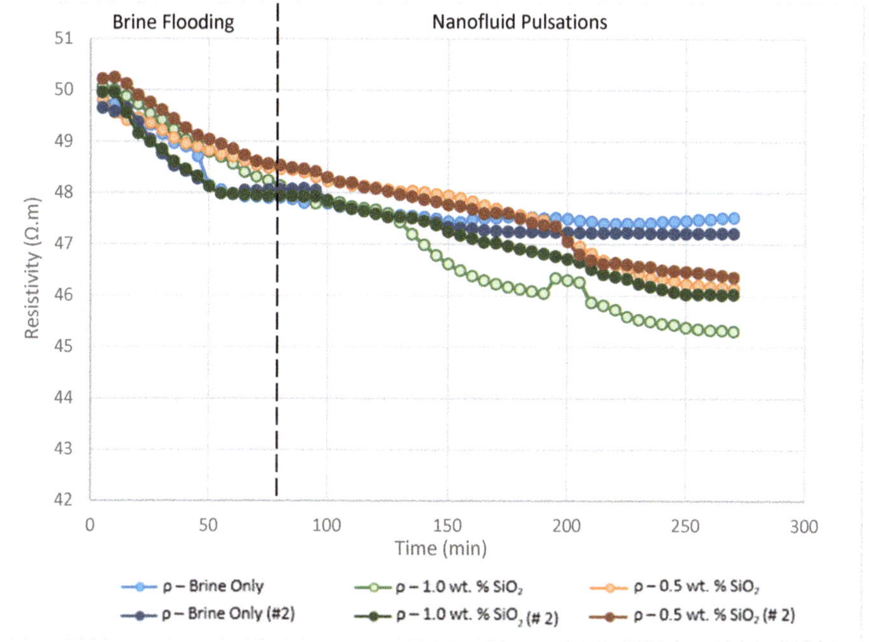


**Figure 9.** Plot of average resistivity vs oil recovered as function of time for runs no. 1 and 2 for brine flooding and both 0.5 wt% and 1.0 wt% $SiO_2$ nanofluid concentrations (see Table 4).

### 3.3. Oil Recovery Rate

The slope function in excel was utilized in order to compare local recovery rate for the various runs. The charts reveal information on the rate of oil recovery since they describe changes in local resistivity, and resistivity in this case is solely a function of oil present in the sensing zones. Since the initial resistivities and oil saturations could not be fixed at exactly the same values, rather than comparing the actual values of resistivity against time, the gradient of the trends in resistivity are compared. Figures 10–13 to show the effects of nanofluids on the local recovery rate when compared with brine-only runs. The slope was calculated between each data point (every 5 min) and presented in the form of a bar chart so that both negative and positive fluctuations could be observed, thus allowing for oil entering and leaving the sensing zone to be accounted for. Figures 10 and 11 reveal the oil recovery rate for both the brine-only runs to stagnate after approximately 170 min. On the contrary the runs involving nanofluids reveal large fluctuations in gradients after 170 min highlighting the ability of SiO$_2$ nanofluids to release residual oil zones and prolong the recovery process.

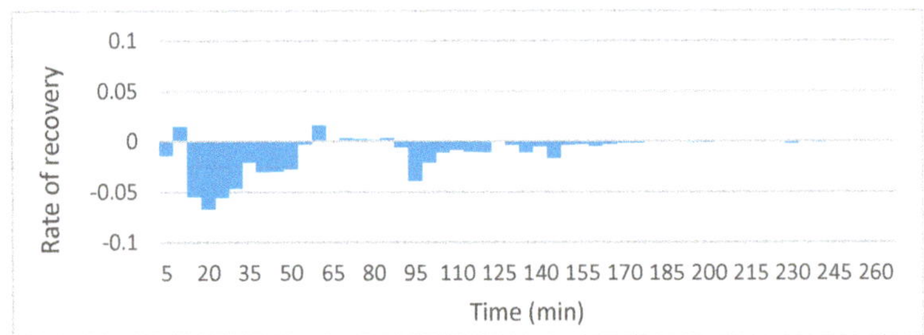

**Figure 10.** The slope function method to reveal the trend in local resistivity during brine-only run no.1.

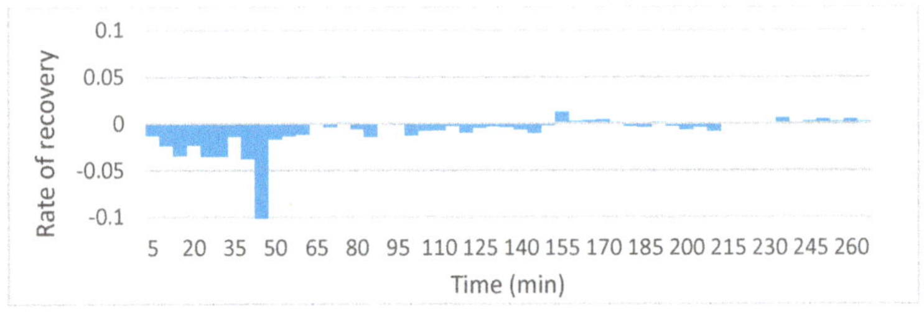

**Figure 11.** The slope function method to reveal the trend in local resistivity during brine-only run no. 2.

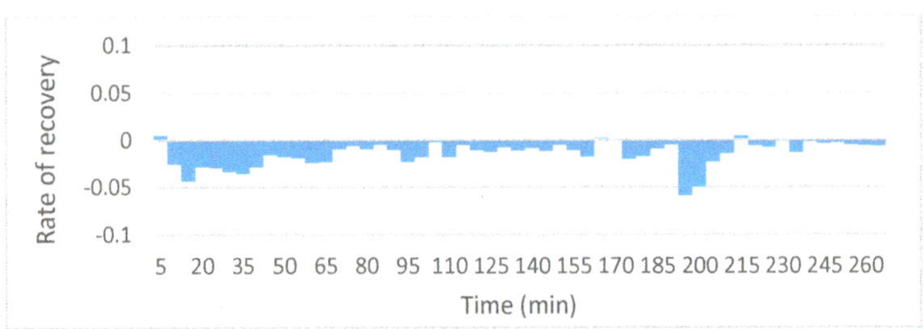

**Figure 12.** The slope function method to reveal the trend in local resistivity during the 0.5 wt% SiO$_2$ run no.1.

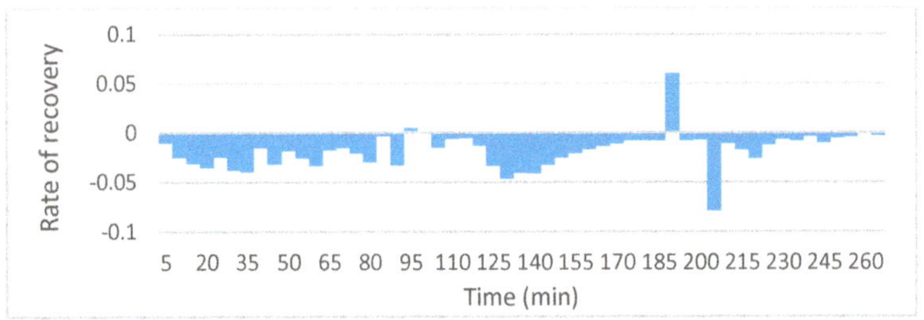

**Figure 13.** The slope function method to reveal the trend in local resistivity during the 1.0 wt% SiO$_2$ run no.1.

### 3.4. Cumulative Oil Recovery

The oil recovery for each run is plotted in Figure 14 as a function of injected brine volume (with and without nanoparticles) related to percentage of pore volumes. The results reveal the nanofluids to be slightly more efficient than the other runs using only brine, with the 0.5 wt% $SiO_2$ nanofluid resulting in an additional 2.4% (4.6 mL) oil recovery, and the 1.0 wt% run allowing for an additional 4.3% enhanced recovery (8.3 mL) (Table 4). The nanofluid pulsations were introduced after approximately 0.8 PV of brine flooding, and the nano-EOR effect was observed after approximately 1.35 PV of nanofluid pulsations in both cases. In contrast, the cumulative oil recovery during the brine-only runs were seen to plateau after around 1.2 PV and became stagnant thereafter. Both the resistivity plots and tomograms confirm the results observed in the cumulative oil recovery plot by revealing rapid changes at the same period (180–240 min) and are related to enhanced oil recovery. Therefore, ERT was found to be successful in relating changes in local resistivity to the overall oil recovery. The underlying reasons for $SiO_2$ nano-EOR effects are widely studied in literature and are believed to be a combination of viscosity enhancement of the displacing phase, IFT reduction, and temporary log-jamming.

An uncertainty analysis was also conducted to account for the degree of uncertainty during the experimental runs and to highlight the repeatability of the results for this novel setup. The vertical error bars for cumulative oil recovery represent the standard deviation of the data set across three experimental runs for each scenario and reveal the sand-pack system to yield very repeatable results with errors of less than 2%. All the runs with no nanoparticles showed recovery to stagnate after around 1.2 PV, which is in line with resistivity plots revealing the same stagnation. The horizontal error bars account for the small change in pressure drop over time as the fluid leaves the drainage bags. The fluid level was found to decrease by 7 cm which corresponded to a negligible change of 0.9 psi or an error of 4% over the duration of the experiments. An error was also attributed to the oil residue left behind in the tubing at the outlet of the sand-pack which was quantified by weighing the tubing section before and after each run. The error was found to be negligible with an average of less than 0.5 g of oil residue left behind, corresponding to a negligible error of 0.06%.

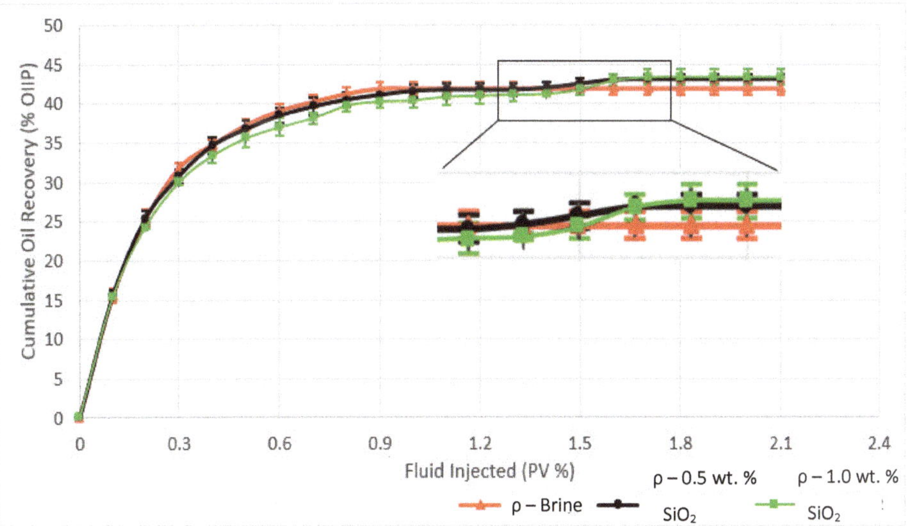


**Figure 14.** Cumulative oil recovery as a function of injected brine volume (with and without nanoparticles) related to pore volume percentage for the 3 flooding scenarios. The highlighted section reveals the duration of the flood where the nano-EOR effects can be observed.

**Table 4.** Summary of sand-pack experiments where the initial water and oil volumes are fixed at 93.2 mL and the initial oil volume is 186.4 mL, respectively. Where %PV is the percentage of pore volume, OIIP is the oil initially in place, WF1 is the water-flooding stage, and NF represents the nanofluid pulsations.

| Fluid Type | Initial Oil Saturation, % PV | Run | Oil Recovery, % OIIP | | Residual Oil Saturation, % PV | | Total Recovery, % OIIP |
|---|---|---|---|---|---|---|---|
| | | | WF1 | NF Pulsations | WF1 | NF Pulsations | |
| Brine | 66 | 1 | 42.6 | - | 37.8 | - | 42.6 |
| | | 2 | 40.9 | - | 39.0 | - | 40.9 |
| | | 3 | 42.0 | - | 38.3 | - | 42.0 |
| | | Mean | 41.8 | - | 38.4 | - | 41.8 |
| 0.5 wt% $SiO_2$ | 66 | 1 | 41.0 | 2.8 | 38.9 | 37.0 | 43.8 |
| | | 2 | 42.0 | 2.0 | 38.3 | 36.9 | 44.0 |
| | | 3 | 42.8 | 2.3 | 37.8 | 36.2 | 45.1 |
| | | Mean | 41.9 | 2.4 | 38.3 | 36.7 | 44.3 |
| 1.0 wt% $SiO_2$ | 66 | 1 | 40.7 | 4.1 | 39.1 | 36.4 | 44.8 |
| | | 2 | 40.8 | 3.8 | 39.1 | 36.5 | 43.0 |
| | | 3 | 41.2 | 5.2 | 38.9 | 35.4 | 45.4 |
| | | Mean | 41.4 | 4.3 | 39.0 | 36.1 | 44.7 |

### 3.5. Recovery Mechanisms

The effective viscosity was measured using an Anton Paar MCR 301 rheometer at 25 °C, and the interfacial tensions were measured using a KSV CAM 200 optical tensiometer. An IFT analysis was done in order to determine its influence as a recovery mechanism. The IFT was measured for the brine-mineral oil system at room temperature with a base value of 41 mN/m. The presence of $SiO_2$ nanoparticles was found to reduce the interfacial tension between the oil and aqueous phases with an increase in concentration resulting in a reduction in interfacial tension, a trend similar to those found in literature. The sand-pack flooding results also reveal the recovery efficiency to increase slightly with nanoparticle concentration as seen in Figure 14 and IFT reduction seems to be a good candidate as the dominant recovery mechanism (Figure 15), although log-jamming may account for some additional recovery. However, quantifying the extent of recovery from log-jamming is difficult as the nanoparticles or pore throats are not visible by ERT, but the results would suggest log-jamming played a part as the IFT was approximately the same when going from 0.5 wt.% to 1.0 wt.%, however there was small trend in oil recovery observed showing 1.0 wt.% $SiO_2$ to be slightly more effective.

The dynamic viscosities of brine and different $SiO_2$ nanoparticle concentrations were considered in Figure 16. As predicted, the concentration of nanoparticles had an ignorable effect on viscosity. A slight increase in viscosity can be observed after the addition of nanoparticles to the brine solution from 0.818 mPa·s to 1.08 mPa·s after the addition of 1 wt.% $SiO_2$. However, such a small change cannot be attributed to any substantial EOR effects.

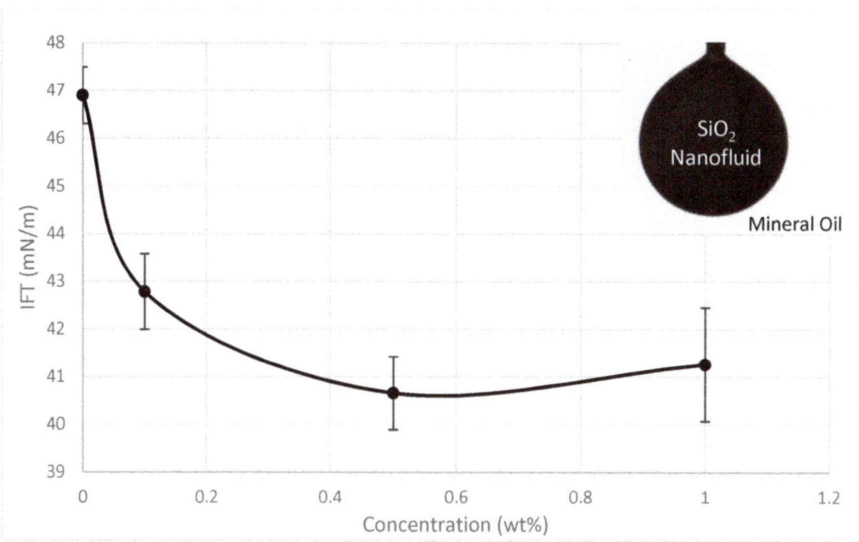



**Figure 15.** Interfacial tension of brine and various SiO$_2$ nanofluid concentrations.

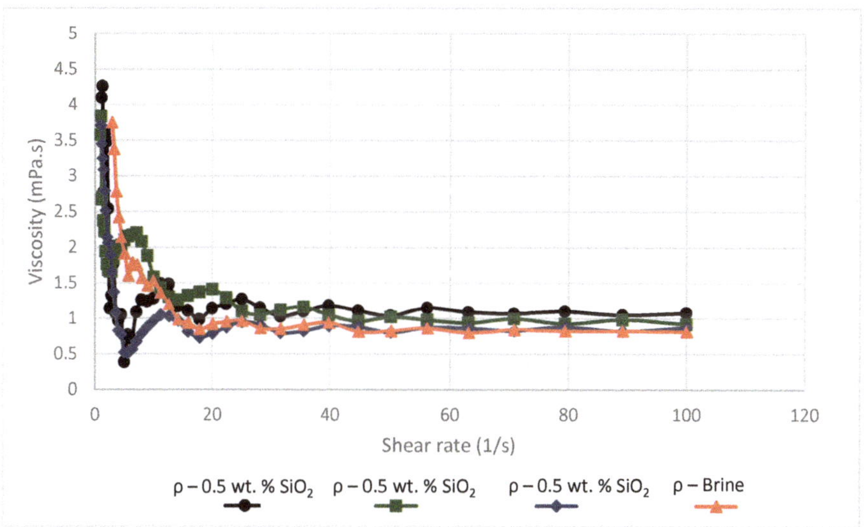



**Figure 16.** Viscosity against shear rate for brine and various SiO$_2$ nanofluid concentrations.

## 4. Conclusions

A novel low-budget sand-pack setup and effluent collection system was developed during the scope of this research and showed to yield similar results to those in literature. Tomography (ERT) showed to be an effective tool in monitoring multiphase flow behavior during flooding experiments to investigate oil recovery. The tomograms also crucially reveal the effect of SiO$_2$ nanofluids on residual oil zones and the dynamics of flow behavior. The oil recovery results reveal the presence of SiO$_2$ nanoparticles to enhance recovery during all runs. The difference in recovery from the 0.5 wt% and 1.0 wt% concentrations was not substantial but there was a slight trend observed showing an increase in concentration yielded approximately an extra 2% recovery. This suggests that log-jamming played a role in the recovery process as the concentration increased, because the IFT was almost the same when comparing the 0.5 wt.% and 1.0 wt.% concentrations. However, quantifying the extent of recovery from log-jamming is not possible due to limitations in ERT's spatial resolution. The plots of local resistivity reveal large differences between nanofluid and brine-only runs with the resistivity trends stagnating during the brine-only runs after approximately 170 min. The slope function technique was introduced as a method of comparing the effects of nanoparticles on local oil recovery rate since the observed changes in resistivity during nanofluid runs were related to oil zones being released. Conductivity profiles along the central x-axis against time to show the effects of nanoparticles and highlight the physical understanding of the process. The changes in oil recovery by $SiO_2$ nanofluids are highlighted by the arrows to show the increases in conductivity along the central x-axis of the tomograms as oil displacement efficiency increases, which weren't present during the brine only runs. Furthermore, the flow behaviour was slightly altered with more pronounced peaks in conductivity after the introduction of nanofluids. A recovery analysis was also performed to determine the underlying reasons behind the observed EOR-effects. Further research involves focusing on a comparative study using different nanoparticles and introducing surfactant and polymer blends. Additionally, the ability of ERT as a monitoring tool is being considered for use in a more realistic core-flooding scenario.

**Author Contributions:** Conceptualization, P.N., M.W.; data curation, P.N.; formal analysis, P.N.; funding acquisition, M.W. and D.W.; investigation, P.N., Z.H., D.W., and M.W.; methodology, P.N.; project administration, M.W.; resources, D.W. and M.W.; software, P.N.; supervision, D.W. and M.W.; validation, P.N. and M.W.; visualization, P.N.; writing—original draft, P.N.; writing—review & editing, Z.H., D.W., and M.W.

**Funding:** This research was funded by Engineering and Physical Sciences Research Council (EPSRC), part of UK Research and Innovation (UKRI).

**Acknowledgments:** A special thanks to the lab technician, Robert Harris, from the School of Chemical and Process Engineering, Leeds, UK, who assisted with the experimental setup and materials.

**Conflicts of Interest:** The authors declare no conflict of interest.

## References

1. Binley, A.; Shaw, B.; Henry-Poulter, S. Flow Pathways in Porous Media: Electrical Resistance Tomography and Dye Staining Image Verification. *Meas. Sci. Technol.* **1996**, *7*, 384–390. [CrossRef]

2. Gnecchi, J.G.; Chávez, A.G.T.; Campos, G.C.; Peregrino, V.O.; Pineda, E.M. Soil Water Infiltration Measurements Using Electrical Impedance Tomography. *Chem. Eng. J.* **2012**, *191*, 13–21. [CrossRef]

3. Kaipo, P. Measuring Water Flow in Soil Using Electrical Impedance Tomography. Ph.D. Thesis, Savonia University of Applied Sciences, Kuopio, Finland, 2014.

4. Singh, S.; Ahmed, R.M.; Growcock, F. Vital role of nanopolymers in drilling and stimulations fluid applications. In Proceedings of the SPE 130413, SPE Annual Technical Conference and Exhibition, Florence, Italy, 19–22 September 2010.

5. Ogolo, N.; Olafuyi, O.; Onyekonwu, M. Enhanced Oil Recovery Using Nanoparticles. In Proceedings of the SPE Saudi Arabia Section Technical Symposium and Exhibition, Al-Khobar, Saudi Arabia, 8–11 April 2016.

6. Rodriguez, E.; Roberts, M.R.; Yu, H.; Huh, C.; Bryant, S.L. Enhanced Migration of Surface-Treated Nanoparticles in Sedimentary Rocks. In Proceedings of the SPE 124418 Annual Technical Conference and Exhibition, New Orleans, LA, USA, 4–7 October 2009.

7. Hendraningrat, L.; Zhang, J. Polymeric Nanospheres as a Displacement Fluid in Enhanced Oil Recovery. *Appl. Nanosci.* **2015**, *5*, 1009–1016. [CrossRef]

8. Elsayed, R.; Fattah, A. A Comparative Study between Nanoparticles Method and the Other Methods for Increasing Heavy Oil Recovery. Master's Thesis, Suez University, Suez, Egypt, 2014.

9. Ju, B.; Fan, T.; Li, Z. Improving Water Injectivity and Enhancing Oil Recovery by Wettability Control Using Nanopowders. *J. Pet. Sci. Eng.* **2012**, *86*, 206–216. [CrossRef]

10. El-Diasty, A.I. The Potential of Nanoparticles to Improve Oil Recovery in Bahariya Formation, Egypt: An Experimental Study. In Proceedings of the SPE 174599 Annual Technical Conference and Exhibition, Cairo, Egypt, 11–13 August 2015.

11. Wang, M.; Yin, W.; Holliday, N. A highly adaptive electrical impedance sensing system for flow measurement. *Meas. Sci. Technol.* **2002**, *13*, 1884–1889. [CrossRef]
12. Baldygin, A.; Nobes, D.S.; Mitra, S.K. New laboratory core flooding experimental system. *Ind. Eng. Chem. Res.* **2014**, *53*, 13497–13505. [CrossRef]

 *sensors*

*Article*

# Monitoring Surface Defects Deformations and Displacements in Hot Steel Using Magnetic Induction Tomography [†]

**Fang Li [1], Stefano Spagnul [2], Victor Odedo [1] and Manuchehr Soleimani [1,]***

[1]   Engineering Tomography Lab (ETL), University of Bath, Bath, BA2 7AY, UK
[2]   Product Division, Ergolines lab s.r.l., Area Science Park, Bldg. R3 Padriciano, Trieste 34149, Italy
*   Correspondence: m.soleimani@bath.ac.uk
†   This paper is an extended version of paper Odedo, V.; Soleimani, M.; Spagnul, S. Magnetic Induction Tomography for Imaging Defects and Deformations of External Surfaces, In proceedings of the 9th World Congress on Industrial Process Tomography, Bath, UK, 2–6 September 2018.

Received: 29 May 2019; Accepted: 5 July 2019; Published: 8 July 2019

**Abstract:** Magnetic Induction Tomography (MIT) is a non-invasive imaging technique that has been widely applied for imaging materials with high electrical conductivity contrasts. Steel production is among an increasing number of applications that require a contactless method for monitoring the casting process due to the high temperature of hot steel. In this paper, an MIT technique is proposed for detecting defects and deformations in the external surfaces of metal, which has the potential to be used to monitor the external surface of hot steel during the continuous casting process. The Total Variation (TV) reconstruction algorithm was developed to image the conductivity distributions. Nonetheless, the reconstructed image of the deformed square metal obtained using the TV algorithm directly does not yield resonable images of the surface deformation. However, differential images obtained by subtracting the image of a perfect square metal with no deformations from the image obtained for a deformed square metal does provide accurate and repeatable deformation information. It is possible to obtain a more precise image of surface deformation by thresholding the differential image. This TV-based threshold-differencing method has been analysed and verified from both simulation and experimental tests. The simulation results reported that 0.92% of the image region can be detected, and the experimental results indicated a 0.57% detectability. Use of the proposed method was demonstareted in a MIT device which was used in continuous casting set up. The paper shows results from computer simulation, lab based cold tests, and real life data from continoeus cating demonstating the effectiveness of the proposed method.

**Keywords:** Magnetic Induction Tomography; imaging defects; imaging deformations; total variation algorithms; threshold-differencing algorithms; continuous casting

---

## 1. Introduction

Magnetic Induction Tomography (MIT) is a non-invasive and contactless imaging technique that is used to display the images describing passive electromagnetic properties (PEP), i.e., conductivity, permeability and permittivity [1,2]. The fundamental principle of MIT can be explained by the mutual inductance and eddy current theory: injecting an alternating current through excitation coils to generate a primary magnetic field, which interact with a conductive sample to induce a secondary magnetic field that can be detected by the sensor coil [3,4]. MIT has typically focused on imaging conductivity distributions of materials with high electrical conductivity by modelling the eddy currents in the forward model and then acquiring the conductivity distribution by solving the inverse problem [5]. The inverse problem in MIT is a major challenge, but this has been solved conventionally by The Tikhonov [3] and Total Variation (TV) [6] regularization algorithms. However, previous works have demonstrated that the Tikhonov regularization method may produce overly smoothed reconstructed images with blurred boundaries between different materials. The quality of the reconstructed images can be improved by an enhanced inverse solver, such as the Total Variation regularization technique [7,8].

In recent years, MIT has been widely proposed as an appropriate imaging technique in both industrial and medical applications, such as brain stroke detection [5], molten metal flow monitoring [9–11], pipelines inspection [12,13], multi-phase flow imaging [14] and non-destructive testing (NDT) for material characterization [15]. It's well known that steel production has been attracting an increasing level of interest for potential use in various applications that require a contactless method for remotely monitoring the casting process. Many previous works [9–11] have sought to monitor the molten steel flow by reconstructing the conductivity distributions of the internal area of interest. However, MIT presents difficulties in detecting and imaging external surface defects and deformations due to its low spatial resolution. The inspection of the outer surface of pipelines using MIT was analyzed in [12] through a traditional MIT pixel-based reconstruction method (PBRM) and a narrowband pass filtering method (NPFM). Metal samples in a cylindrical shape with different types of damage at the outer surface have been investigated as defected pipelines using an 8-channel MIT system, which demonstrated that PBRM can retrieve information with a detectability of 10%, while NPFM can achieve a resolution of 2%.

This paper focuses on detecting the defects or deformations on the external surface of metallic targets using a Total variation algorithm-based MIT. This novel algorithm is described in [16]. Here, several simulation and experimental tests were carried out to investigate the ability of the novel algorithm to identify and image deformations and defects in metal. The proposed TV-based threshold-differencing algorithm was compared to the TV reconstruction method, and significant improvements were observed, which indicates that the novel algorithm for MIT may provide a suitable and valuable method for monitoring surface defects, deformations and displacements in hot steel continuous casting processes.

## 2. Methodology

The forward problem in MIT for surface defect application is based on the laws of induction and the eddy currents which are induced in the magnetic field with an alternating current [3]. The governing equation can be written as Equation (1):

$$\nabla \times \frac{1}{\mu} \nabla \times A + j\omega\sigma A = J_s \tag{1}$$

where $\mu$ is the permeability, $\omega$ is the angular frequency, $A$ is the total magnetic vector potential as a result of the current source $J_s$ and the effect of eddy current induced by the electrical conductivity $\sigma$. The current density can be determined by the magnetic vector potential according to the Biot-Savart Law. Equation (1) is solved by approximating the system as a combination of linear equations in small elements with appropriate boundary conditions using the Galerkin's approximation [4,7]:

$$\int_{\Omega_{all}} \left( \nabla \times N_i \cdot \frac{1}{\mu} \nabla \times A \right) dv + \int_{\Omega_{all}} (j\omega\sigma N_i \cdot A) dv = \int_{\Omega_s} (\nabla \times N_i \cdot T_s) dv \tag{2}$$

where $N_i$ is the linear combination of edge shape functions, $\Omega_S$ is the current source region (excitation coil), $\Omega_{all}$ is the entire region (current source region and eddy current region), $T_s$ is the electric vector potential, which is defined as:

$$J_s = \nabla \times T_s \tag{3}$$

Then, after applying the volume integration equation, the induced voltage in the measuring coil can be calculated:

$$V_{mn} = -j\omega \int_{\Omega_s} (A \cdot J_0) dv \tag{4}$$

where $J_0$ is the unit current density passing through the coil.

The Jacobian matrix $J$ can be expressed by the relationship between the induced voltage in the sensing coil and the conductivity:

$$J = \frac{\partial V_{mn}}{\partial \sigma_x} = -\omega^2 \frac{\int_{\Omega_x} A_m \cdot A_n dv}{I} \tag{5}$$

where $\sigma_x$ is the conductivity of pixel $x$, $\Omega_x$ is the volume of the perturbation, $A_m$ is the forward solver of excitation coil $m$ excited by $I$ and $A_n$ is the forward solver of sensor coil excited by unit current.

The inverse problem in MIT is defined as the retrieval of the unknown conductivity distributions $\sigma$ of targets from the measured voltage $V_{measured}$, expressed by the linear equation:

$$\Delta v = J \Delta \sigma \tag{6}$$

where $\Delta \sigma = \sigma - \sigma_0$, $\Delta v = V_{measured} - F(\sigma_0)$, $F$ is the forward operator, $F(\sigma_0)$ means the initial estimate voltage obtained from forward problem, $\sigma_0$ is the initial estimate conductivity, $J$ is the Jacobian matrix obtained from forward problem. The resolution of the reconstructed images can be improved by increasing the size of Jacobian matrix.

The inverse problem can then be represented by solving the least-square problem:

$$\sigma_\alpha = argmin_{\Delta\sigma} (\|J\Delta\sigma - \Delta v\|^2) \tag{7}$$

The theory and application of both the Tikhinov and TV inverse solvers have been investigated and explained detailed in [7]. The total variation method can be defined by adding a penalty term to Equation (7):

$$\sigma_\alpha = argmin_{\Delta\sigma} (\|J\Delta\sigma - \Delta v\|^2 + \alpha \|\nabla \Delta\sigma\|_1) \tag{8}$$

where $\alpha$ is the regularization parameter, $\nabla$ is the gradient and $\|\cdot\|_1$ is the $l_1 - norm$. The anisotropic version of the discrete TV functional [6,8] is adopted in this paper:

$$\alpha \|\nabla \Delta\sigma\|_1 = \alpha_x \|\nabla_x \Delta\sigma\|_1 + \alpha_y \|\nabla_y \Delta\sigma\|_1 + \alpha_z \|\nabla_z \Delta\sigma\|_1 \tag{9}$$

It should be mentioned that in this study, $\alpha_x = \alpha_y = \alpha_z = 0.1$.

Then, the difficulty is to solve the constrained optimization problem:

$$x_\alpha = argmin_{\Delta\sigma} \alpha \|\nabla \Delta\sigma\|_1 \; such \; that \; \|J\Delta\sigma - \Delta v\|^2 < \rho \tag{10}$$

The above constrained optimization problem can be controlled by applying the Split Bregman iteration, which is an iterative method based on Bregman distance [8]. Then, the problem may be transferred to solve the Split Bregman equations [7]:

$$\Delta\sigma^{k+1} = argmin_{\Delta\sigma} \frac{1}{2} \|J\Delta\sigma - \Delta v^k\|^2 + \frac{\beta}{2} \|d^k - \nabla\Delta\sigma - b_d^k\|^2 \tag{11}$$

$$d^{k+1} = argmin_d \alpha \|d\|_1 + \frac{\beta}{2} \|d - \nabla\Delta\sigma^{k+1} - b_d^k\|^2 \tag{12}$$

$$b_d^{k+1} = b_d^k + \nabla\Delta\sigma^{k+1} - d^{k+1} \tag{13}$$

where $k$ is the $k^{th}$ iteration and $d$ is an auxiliary variable.

Based on [16], the TV algorithm seeks to achieve the conductivity distribution of a perfect target $\Delta\sigma_{ref}$ which retains as reference data. Consequently, the conductivity distribution of the target contain a deformation, $\Delta\sigma_{def}$, which can be obtained using the TV method as well. Subtracting the reference conductivity distribution from the target conductivity distribution gives the deformed target conductivity as:

$$\Delta\sigma_{diff} = \Delta\sigma_{def} - \Delta\sigma_{ref} \tag{14}$$

The image of the deformed target conductivity distribution was expected to indicate the location and the deformation information. And it's obvious that the negative differential conductivity values represent the depression deformation part, and that the positive differential conductivity values represent the bulging. However, the reconstructed image of the deformed target is inadequate. For instance, the differential reconstructed image for a metal square with a depression deformation on one side may also show a bulging deformation on the opposite side due to the ill-posed nature of the inverse problem. This mirror effect can be tackled by comparing the absolute value of the maximum differential conductivity of deformed target $|\max\{\Delta\sigma_{def}\}|$ to the absolute value of the minimum differential conductivity $|\min\{\Delta\sigma_{def}\}|$, and then eliminating the minimum to obtain the threshold-differential conductivity distribution, defined as [16]

$$\Delta\sigma_{thres} = \begin{cases} \begin{cases} \Delta\sigma_{def} & \text{when} & \Delta\sigma_{def} > 0 \\ 0 & \text{when} & \Delta\sigma_{def} \leq 0 \end{cases} & \text{for} & \left|\max\{\Delta\sigma_{def}\}\right| > \left|\min\{\Delta\sigma_{def}\}\right| \\ \begin{cases} \Delta\sigma_{def} & \text{when} & \Delta\sigma_{def} < 0 \\ 0 & \text{when} & \Delta\sigma_{def} \geq 0 \end{cases} & \text{for} & \left|\max\{\Delta\sigma_{def}\}\right| < \left|\min\{\Delta\sigma_{def}\}\right| \end{cases} \tag{15}$$

In fact, the quality and accuracy of deformed image can be further improved by forcing values below or above a certain threshold $\gamma$ to zero, such that:

$$\Delta\sigma_{thres} = \begin{cases} \begin{cases} \Delta\sigma_{def} & \text{when} & \Delta\sigma_{def} > \gamma.\max\{\Delta\sigma_{def}\} \\ 0 & \text{when} & \Delta\sigma_{def} \leq \gamma.\max\{\Delta\sigma_{def}\} \end{cases} & \text{for} & abs(\max\{\Delta\sigma_{def}\}) > abs(\min\{\Delta\sigma_{def}\}) \\ \begin{cases} \Delta\sigma_{def} & \text{when} & \Delta\sigma_{def} < \gamma.\min\{\Delta\sigma_{def}\} \\ 0 & \text{when} & \Delta\sigma_{def} \geq \gamma.\min\{\Delta\sigma_{def}\} \end{cases} & \text{for} & abs(\max\{\Delta\sigma_{def}\}) < abs(\min\{\Delta\sigma_{def}\}) \end{cases} \tag{16}$$

where $0 \leq \gamma \leq 1$. Introducing the $\gamma$ will remove some amount of the noise data to make the results more stable and precise. After applying different values of $\gamma$ during the research, we discovered that $0.3 \leq \gamma \leq 0.5$ gave the optimal reconstructed images for the scenarios considered here. Appropriate values for $\gamma$ give threshold-differential images that display the position of deformation individually. The complete images with deformation of the target can be obtained simply by adding the reference conductivity distribution to the threshold-differential conductivity distribution, such as:

$$\Delta\sigma_{tot} = \Delta\sigma_{ref} + \Delta\sigma_{thres} \tag{17}$$

where $\Delta\sigma_{tot}$ is the total conductivity distribution.

## 3. Simulation Results

In this section, we displayed the results in [16], which investigated the ability of the TV-based threshold-differential reconstruction method by reconstructing a $74 \times 74$ mm$^2$ square sample with surface deformations using simulation data based on the laboratory prototype ( Figure 1A). The metal square sample was located at the center of a 16 coil array MIT system, as illustrated in Figure 1B, with an operational frequency of 130 Hz.

Figure 2 shows the conductivity distribution obtained for a perfect square metal without any deformation using the TV reconstruction algorithm, which indicates that the algorithm can produce reconstructed images representing the square metal with reasonable accuracy. Then, bulging and depression deformation were considered.

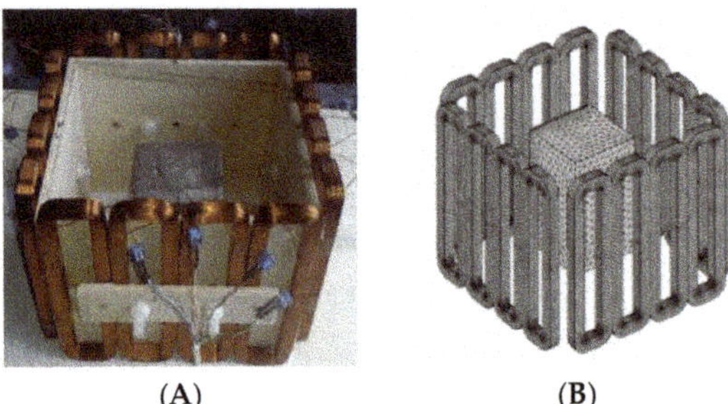

**Figure 1.** (**A**) An illustration of the laboratory prototype and (**B**) the simulation scenario.

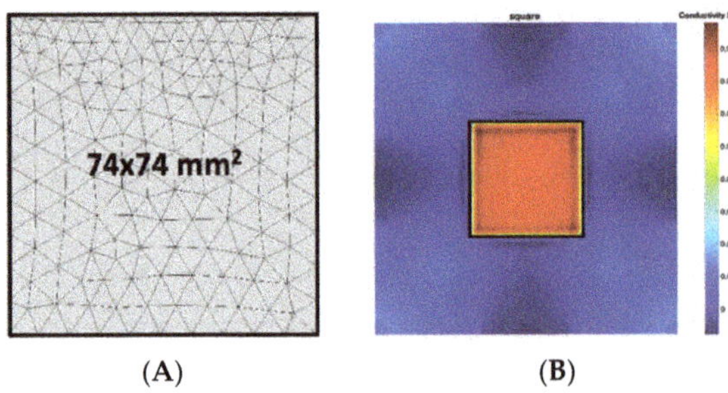


**Figure 2.** (**A**) Simulation setting of a perfect metal sample without any deformation and (**B**) Corresponding reconstructed image obtained by the simulation data.

Figure 3 illustrates the simulation scenarios and the corrosponding reconstructed conductivity distribution images using TV algorithm for (A) a metal with a surface bulge on top, and (B) a metal with a surface depression on top, where the black lines represent the location of perfect square, the red line represents the boundaries of simulation setting of sample with bulging and the blue line represents the boundaries of sample with depression. The two scenarios were deformed by 9.25 mm at the midsection of one external surface. It's obvious that although the bulge deformation in Figure 3A can be detected, the image may also be interpreted to include a depression at the bottom, as shown in Figure 3B.

Figure 4 illustrates the image obtained for the deformed target conductivity from applying Equation (14) for (A) the scenario with a bulge deformation on top, and (B) the scenario with a depression deformation on top, which therefore show only the effects of the deformation. It is obvious that a bulge (red) represented by positive numbers on the colorbar. A depression (blue) represented by negative numbers on the colorbar in both images can also be observed. However, the maximum and minimum values of $\Delta\sigma_{def}$ in the colorbar of the images shown in Figure 4 give a clue about the actual deformation in the considered scenario. In Figure 4A, the absolute value of the maximum deformed target conductivity, which represents a bulge, is greater than the absolute value of the minimum deformed target conductivity that represents a depression. Similarly, Figure 4B shows that the the absolute value of the minimum deformed target conductivity, which represents a depression, is greater than the absolute value of the maximum deformed target conductivity that represents a bulge.

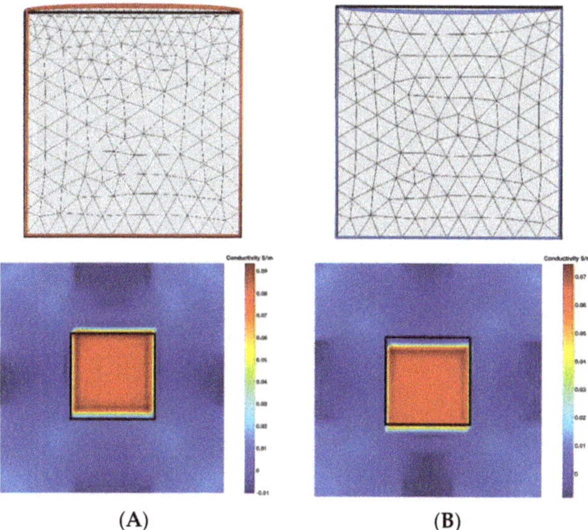

**Figure 3.** The simulation scenarios and the corresponding reconstructed images for (**A**) the metal with a surface bulge on top and (**B**) the metal with a surface depression on top, where the black line represents the actual location of a perfect square metal.

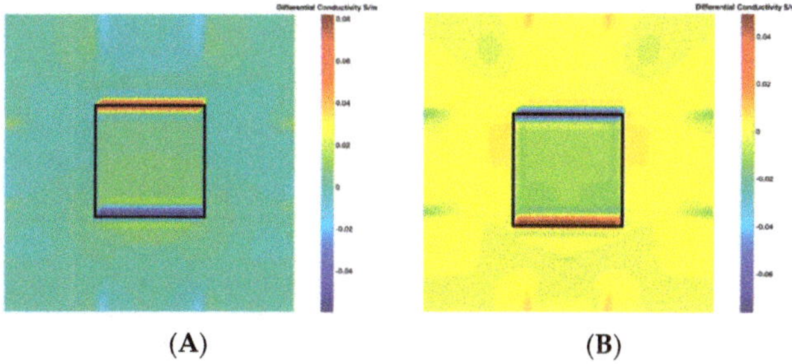

**Figure 4.** The differential conductivity distribution images obtained for a metal with (**A**) a bulge on top and (**B**) a depression on top.

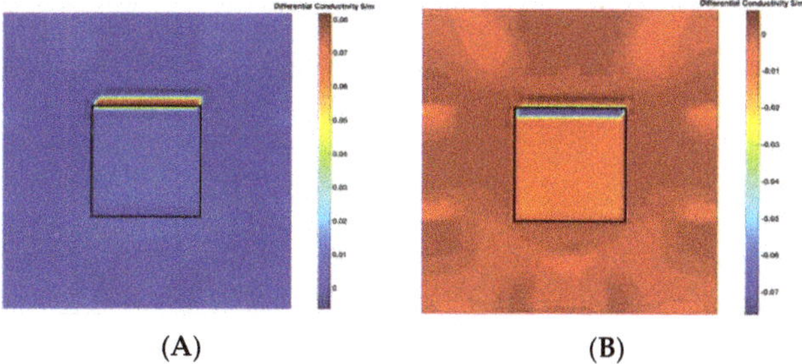

**Figure 5.** The threshold-differential conductivity distribution images obtained for (**A**) the metal with a bulge on top and (**B**) the metal with a depression on top.

The threshold-differential conductivity distribution can be plotted from Equation (15), obtained for the scenarios with deformation to get a more accurate representation of the particular deformation. Figure 5 shows the threshold-differential conductivity distribution obtained for (A) the metal with a bulge, and (B) the metal with a depression. Hence, a bulge represented by positive numbers on the colorbar in Figure 5A and a depression represented by negative numbers on the colorbar in Figure 5B can be observed, which indicates that we can obtain a better representation of the deformation in the scenario from $\Delta\sigma_{thres}$ than from $\Delta\sigma_{def}$.

By applying an appropriate value of $\gamma$ to the Equation (16), the threshold-differential conductivity distribution images illustrate the location of the deformation alone. Figure 6 shows the threshold-differential conductivity distribution images obtained for (A) the metal with a bulge, and (B) the metal with a depression deformation when $\gamma = 0.4$. The threshold-differential conductivity distribution images located the deformation more clearly by applying $\gamma$ in Figure 6 than in Figure 5.

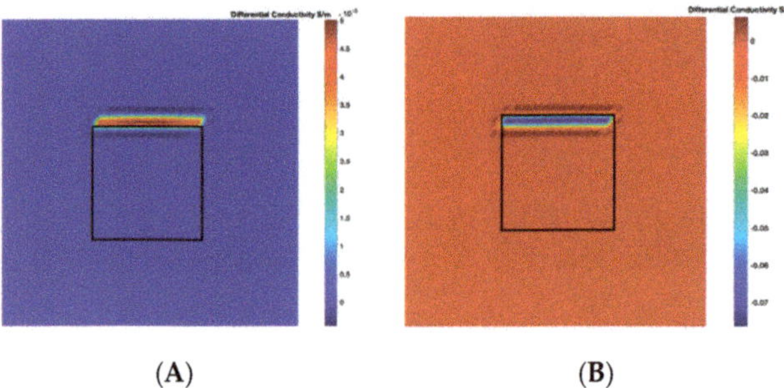

(**A**)  (**B**)

**Figure 6.** The threshold-differential conductivity distribution images obtained when $\gamma = 0.4$ for (**A**) the metal with a bulge deformation on top and (**B**) the metal with a depression deformation on top.

The TV-based threshold-differential image reconstruction algorithm with reference sample was analyzed for the outer shape reconstruction with surface deformations (bulging and surface depression defects), showing potential. Therefore, further research was conducted based on this method to analyze the image reconstruction of shapes with surface deformations. Supplementary simulation data were produced and analyzed: bulging or depression with different depth of the defect; bulging and depression at the same time on different surfaces and on the same surface.

Figure 7 shows the reconstructed image obtained for a square metal ($74 \times 74$ mm$^2$) without any deformation using the TV reconstruction algorithm and larger Jacobian matrix ($100 \times 100$). The resolution in this dissertation can be defined by a pixel resolution in pixel per mm. The size of the square in the reconstructed image is $64 \times 64$ pixels, which means that 1 pixel $= 1.156$ mm; the resolution of this simulation scenario is $1/1.156 = 0.865$ pixel/mm.

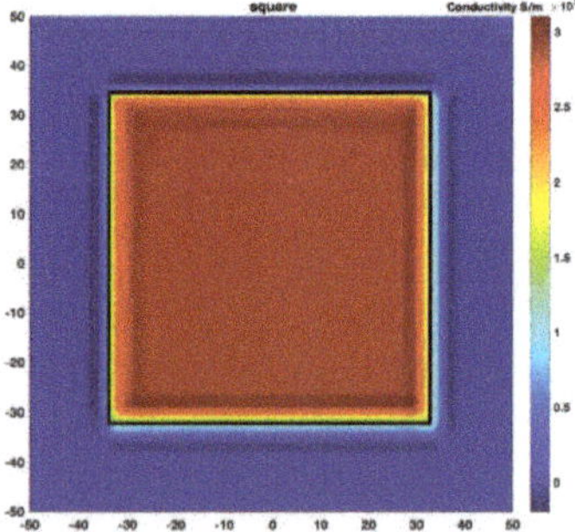

**Figure 7.** Reconstructed image for a square metal without any deformation.

Samples with a bulging or depression defect on the top surface with a length of 74 mm (i.e., in the whole side) and with the 5 mm deformation depths were analyzed.

Table 1 indicates the threshold differential reconstructed images and their contour images obtained for the different forms of defects. The second column is the threshold differential image obtained for the sample with a bulging deformation on top with different defect depths. The fourth column is the threshold differential image obtained for the sample with depression deformation on top. The third and fifth columns are the corresponding contour images of reconstructed images. The blue line is the sample without any deformation, while the black lines are deformation parts. The threshold factor $\gamma$ was set to 0.5. The iteration number used for solving TV function equals 500, in reference to Equations (7)–(9), which can provide reconstructed images with sharp boundaries.

**Table 1.** Threshold Differential reconstructed images for bulging and depression defects.

| Defect Depth (Simulated) | Threshold Differential Image Obtained for the Square with a Bulge Deformation on Top | Defect Depth (Reconstructed) | Threshold Differential Image Obtained for the Square with a Depression Deformation on Top | Defect Depth (Reconstructed) |
|---|---|---|---|---|
| 5 mm | | 3.5 pixels = 4.046 mm | | 4 pixels = 4.624 mm |

The deformation detectability can be defined as:

$$Deformation\ detectability = \frac{area\ of\ the\ detectable\ deformation}{area\ of\ the\ no\ deformed\ sample} \tag{18}$$

So, the deformation detectability of the proposed simulation system is $74 \times 5/\left(200^2\right) = 0.92\%$. In order to evaluate the accuracy of the proposed method applied to detect the surface defect, the normalized mean square error can be defined as [17]:

$$err = \frac{\|X_{estimated} - X_{actual}\|^2}{\|X_{actual}\|^2} \tag{19}$$

where $X_{estimated}$ is the reconstructed defect depth value and $X_{actual}$ is the actual one. For the case in which bulging and depression deformations occur at the same time:

$$err_{B+D} = \left(err_{Bulge} + err_{Depression}\right)/2 \tag{20}$$

By referring to these quantitative metrics and the above results, the conclusion can be drawn that the value of the defect depth can be detected and acquired accurately with an error ($err_{Bulge\ deformation} = 3.6\%$ and $err_{Depression\ deformation} = 0.57\%$) through the proposed method.

Table 2 illustrates the simulation scenario with two deformations at the same time (one bulging on the top surface and one depression at the bottom surface of the sample, both with a depth of 9.25 mm), as well as the differential reconstructed images.

**Table 2.** Differential reconstructed images of two deformations at same time (Pixel = 100).

| Scenarios | Differential Reconstructed Images | Contour Images | Defect Depth(Reconstructed) |
|---|---|---|---|
| 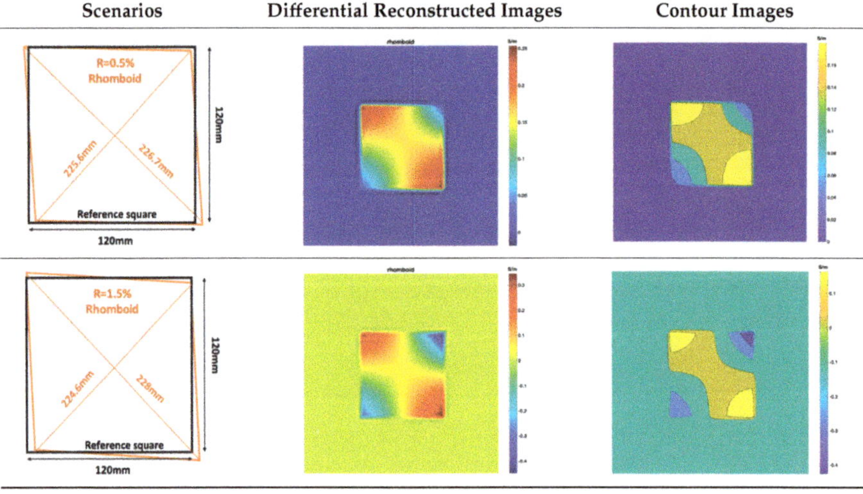 | | | Bulging (top):<br>7 pixels = 8.092 mm<br>$err_{Bulge} = 1.6\%$<br>Depression (bottom):<br>7.1 pixels = 8.208 mm<br>$err_{Depression} = 1.3\%$ |

The results show that the TV-based differencing algorithm has the ability to detect two different types of deformation at the same time. By referring to Equations (19) and (20), the quantitative metrics *err* and the reconstructed defect depth can be acquired with an average error level of 1.45%.

Further simulation work was conducted on bulging and depressions at the same surface, which was achieved by transferring the square sample ($120 \times 120$ mm$^2$) to a rhomboid one. The rhomboid samples can be defined by the parameter $R = \frac{Longest\ Diagonal\ length\ -\ Shortest\ Diagonal\ length}{Shortest\ Diagonal\ length}$. Table 3 illustrates the simulation scenario and the results of the rhomboid problem with $R = 0.5\%$ and $R = 1.5\%$.

**Table 3.** Differential reconstructed images of two deformations at same surface (Pixel = 50).

| Scenarios | Differential Reconstructed Images | Contour Images |
|---|---|---|
| R=0.5% Rhomboid; 225.6mm, 226.7mm; Reference square; 120mm; 120mm | | |
| R=1.5% Rhomboid; 224.6mm, 228mm; Reference square; 120mm; 120mm | | |

The positive values on the color bar represent bulging as well as the negative value mean depression. The conductivity values also give a clue of the amount of deformation. The results shows that the TV-based differencing algorithm has the ability to detect two deformations on the same surface.

## 4. Experimental Tests

Regarding the inspection of pipelines, according to [12], similar results can be obtained when testing the same damaged cylinder samples using the proposed algorithms. In this section, three sets of MIT machine experimental tests were conducted to investigate and display the ability of the threshold-differential algorithm to utilise the Total Variation (TV) reconstruction method to image a $50 \times 50$ mm$^2$ metal cube with surface deformations located at the center of an 8-coil array, as illustrated in Figure 8. The MIT system in this research consists of (i) a host computer (ii) an National Instrument data acquisation system, and (iii) an equally-spaced eight x 50 turns coils with 40 mm diameter circular-array forming a 110 mm diameter imaging-region. The working frequency is 180 kHz. One metal cuboid represents the surface deformation. The demonstrated experiment system detectability is $6 \times 9 / \left( \pi \times 55^2 \right) = 0.57\%$. In this case:

$$\Delta\sigma_{def} = \Delta\sigma_{cube+cuboid} \tag{21}$$

$$\Delta\sigma_{ref} = \Delta\sigma_{cube} \tag{22}$$

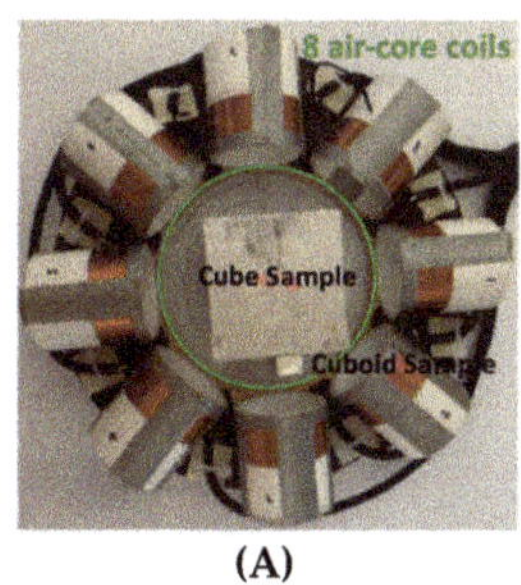

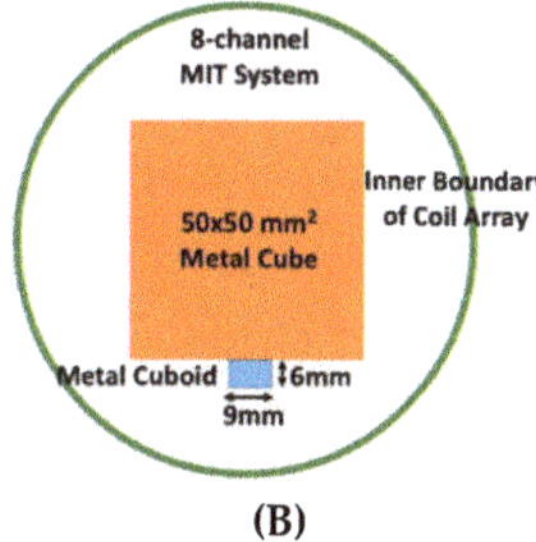


(A)        (B)

**Figure 8.** (**A**) Real experimental scenario of a metal cube and metal cuboid located at the center; (**B**) Illustration of the samples.

Then, Equation (14) turns out to be:

$$\Delta\sigma_{diff} = \Delta\sigma_{cube+cuboid} - \Delta\sigma_{cube} \tag{23}$$

Table 4 indicates the differential and threshold differential reconstructed images obtained for the sample with a bulging deformation at the center of different surfaces with different defect depths. The metal cuboid was placed at different positions which were close enough to the center of the metal cube surface. The two side lengths of the metal cuboid represent different defect depths (6 mm and 9 mm) according to different orientations. The threshold factor $\gamma$ was set to 0.5. It can be determined that 1 pixel = 2.08 mm, so the resolution of this experimental scenario is $1/2.08 = 0.48$ pixel/mm.

The reconstructed images obtained by differential and threshold differential TV algorithms both precisely indicate the position of the defect. The threshold differential reconstructed images and the corresponding contour images shown in Column 3 and Column 4 demonstrate better image performance than the differential images displayed in Column 2. When compared to the results of Scenario 1 and 2, the threshold differential image shown in Row 2 has a longer reconstructed length of the bulging defect than that of the image shown in Row 1, which is consistent with the real experimental scenario. So, the results displayed above prove that the TV-based differencing algorithm has the ability to locate and reconstruct the bulging defect.

A second stage threshold can be adopted to refine the accuracy of position locating and depth detection by applying a larger number of $\gamma$ to focus on the values around the maximum or minimum values and force the other values to zero. $\gamma$ was set at 0.8 in this scenario. The metal cuboid sample was located at a different corner of the metal cube with a different orientation to achieve different positions and variable defect depths (6 mm and 9 mm). Table 5 displays the second stage threshold differential reconstructed images obtained for the sample with a bulging deformation in different corners with variable defect depths.

**Table 4.** Differential and Threshold Differential reconstructed images for a defect at the center of different surfaces with different depths (Pixel = 50).

| Scenarios | Differential Reconstructed Images | Threshold Differential Reconstructed Images | Contour Images |
|---|---|---|---|

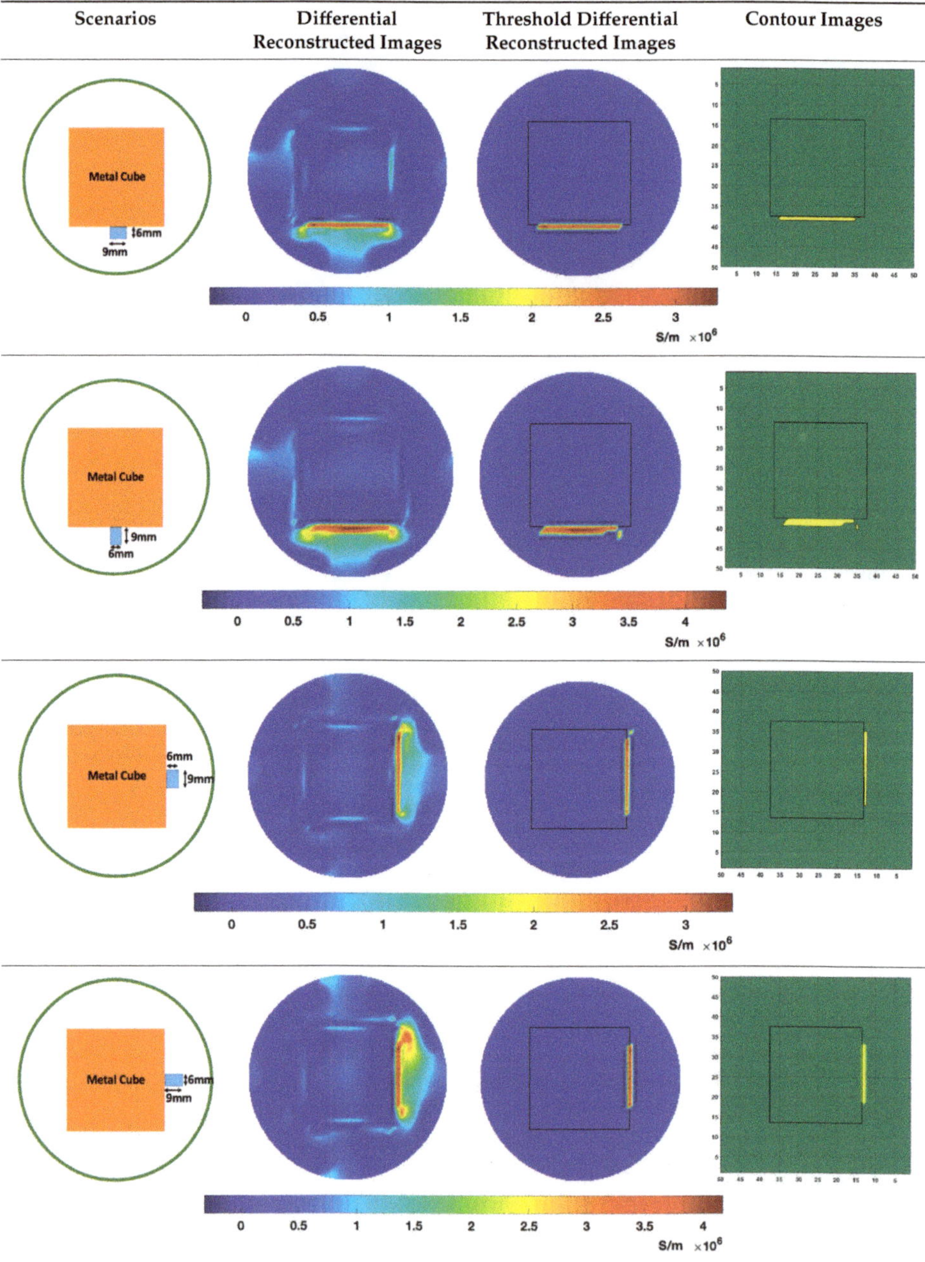

**Table 5.** Differential reconstructed images refined by A second stage threshold for defects at different positions with different depths (Pixel = 50).

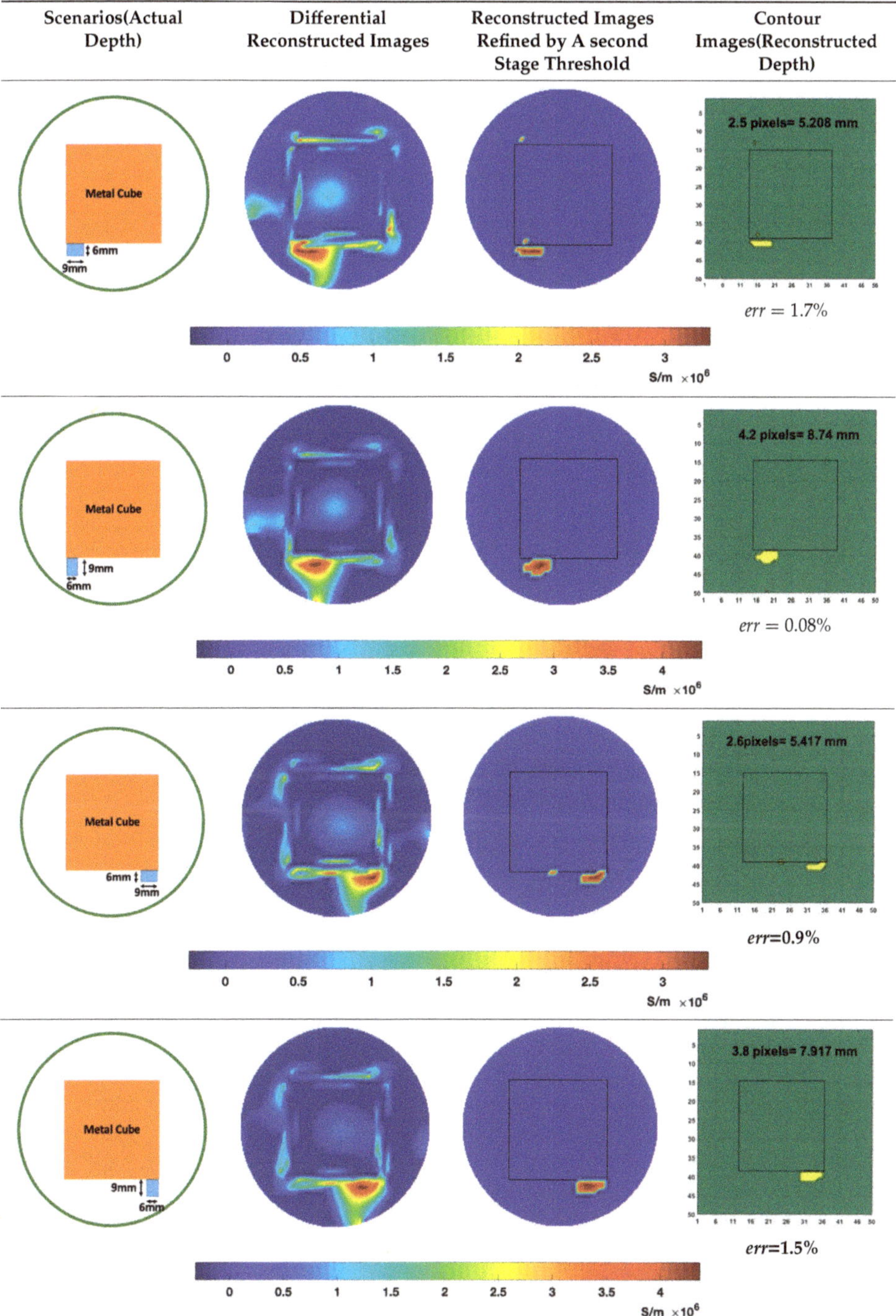

It's obviously that the conductivity distribution images from the further threshold differential algorithm shown in Column 3 can locate and image the bulge more accurately than those obtained directly from the differential algorithm shown in Column 2. The defect depth can be acquired from the contour images displayed in Column 4 with an acceptable degree of error (1.7%, 0.08%, 0.9% and 1.5%).

Based on the simulation scenario with two deformation defects at the same time displayed in Table 2, the corresponding experimental tests were implemented by the displacement of the square sample. In this case, Equation (14) turns out to be:

$$\Delta\sigma_{diff} = \Delta\sigma_{cube\ moved\ left\ or\ up} - \Delta\sigma_{cube} \tag{24}$$

Table 6 indicates the differential reconstructed images obtained for the sample with bulging and depression deformations at same time, which were achieved by moving the metal cube sample and then applying the abstraction. The red color in the differential reconstructed images, formed by positive numbers in the colorbar, and blue color parts, produced by negative numbers in the colorbar, represent the bulging and depression defect, respectively.

**Table 6.** Differential reconstructed images for bulging and depression defects at same time (Pixel = 50).

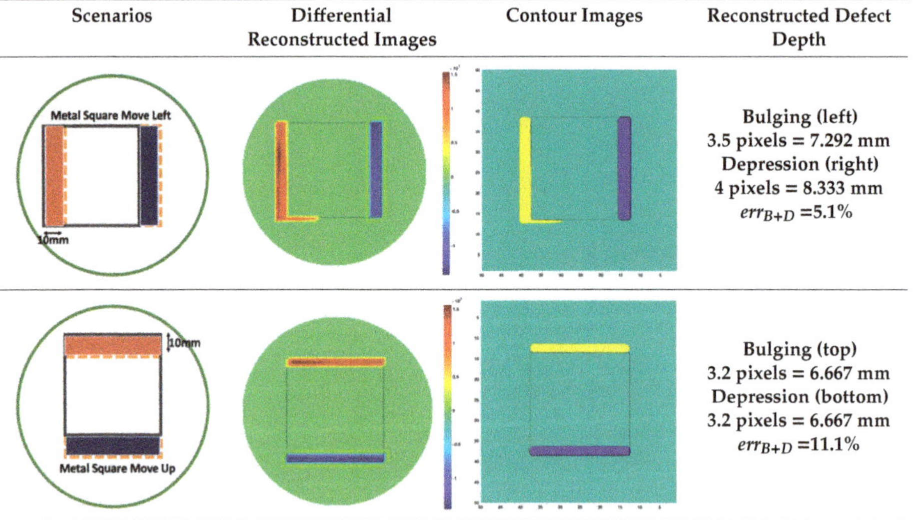

| Scenarios | Differential Reconstructed Images | Contour Images | Reconstructed Defect Depth |
|---|---|---|---|
|  | | | Bulging (left)<br>3.5 pixels = 7.292 mm<br>Depression (right)<br>4 pixels = 8.333 mm<br>$err_{B+D}$ =5.1% |
|  | | | Bulging (top)<br>3.2 pixels = 6.667 mm<br>Depression (bottom)<br>3.2 pixels = 6.667 mm<br>$err_{B+D}$ =11.1% |

The differential reconstructed images in Column 2 prove that the TV-based differencing algorithm can locate bulging and depression defects at the same time, as predicted from the simulation results shown in Table 2. It can be seen from the colorbar that the bulging amount is approximately the same as the depression amount, as it should be. Meanwhile, the reconstructed depth can be acquired by the corresponding contour images shown in Column 3, with a low level of error (5.1% and 11.1%). Moreover, the dispalcements of the sample can be detected by analyzing the differential resonstructed images.

Based on the simulation scenario of the rhomboid problem displayed in Table 3, the corresponding experimental tests were implemented by rotating the square sample with 45°. Table 7 indicates the differential reconstructed images obtained for the sample with bulging and depression deformations on the same surface, which were achieved by rotating the metal cube sample and then applying the subtraction.

**Table 7.** Differential reconstructed images for bulging and depression defects on the same surface (Pixel = 50).

| Scenarios | Differential Reconstructed Images | Contour Images |
| --- | --- | --- |
| 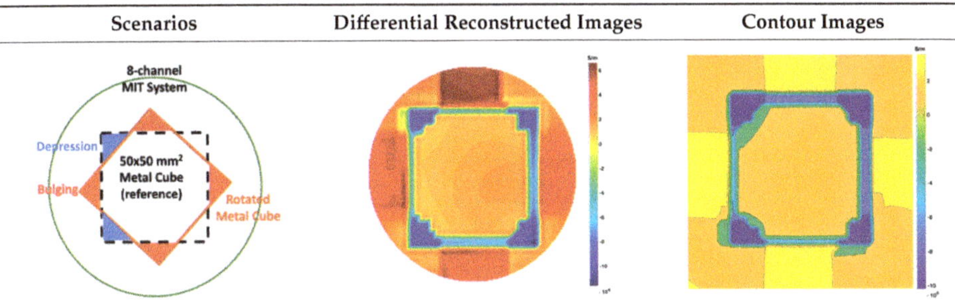 | | |



The red and blue color parts in differential reconstructed images represent the bulging and depression defects respectively. The results indicates that the TV-based differencing algorithm has the potential to reconstruct the two deformations at the same surface, but further research should be conducted to obtain more accurate results.

## 5. Validation and Quantitative Evaluation of the Method

The TV-based threshold-differential algorithm in MIT was adopted into real factory hot test data to track the movement of the strand during a continuous casting process. In this case:

$$\Delta\sigma_{diff} = \Delta\sigma_{Time\_i} - \Delta\sigma_{referece} \tag{25}$$

where is $\Delta\sigma_{referece}$ is the conductivity distribution of the strand at one fixed time and $\Delta\sigma_{Time\_i}$ is any random time selected during the casting process.

Figure 9 shows the test set up for the 16 channel MTI system in the secondary stage of continuous casting. The system model is the same as coil geometry for Figure 1.

**Figure 9.** The MIT system in continuous casting set up.

Figure 10 illustrates the movement distance of the strand in X- and Y-directions with its corrosponding differential reconstructed images. The positive number in the colorbar form the red colour part of the diffrential conductivity distribution, which indicated bulging and also the moving direction of the strand.

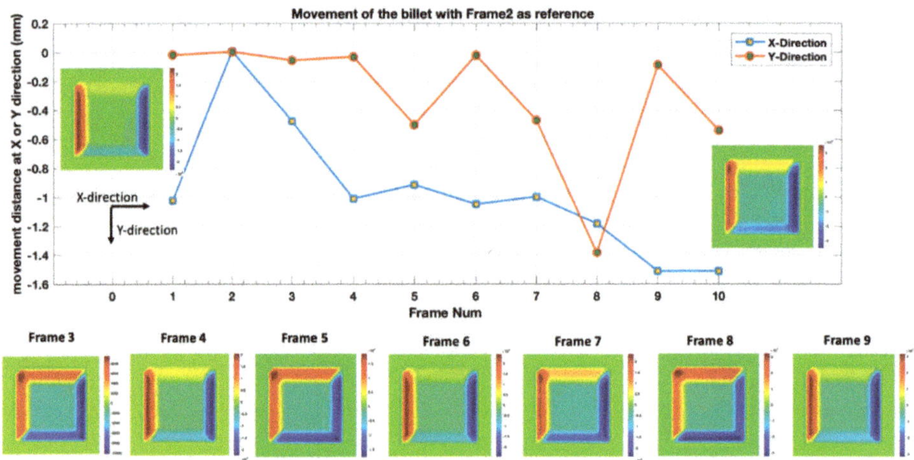

**Figure 10.** The movement direction and distance of the strand and its corresponding differential reconstructed images.

Moreover, two factors were defined to evaluate and validate the performance of the proposed method for monitoring the movement of the strand during a continuous casting process:

$$pp = \frac{Sum\ of\ the\ differential\ conductivity\ distribution}{Sum\ of\ the\ conductivity\ distribution\ of\ reference\ Time} \times 100 \quad (\%) \tag{26}$$

$$dis = \sqrt{\left(Pos\_X(i) - Pos\_X_{reference}\right)^2 + \left(Pos\_Y(i) - Pos\_Y_{reference}\right)^2} \quad (mm) \tag{27}$$

Figure 11 displays the relationship between the sum value of bulging and depression with respect to the reference, and the movement distance of the strand to the reference point.

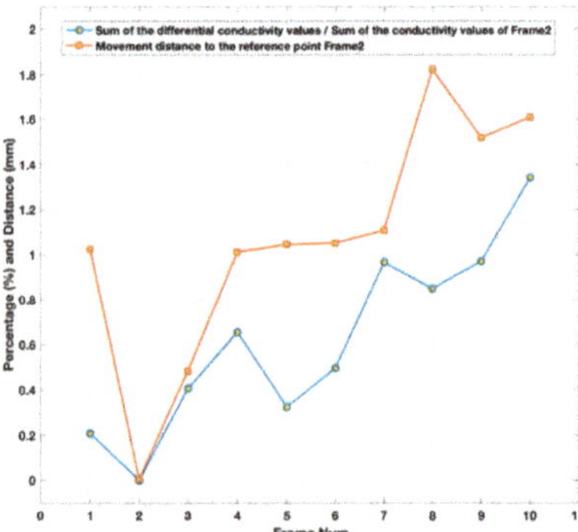

**Figure 11.** The movement distance of the strand to the reference point versus the sum values of bulging and depression with respect to the reference.

It is clear that the trend of these two plots is nearly the same, which indicates that the movement pattern of the strand is related to the sum conductivity distribution values of the differential reconstructed images. Therefore, the TV-based threshold-differential algorithm can be treated as a suitable method for monitoring surface defects, deformations and displacements.

## 6. Conclusions

A TV-based threshold-differential algorithm in MIT for surface defects and deformations is presented and validated using simulation and lab experimental tests as well as hot test in real continuous casting. The total conductivity distribution image obtained by simulations and experiments shows that the threshold-differential algorithm gives much better representation of bulge or depression deformation than that obtained from the TV method directly. The proposed simulation scenario with a resolution of 0.865 pixel/mm can achieve a deformation detectability of 0.92%. The proposed MIT experimental system can achieve a detectability of 0.57% with a resolution of 0.48 pixel/mm. The displacements of the sample can also be detected and located using the proposed method. The results presented in this paper confirm and prove the TV-based threshold-differencing method applied in MIT as a viable tool for monitoring and detecting surface defects, deformation and displacements in hot steel tests. The lab sample shown for the laboratory-based experiments conducted in this paper are the combination of two samples several more lab based experiments are conducted with success not included in here. The real process based data and results shows demonstartion of this new imaging method and system for contneous casting application.

**Author Contributions:** M.S. conceived the idea and led the project. S.S. produced simulated and hot test data, V.O. conducted simulation analysis. F.L. conducted lab based tests and analysis of all data and led the writing of the article. All authors read the paper and commented and participated in data analysis.

**Funding:** This work is part of Improvement of the continuous casting through a new system for the real-time measurement of SHELL THICKness in several locations of the casting strand project between the University of Bath, Ergolines, Tecnalia and Ferriere Nord. The funding is from EU commission on 709830-RFCS-2015 project.

**Acknowledgments:** We thank the steel manufacturer Ferriere Nord for helping us to set up our MIT system and collect long term data in continuous casting, also thank Tecnalia for several insightful project disucssions.

**Conflicts of Interest:** The authors declare no conflict of interest.

## References

1. Wei, H.Y.; Soleimani, M. A Magnetic Induction Tomography System for Prospective Industrial Processing Applications. *Chin. J. Chem. Eng.* **2012**, *20*, 406–410. [CrossRef]
2. Ma, X.; Peyton, A.J.; Higson, S.R.; Lyons, A.; Dickinson, S.J. Hardware and software design for an electromagnetic induction tomography (EMT) system for high contrast metal process applications. *Meas. Sci. Technol.* **2005**, *17*, 111. [CrossRef]
3. Vauhkonen, M.; Vadasz, D.; Karjalainen, P.A.; Somersalo, E.; Kaipio, J.P. Tikhonov regularization and prior information in electrical impedance tomography. *IEEE Trans. Med. Imaging* **1998**, *17*, 285–293. [CrossRef] [PubMed]
4. Biro, O. Edge element formulations of eddy current problems. *J. Comput. Methods Appl. Mech. Eng.* **1999**, *169*, 391–405. [CrossRef]
5. Park, G.S.; Kim, D.S. Development of a magnetic inductance tomography system. *IEEE Trans. Magn.* **2005**, *41*, 1932–1935. [CrossRef]
6. Osher, S.; Burger, M.; Goldfarb, D.; Xu, J.J.; Yin, W.T. An iterative regularization method for total variation-based image restoration. *Multiscale Model. Simul.* **2005**, *4*, 460–489. [CrossRef]
7. Li, F.; Abascal, J.F.P.J.; Desco, M.; Soleimani, M. Total Variation Regularization with Split Bregman-Based Method in Magnetic Induction Tomography Using Experimental Data. *IEEE Sens. J.* **2017**, *17*, 976–985. [CrossRef]
8. Abascal, J.F.P.J.; Chamorro-Servent, J.; Aguirre, J.; Arridge, J.; Correia, T.; Ripoll, J.; Vaquero, J.J.; Desco, M. Fluorescence diffuses optical tomography using the split Bregman method. *J. Med. Phys.* **2011**, *38*, 6258. [CrossRef] [PubMed]
9. Ma, X.; Peyton, A.J.; Binns, R.; Higson, S.R. Electromagnetic techniques for imaging the cross-section distribution of molten steel flow in the continuous casting nozzle. *IEEE Sens. J.* **2005**, *5*, 224–232.
10. Wei, H.-Y.; Wilkinson, A.J. Design of a Sensor Coil and Measurement Electronics for Magnetic Induction Tomography. *IEEE Trans. Instrum. Meas.* **2011**, *60*, 3853–3859. [CrossRef]

11. Peyton, A.J.; Yu, Z.Z.; Al-Zeiback, S.; Saunders, N.H. Electromagnetic Imaging Using Mutual Inductance Tomography: Potential for process applications. *Part. Part. Syst. Charact.* **1995**, *12*, 68–74. [CrossRef]

12. Ma, L.; Wei, H.-Y.; Soleimani, M. Pipelines inspection using magnetic induction tomography based on a narrowband pass filtering method. *Prog. Electromagn. Res. M* **2012**, *23*, 65–78. [CrossRef]

13. Ma, L.; Soleimani, M. Electromagnetic imaging for internal and external inspection of metallic pipes. *Insight Non Destr. Test. Cond. Monit.* **2012**, *54*, 493–495. [CrossRef]

14. Ma, L.; Hunt, A.; Soleimani, M. Experimental evaluation of conductive flow imaging using magnetic induction tomography. *Int. J. Multiph. Flow* **2015**, *72*, 198–209. [CrossRef]

15. Borges, A.R.; De Oliveira, J.E.; Velez, J.; Tavares, C.; Linhares, F.; Peyton, A.J. Development of electromagnetic tomography (EMT) for industrial applications. Part 2: Image reconstruction and software framework. In Proceedings of the 1st World Congress on Industrial Process Tomography, Buxton, UK, 14–17 April 1999; pp. 219–225.

16. Odedo, V.; Soleimani, M.; Spagnul, S. Magnetic Induction Tomography for imaging defects and deformations on external surfaces. In Proceedings of the 9th World Congress in Industrial Process Tomography, WCIPT9, Bath, UK, 2–6 September 2018; pp. 211–217.

17. Bevacqua, M.; Palmeri, R.; Di Donato, L.; Crocco, L.; Isernia, T. Microwave imaging via iterated virtual experiments. In Proceedings of the 10th European Conference on Antennas and Propagation (EuCAP), Davos, Switzerland, 10–15 April 2016; pp. 1–5.

*Article*

# Assessment of the Spatial Distribution of Moisture Content in Granular Material Using Electrical Impedance Tomography [†]

Jan Porzuczek

Faculty of Environmental Engineering, Cracow University of Technology, Warszawska 24,
31-155 Krakow, Poland; jan.porzuczek@pk.edu.pl

† This paper is an extended version of the paper: Porzuczek, J. Monitoring of the drying process using EIT method. In Proceedings of the 9th World Congress on Industrial Process Tomography, Bath, UK, 2–6 September 2018.

Received: 28 May 2019; Accepted: 20 June 2019; Published: 23 June 2019

**Abstract:** This paper presents a method for the online determination of the spatial distribution of the moisture content in granular material. It might be essential for the monitoring and optimal control of, for example, drying processes. The proposed method utilizes Electrical Impedance Tomography (EIT). As an exemplary material for experimental research, the black chokeberry (Aronia melanocarpa) was used. The relationship between the electrical impedance of the chokeberry and its moisture content was determined for a wide range of frequencies (20 Hz–200 kHz). The EIT research consisted of both simulation and experimental investigation. Experimental studies of the spatial distribution of the moisture content were performed in a cylindrical vessel equipped with 8 electrodes circumferentially arranged. The voltage signal from the electrodes was acquired simultaneously using the data acquisition module. Due to the high impedance of the chokeberries, exceeding $10^9$ $\Omega$ for the dried matter, extraordinary instrumentation was necessary to be applied. On the other hand, raw chokeberry was characterized by a several orders of magnitude lower impedance ($10^3$–$10^4$ $\Omega$), especially for high frequencies. The wide range of the observed impedance was able to be measured owing to its use of the voltage stimulation instead of the current stimulation (which is most common for EIT). The image reconstruction problem was solved using an iterative Gauss–Newton algorithm and the EIDORS (Electrical Impedance Tomography and Diffuse Optical Tomography Reconstruction Software) package. The obtained results showed a satisfactory ability to localize an insufficiently dried part of the material. Prospective ways to improve the imaging quality are also discussed.

**Keywords:** drying process; Electrical Impedance Tomography; impedance spectroscopy; chokeberry; EIDORS

---

## 1. Introduction

Measurement of moisture content in granular or solid materials is vital for a number of technological processes (e.g., drying or storage in silo). This measurement might be also be an essential tool for the monitoring the states of buildings or other works of construction. The inspection of dampness in building foundations and the leakage of flood embankments are possible examples. This paper presents research focused on the assessment of the spatial distribution of moisture content in granular material during its drying. However, the results acquired in this study might also be adapted to other applications that require similar measurements.

Drying processes (e.g., agricultural crops processing or fertilizers production) are known to be one of the most energy consuming processes. Therefore, any improvement in such processes may lead to significantly reduced energy consumption and, thus, ecologic as well as economic benefits. One of possibilities for optimizing the drying process is a proper process control strategy taking advantage of modern measurement techniques and instrumentation. The main purpose of this paper is to study the prospect of applying Electrical Impedance Tomography (EIT) is a non-intrusive measurement technique that could allow the drying of granular materials to be monitored online. The use of such a method would give better control to a process and may contribute in the improvement of the product's quality as well as reducing its energy consumption. As an exemplary material for experimental research, the black chokeberry (Aronia melanocarpa) [1] was used, but it appears possible to use this technique for a range of other agricultural crops (e.g., currants and cranberries) or other conductive granular materials (e.g., fertilizers and biomass).

The term 'tomography' is common for all measurement and visualization methods that provide information about the spatial distribution of certain material properties within an examined object, using measurements performed at the border of the object [2,3]. The graphical representation of those properties' distributions can be achieved as a result of measurement. Originally, this technique was used in medicine, but since the early 1990s, the first engineering applications of tomography appeared, leading to the development of the modern measurement technique known as 'process tomography'. The interest in the utilization of electrical phenomena in tomography contributed to the origin of electrical tomography, which was relatively inexpensive in application. A main advantage of all the tomographic methods may also include non-intrusiveness. Depending on the electrical and magnetic properties of the examined object, different measurement methods can be applied [4,5]:

- Electrical Impedance Tomography (EIT): The distribution of electrical conductivity is determined within the examined object.
- Electrical Capacitance Tomography (ECT): ECT allows one to determine the permittivity distribution in an area filled with dielectrics.
- Electromagnetic Tomography (EMT) or Magnetic Induction Tomography (MIT). These techniques are used to determine the permeability distribution in the examined object.

Application of tomography for the assessment of moisture content in the material-of-interest is not a new idea. However, the most common applications utilize ionizing radiation tomography [6] or ultrasound tomography [7]. In the literature, one can find a few examples of the use of electrical tomographic methods for monitoring of the drying processes [8,9], but they relate mainly to applications in pharmacy and use capacitance tomography instead of EIT, as proposed in this paper. Application of the Electrical Capacitance Tomography is suitable in the case of highly dielectric materials or processes in which the contact between electrodes and materials is inadvisable (e.g., fluidized bed processes, due to electrode erosion) [3,10]. Impedance tomography seems to be more adequate than ECT for the considered purpose. This is due to the relatively high conductivity of the wet material, which is typical for plant materials and many chemical substances [11], as well as a meaningful relationship between a material's conductivity and its moisture content. This feature was already implemented in research on building wall dampness [12] and the monitoring of waste landfills [13]. So far, no studies have been carried out to determine the moisture content of granular materials using EIT. On the other hand, in the state of low moisture content, a significant reduction of material conductivity might be expected. This results in the increased complexity of measurement appliance. In particular, utilizing the current stimulation commonly used in EIT is, in this case, virtually impossible. Instead of this, a voltage source should be used. The high impedance of the dry material shows the need to accurately measure very low current (a few μA). Particular attention must be given to ensuring the best possible contact between the material and the sensor electrodes. For future research, the use of multifrequency impedance tomography or spectro-tomography [14–16] seems to be promising.

The program of the research presented in this work consisted of three stages. In the first stage, the dependence between the complex impedance and the moisture content of the tested material was experimentally determined for a wide frequency range. In a later part of the research, a series of computer simulations were carried out. The aim of these simulation was analysing the imaging quality obtainable for the given measurement system and the selected algorithm for image reconstruction (Gauss-Newton algorithm [2]). The last part of the research was the experimental verification of the ability to identify the material with increased humidity in the sensor area. Experimental research was carried out for three frequency values of the stimulating voltage.

The experimental investigation proved the significant dependence between the absolute impedance of the chokeberry and its moisture content. This dependence showed a good potential to use it for an EIT purpose. The simulation analysis showed that even if some limitations might be expected, the 8-electrode EIT system with larger electrodes can provide a rewarding imaging quality, especially for process control purposes. The validation of the imaging quality of the spatial distribution of moisture content using the 8-electrode EIT system, conducted experimentally, confirmed the simulation results. The artefacts that appeared in the reconstructed images tended to decrease with a raise in frequency.

## 2. Materials and Methods

### 2.1. Impedance Spectrum of the Chokeberry and Its Dependence on Humidity—The Method of Examination

Impedance spectroscopy is the research method that allows one to obtain the complex impedance of the investigated system as a function of frequency [11]. After the insertion of the conductive material between two electrodes, such a system can be regarded as a combination of several passive elements, such as resistors and capacitors. Depending on the adopted structure, the model will approximate the experimental results with a certain level of accuracy. For the material originating from plants, the first model, consisting of three elements, was proposed by Hayden at the turn of the 1960s and 1970s [17]. This model was then widely used by other researchers [18]. However, examples utilizing more sophisticated models can be found. For instance, in [19], the authors proposed a model consisting of five elements. According to the authors, this model better represents the electrical properties of the plant cells. The model proposed by Zhang was subsequently used during the impedance spectroscopic investigation of kiwi fruit [20]. In the present paper, the simplification of the Hayden model was used. This system is modelled as a resistor and capacitor connected in parallel. The reason for such a simplification is the relatively narrow frequency range in comparison to the above-mentioned research.

As mentioned, after the insertion of the conductive material (in this case, chokeberry) between two electrodes, such a system can be regarded (as a simplification) as a parallel combination of a resistor $R$ and a capacitor $C$ [11]. The reactance $X$ of the capacitor shows the dependence of angular frequency $\omega$ as shown in Equation (1):

$$X = \frac{1}{\omega C},\tag{1}$$

Likewise, complex impedance of the circuit, Equation (2), is a function of angular frequency $\omega$:

$$Z = \frac{RX^2}{R^2 + X^2} + j\frac{R^2 X}{R^2 + X^2},\tag{2}$$

Complex impedance can be equivalently represented as in Equation (3), using the absolute impedance $|Z|$ and the phase angle $\varphi$:

$$Z = |Z| \cdot e^{-j\varphi},\tag{3}$$

For a given measurement system, both the absolute impedance $|Z|$ and the phase angle $\varphi$ are related to the condition of a material being investigated (temperature, humidity) and that relationship might be obtained experimentally. On the other hand, knowledge about that relationship leads to the conclusion of the state of the material by measuring the above-mentioned values.

According to the literature data [1], initial humidity of fresh harvested chokeberry is typically 71.2%–84.8%. The chokeberry frozen immediately after harvest was used for the experimental research. The material contains whole fruits, without any mechanical processing and damages This is noteworthy, since in the case of processed chokeberry (e.g., chopped or pressed), a peel of the fruit gets damaged.

Thus, in such a situation, the results described below cannot be reliable [21]. After defrosting and reaching room temperature, a thin layer of water on the surface of the peel still existed. This state of berries was assumed to be the initial state. In a further step, the chokeberry was dried in the open air at a temperature of 20 °C and relative air humidity of 35%–40%. The sample of material (approximately 0.75 dm$^3$) was gathered periodically and the measurement of absolute impedance and phase angle was taken using the procedure described below. Using the moisture analyser Radwag MAC 210, the humidity of the chokeberry sample was designated afterward. A measurement was taken for the chokeberry's humidity range of 80.0%–12.6%. In the case of low water content (<20%), the impedance exceeded the measuring range of the instrument reaching more than $10^9$ $\Omega$ (for low frequency).

The study was conducted in the measurement vessel, as detailed in Section 2.3, at a fixed temperature of 20 °C. For the purpose of verification, the procedure was repeated several times for each sample of the chokeberry. This repetition involved both changing the measuring electrodes and removing and re-inserting the material into the area of the sensor. Each time the observed differences did not exceeded a few percent, and this is most likely the result of differences in the contact impedance between the electrode and the berries, as well as between the berries themselves at different arrangements of the material.

The impedance of the chokeberry was measured using a programmable LCR bridge Rhode-Schwarz Hameg HM 8118. This device allows one to designate selected parameters of the circuit (in this case absolute impedance and phase angle) for 69 frequency values in the range of 20 Hz–200 kHz. The frequency range as well as the number of measurement points were limited by the LCR bridge specification. Since the LCR bridge is equipped with both RS232 and USB communication interfaces, allowing one to control the device using a computer, it was possible to automate the measurement procedure for the entire frequency range.

*2.2. Simulation Analysis of Projected EIT Image Quality*

Determination of the conductivity distribution in the cross-section of a sensor, based on the measured voltage, requires a two-step process. The aim of the first stage is to solve the forward problem, leading to a designation of the transformation matrix of the measurement system. This step is based on the Laplace Equation (4), which defines the distribution of electrical potential $V$ in the studied area with the material of conductivity $\sigma$ [2]:

$$\nabla \cdot \sigma \nabla V = 0, \tag{4}$$

The solution of Equation (4) for the given measurement system is usually carried out with the Finite Elements Method. Introducing $v$ as the vector of measured potentials at the border of the sensor leads to the designation of the transformation matrix $\mathbf{T}$, satisfying the Equation (5):

$$v = \mathbf{T} \cdot \sigma, \tag{5}$$

Reconstruction of the conductivity image is the second stage of the analysis of acquired data. The image is constructed by solving the inverse problem defined by Equation (6).

$$\sigma = \mathbf{T}^{-1} \cdot v \tag{6}$$

However, the T matrix is not a square one, so it is not possible to use conventional methods for determining the inverse matrix. This problem has resulted in a number of algorithms for image reconstruction, approximating the solution of the inverse problem. Starting from linear methods (e.g., Linear Back-Projection) through nonlinear and iterative methods (e.g., Gauss–Newton) [2,3,5], the approach to the inverse problem has evolved into a dozen modern algorithms that utilize so-called artificial intelligence methods. For instance, in [22], the authors reviewed almost 70 papers describing novel methods for image reconstruction. Some of the modern, advanced algorithms, e.g., Sparse Bayesian Learning [23], are thought to offer a better resolution for imaging in comparison to classical methods.

Due to the ill-conditioning of the inverse problem, the solution obtained may not correctly reflect the distribution of the analysed properties in the area of the sensor. The feasibility of the correct imaging and the projected image quality should by verified using simulation studies. For this reason, the experimental studies were preceded by conducting a series of simulation analyses that gave an overview of the predicted quality of imaging in a considered EIT system. The use of specialized software, for example EIDORS (Electrical Impedance Tomography and Diffuse Optical Tomography Reconstruction Software) [24,25], allows both the simulation test of impedance tomography system and the reconstruction of images based on experimental data. Reconstruction of the image in the software package is possible using a number of algorithms. A broad overview of the application of the EIDORS software was presented by Holder [2]. Commercially available EIT systems usually come with proprietary software, e.g., the ITS Reconstruction Toolsuite [26], which may be used alternatively.

The simulation research consisted of two main stages. In the first stage, the investigation was focused on the possibility of identifying the region that is characterized by a different moisture content than the rest of the material. In the scope of this research, the impact of the diameter of the area with increased humidity, as well as its position inside the sensor area, on detection capability was investigated. In this study, the circular EIT sensor with an internal diameter of 11 cm was used. A simulation study was conducted for the situation in which an area of higher humidity exists inside the sensor filled with the chokeberry. That area had a circular shape with a diameter 60 mm or 30 mm, located in the centre of the sensor or near its border. Four cases were analysed:

- Area of diameter 30 mm located near the border of the sensor;
- Area of diameter 60 mm located near the border of the sensor;
- Area of diameter 30 mm located in the centre of the sensor;
- Area of diameter 60 mm located in the centre of the sensor.

The rest of the material had homogeneous moisture content. Using the EIDORS software together with Netgen meshing software [27], the model of the measurement vessel, electrodes, and distribution of material inside the vessel was created, and a finite element grid was generated. Each simulation was performed both for the 8-electrode and 16-electrode EIT sensor. In the case of the 8-electrode sensor, each electrode was 30 mm wide, while in the case of the 16-electrode sensor, each electrode width had to be limited to 15 mm. In order to improve the accuracy of the calculation, the grid was thickened near the electrodes. This model was subsequently used to solve the forward problem and determine the potential distribution on the border of the area. As a consequence, a simulated measurement set was obtained, which was used to attempt the image reconstruction afterwards. For the solution of the inverse problem, the iterative Gauss–Newton algorithm was applied [2,28].

In the second part of the research, particular attention was given to an analysis of the distinguishability of the two separate regions of material with a higher moisture content. For this reason, a model of the sensor, with the shape as described above, was created. Subsequently, two areas with a 30 mm diameter, characterized by higher moisture content, were defined inside the sensor. During several simulations, the distance between those areas was gradually decreased until they became one area in the reconstructed image. The simulation research was carried out for the 8-electrode and the 16-electrode systems. Additionally, the influence of grid density on the quality of imaging the calculation for the grid of increased density was also investigated. The Gauss–Newton algorithm was also used in this part of research.

### 2.3. The EIT Experimental Setup

The block diagram of the complete measurement system is shown in the graph (Figure 1). The measurement vessel (the sensor of the tomograph) was constructed using a polycarbonate pipe with an internal diameter of 11 cm and a height of 8 cm. Eight electrodes (sized 7.5 × 3.0 cm), made of copper foil, were circumferentially arranged on the internal side of the vessel. The applied size of the electrodes, unusually large for EIT, results from the need to ensure similar contact impedance between the material and all electrodes. Granular material contacts with the electrodes only on certain points. Providing as many contact points as possible is important, therefore. This results in a drop of contact impedance. Furthermore, an increase of the electrode area in the considered case may contribute to the equalization of the contact impedance on all electrodes.

To allow the sequential switching of the stimulating electrodes, the switching circuit (MUX) was designed and manufactured based on a semiconductor multiplexer and reed relays. This subsystem was controlled by the digital outputs of data acquisition card. In most of the implementations of an impedance tomography, an AC current stimulation (typically several milliamps) is used. This requires the use of a voltage controlled current source (a so-called 'current pump'), typically the Howland circuit [2,29]. However, a current source is characterized by a significantly higher output impedance than the stimulated load. Due to the very high impedance of the chokeberry (at a low moisture content), it is virtually impossible to build a current source that would allow the generation of stimulus signal in the required frequency range [29,30]. These circumstances caused the need to apply, in this research, voltage stimulation and accurate current measurement systems [2,31]. As the voltage source (SRC), the programmable signal generator Agilent 33500B was used. Since the current flowing through the chokeberry reaches very low values (1–4 µA), it was necessary to develop a current-to-voltage converter (TIA, transimpedance amplifier), characterized by a high amplification factor. For this purpose, the feedback converter circuit, based on operational amplifier OPA2107, itself characterized by the very low bias current, was built [11,29]. This, in turn, allowed the amplification factor to reach 1 V/µA, and the resolution of current measurement was better than 0.2 nA.

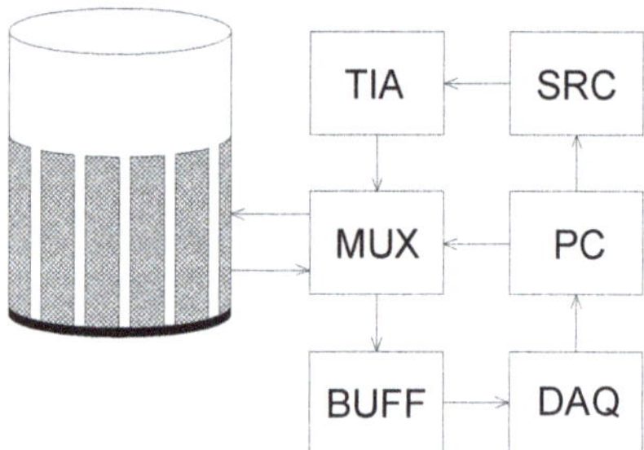


**Figure 1.** Block diagram of the EIT test stand; TIA: transimpedance amplifier, SRC: voltage source, MUX: switching circuit, PC: computer, BUFF: buffering circuit, DAQ: data acquisition card.

The voltage on the electrodes was measured using the high-speed data acquisition card (DAQ), National Instruments PCIe-6351. This is a 16-channel multifunction measurement card with an accurate 16-bit analog-to-digital converter with a maximum sampling rate of 1.25 MHz. The instrument measuring voltage is usually required to have the largest possible input impedance. Although the DAQ card manufacturer reports [32] that the input resistance of the device is very large (>10 GΩ), the significant input capacitance (100 pF) still results in a noteworthy decrease in the input impedance when measuring high frequency voltage. Moreover, the connection cable is characterized by its own capacitance, which is not negligible. In order to reduce the impact of this effect on the achievable measurement accuracy, each input of the DAQ card was equipped with an additional buffering circuit (BUFF) based on the low noise operational amplifier, OPA4228. This allowed for a decrease of approximately 50 times in the input capacitance of the measurement system. The complete stimulus switching and measurement system was controlled using a PC through an application written in LabVIEW [33].

## 3. Results

### *3.1. Impedance Spectrum of the Chokeberry and Its Dependence on Humidity—The Results*

The results of the experimental research showed that the humidity of the chokeberry is significantly dependent on its electrical properties. As could be expected, absolute impedance explicitly increases by decreasing the moisture content in the chokeberry. This is due to a number of phenomena that occur during the drying of the fruit. The key elements are reducing the number of free ions due to the evaporation of the solvent, as well as reducing the contact surface caused by changes in the structure of the peel [21]. Accordingly, the effect should be present for most of the agricultural crops being dried. It is noteworthy that the results obtained for 12.8% moisture content might be partially charged with higher uncertainty. This is caused by exceeding the LCR bridge measurement range ($10^8$ Ω) for frequencies lower than 1 kHz. The obtained results are shown in the graph (Figure 2).

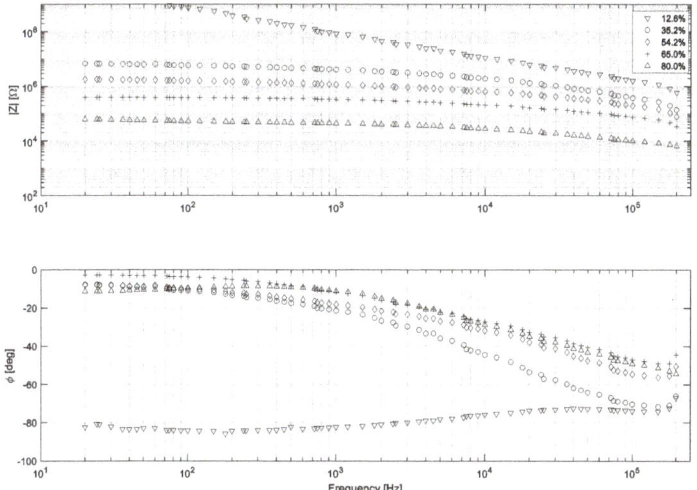

**Figure 2.** Impedance spectra of the chokeberry, with different moisture contents, at 20 °C.

The approximate trends of absolute impedance change with the moisture content in the chokeberry, as visualized in the graph (Figure 3) for selected frequency (100 Hz, 1 kHz, 10 kHz). It could be also stated that the absolute impedance decreases with an increase in frequency. For this reason, utilising a higher frequency makes the measurement easier to conduct, even using low-cost appliances.

The phase angle dependence on humidity of the chokeberry does not show such a clear trend. It is noteworthy that the chokeberry with a low moisture content shows characteristics similar to the dielectric. In this case, particularly at a higher frequency ($10^4$–$10^5$ Hz), the relationship between the phase angle and the moisture content of the material becomes meaningful. This gives the opportunity to use, at a later stage of the research, both the resistive and capacitive properties of the system by proposing a dual modality tomography [34] method or spectro-tomography [14–16]. Further research should obviously include carrying out measurements at different temperatures because, a significant impact on the results might be expected.

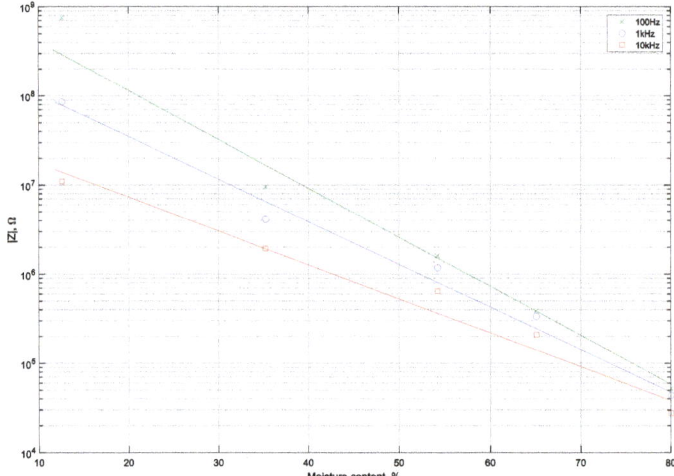


**Figure 3.** Relationship between the absolute impedance of the chokeberry and its moisture content.

During the examination, the phenomenon, which is important in practice, was observed. After the insertion of the chokeberry into the measurement vessel, a gradual decrease in the measured impedance occurred. After approximately 15–50 min (depending on material humidity), the value of the impedance became steady. This phenomenon was stronger the higher the moisture content in the material, and at humidity lower than 30%, the change of impedance was already insignificant. Moreover, the settling time appeared to be influenced by the frequency. For a lower frequency, the settling time reached its minimum, while for higher frequency it increased. The practical importance of this phenomenon should be additionally investigated in future research.

*3.2. Results of the Simulation Analysis*

As shown by the results achieved, described below in detail, identification of the area with the higher humidity is mostly correct. However, even in the case of the simulation data, which is free from noise, material distribution non-uniformity, or inaccuracies in the sensor design, the image reconstruction is not perfect. Outside the area of higher moisture content, the image should be uniform. However, mapping errors (artefacts) can be observed. This indicates a certain limitation of the image reconstruction algorithm.

The results of the first stage of simulation research (Figures 4–7) indicate a good ability to locate a material with higher humidity. Both positions, as well as areas, are obtained precisely enough for the monitoring of the drying processes or control purposes. This confirms the possibility to use the EIT sensor with adopted geometry, as well as the Gauss–Newton algorithm, to achieve the intended purpose. The case where the area of diameter of 30 mm was located in the centre of the sensor (Figure 6) was the exception. In this case, regardless of the sensor used, the designated area was significantly overestimated. Interestingly, in the remaining cases, the results obtained using 8-electrode and 16-electrode sensors did not differ significantly. Obviously, the images obtained with 16-electrode sensor were characterized by a better ability to render the shape of the area. This was particularly noticeable for cases in which the region-of-interest was localized near the border of the sensor (Figures 4 and 5). In the case of area of diameter 60 mm placed in the centre of the sensor (Figure 7), the differences between images acquired using the 8-electrode and 16-electrode system appeared to be negligible.

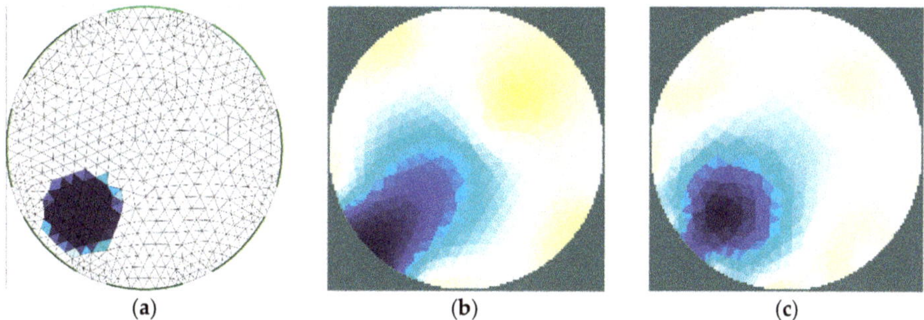

**Figure 4.** Area of diameter 30 mm located near the border of the sensor—finite element grid (**a**) and image reconstructed using simulated data for the 8-electrode (**b**) and 16-electrode (**c**) sensor.

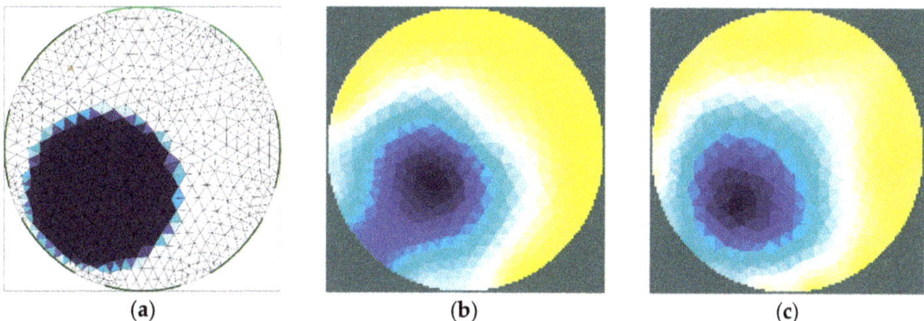

**Figure 5.** Area of diameter 60 mm located near the border of the sensor—finite element grid (**a**) and image reconstructed using simulated data for the 8-electrode (**b**) and 16-electrode (**c**) sensor.

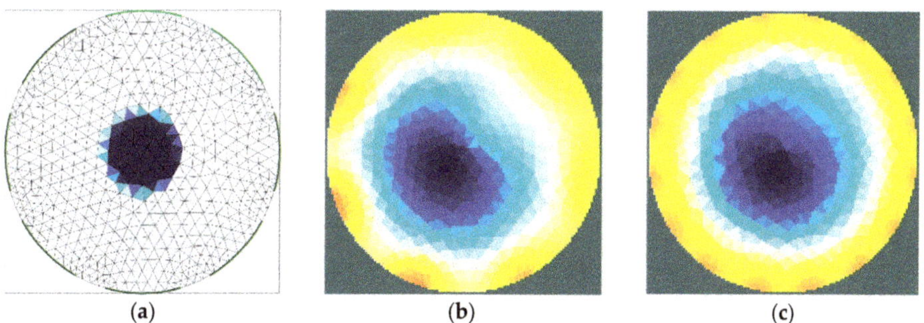

**Figure 6.** Area of diameter 30 mm located in the centre of the sensor—finite element grid (**a**) and image reconstructed using simulated data for the 8-electrode (**b**) and 16-electrode (**c**) sensor.

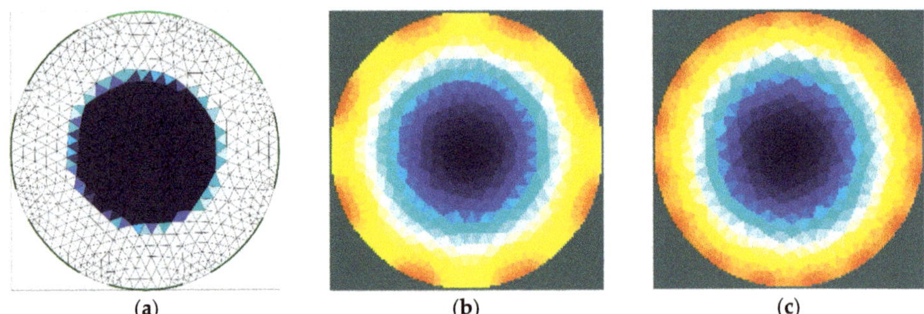

**Figure 7.** Area of diameter 30 mm located in the centre of the sensor—finite element grid (**a**) and image reconstructed using simulated data for the 8-electrode (**b**) and 16-electrode (**c**) sensor.

The second stage of simulation research was aimed to estimate the ability of the EIT sensor to identify two separate areas of material with higher moisture content placed inside the sensor. The importance of this feature is related to the relatively low imaging resolution, which is typical of impedance tomography. The results achieved (Figures 8 and 9) illustrate the minimum distance between the areas for which the areas remain disjoint in the reconstructed image. The obtained minimum limit between the borders of the areas (12 mm) was proven to be similar to the radius of these areas. At this stage of the research, the results obtained for the 16-electrode sensor (Figure 8) turned out to be significantly better than the image obtained using the 8-electrode sensor (Figure 7). In each case, the implementation of the dense grid resulted in improved imaging quality. However, the results obtained cannot justify the significant increase in the computation time needed for image reconstruction. In the subsequent graph (Figures 10 and 11), the results of a similar simulation are shown, but in this case, the distance between the borders of the areas was lower than the minimum limit. In this example, there is no possibility to distinguish the areas. However, indentifying the region-of-interest's existence is possible. To control a drying process, this possibility would also be useful.

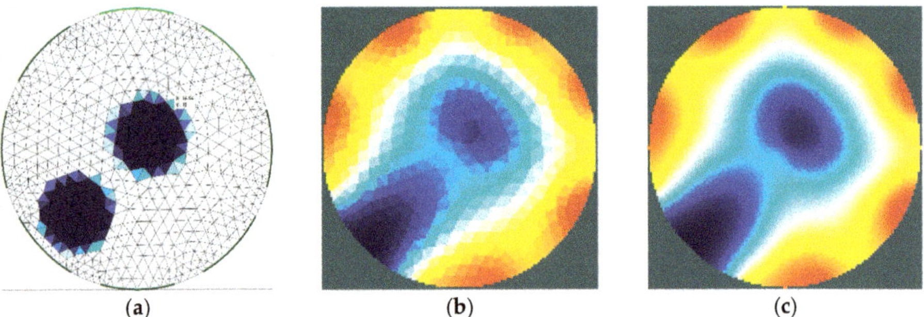

**Figure 8.** Distinguishability analysis: 8-electrode sensor—finite element grid (**a**) and image reconstructed using simulated data for the normal grid (**b**) and dense grid (**c**).

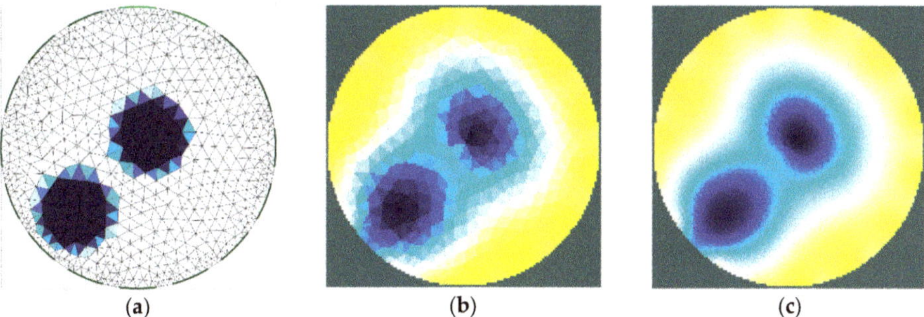

**Figure 9.** Distinguishability analysis: 16-electrode sensor—finite element grid (**a**) and image reconstructed using simulated data for the normal grid (**b**) and dense grid (**c**).

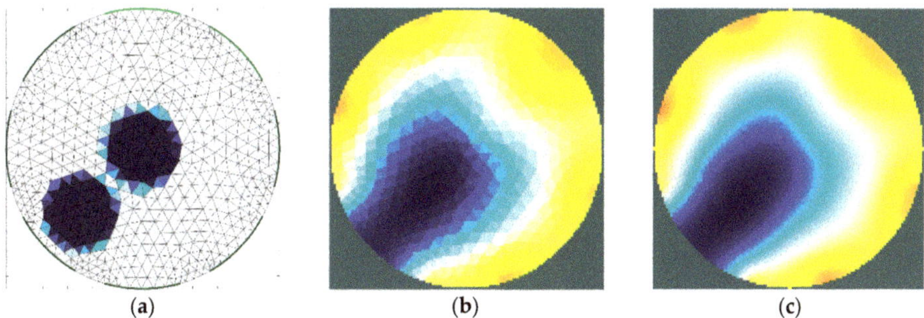

**Figure 10.** Distinguishability analysis: 8-electrode sensor, regions too close—finite element grid (**a**) and image reconstructed using simulated data for the normal grid (**b**) and dense grid (**c**).

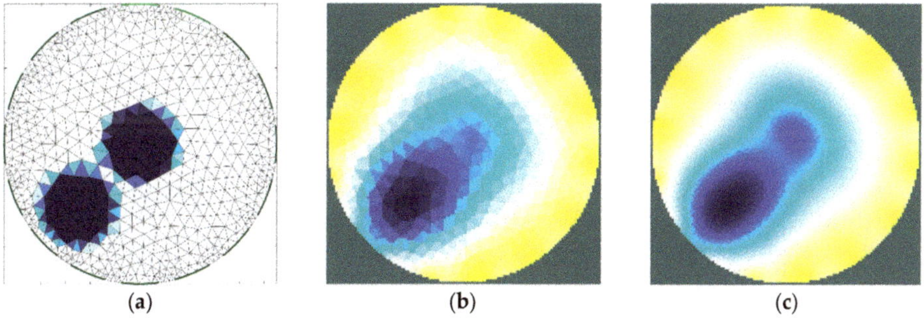

**Figure 11.** Distinguishability analysis: 16-electrode sensor, regions too close—finite element grid (**a**) and image reconstructed using simulated data for the normal grid (**b**) and dense grid (**c**).

*3.3. Imaging of the Spatial Distribution of Moisture Content in the Chokeberry—Image Reconstruction Based on Experimental Data*

Experimental verification of the simulation results was performed using the measurement system described in Section 2.3. The chokeberry was arranged inside the sensor, as in one of the simulation studies (Figure 5). Berries filling the majority of the volume of the sensor had moisture content as follows:

- Case 1: 35.2%;
- Case 2: 54.2%.

However, inside the 'wet' area, the moisture of the material was 65.0%. The large difference in humidity between areas in 'case 1' stems from the need to investigate obtainable imaging contrast. Determining the achievable imaging resolution of the moisture distribution in the material and attempting to improve it will be a further stage of this study.

In the experimental research, a sensing strategy called the 'opposite electrode pair strategy' was applied [31]. This strategy means that the voltage source was connected to an opposite pair of electrodes, e.g. 1–5, 2–6, etc. (for the 8-electrode sensor), while the voltage was measured on every electrode. After completion of the voltage measurement, the sourcing electrode pair was switched to the next one. Current drawn from the source was measured simultaneously and maintained at a constant level. Therefore, the complete measurement set is comprised 64 voltage measurements. A measurement was taken for three values of stimulus voltage frequency: 100 Hz, 1 kHz, and 10 kHz. Similar to the simulation study, an iterative Gauss-Newton algorithm was applied to solve the inverse problem. The results of the image reconstruction for each frequency are shown in Figures 12 and 13.

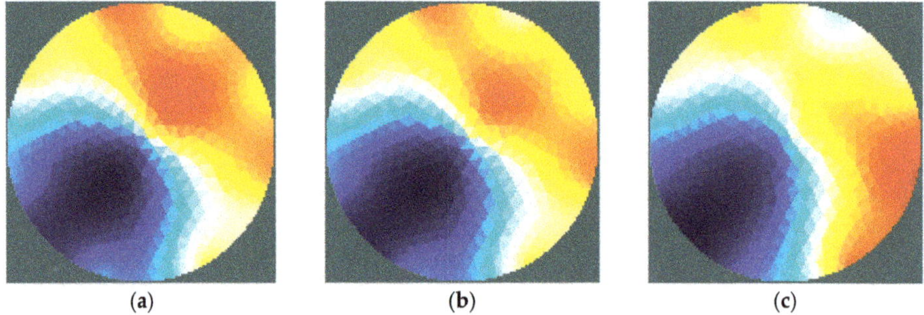

(a)        (b)        (c)

**Figure 12.** Case 1: Chokeberry with water content of 69.0%/35.2%. Reconstructed images for voltage stimulation of 100 Hz (**a**), 1 kHz (**b**), 10 kHz (**c**).

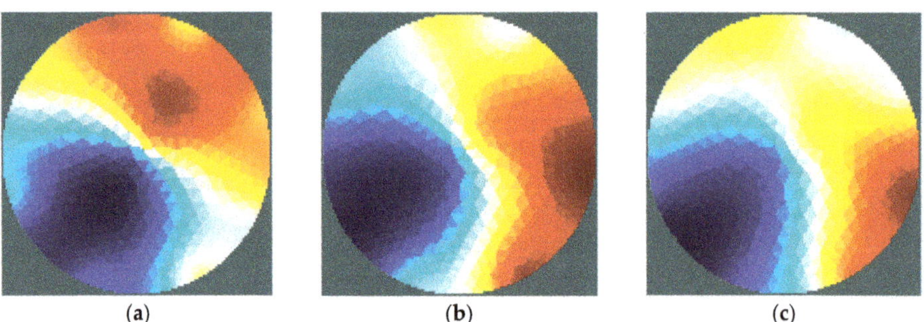

(a)        (b)        (c)

**Figure 13.** Case 2: Chokeberry with water content of 69.0%/54.2%. Reconstructed images for voltage stimulation of 100 Hz (**a**), 1 kHz (**b**), 10 kHz (**c**).

As shown by the results obtained, irrespective of the frequency of the excitation voltage, the location of areas of high humidity can be considered correct. However, it can be seen that all of the reconstructed images are characterized by the artefacts of an intensity significantly greater than in the simulation results. Moreover, the artefacts' influence on image quality is explicitly dependent on frequency. The quality of imaging clearly increases with a raise in the frequency of the stimulus voltage. It might be a prerequisite to extend the frequency range during future research. In 'case 2', the moisture content difference between the areas was lower and led to an increased impact of noise on the measured signals. This, in turn, resulted in an increased presence of artefacts (Figure 13). Presumably, this result also indicates the need to introduce some modifications to the measurement system, especially an increase of the resolution of current measurement. Moreover, a reduction of the noise impact should lead to sensitivity improvement for whole EIT system.

In this stage of the study, the experimental research was focused mainly on proving the feasibility of humidity imaging with the applied method. For this reason, only one material arrangement was tested. Any quantitative indicators of the quality of imaging were not introduced at this stage of the research. Undoubtedly, in further research, such indicators should be introduced, giving a more detailed comparison of the results. Since the method needs to be additionally validated, in the next season of fruit harvest, the research will be continued. Further research should also be concerned with improving measurement accuracy, as well as imaging resolution. Additional humidity ranges, different positions, and sizes of the region of interest and multiple-region detection have been shown to be particularly important for further investigation.

## 4. Discussion

The study showed a significant correlation between the absolute impedance of the chokeberry and its moisture content in a wide frequency range. This indicates the possibility to use the impedance measurement to assess moisture content. In practice, the condition for the correct interpretation of the interdependence of impedance and humidity should take into account temperature compensation—in this study, it was not analysed. Attention was drawn to the phenomenon of long-term stabilization of the measurement result after insertion of the chokeberry into the sensor. That phenomenon limits, but does not exclude, the application of the presented method to dryers characterized by periodic mixing of the material. The high impedance of the chokeberry requires an unusual approach to design the measurement system of the EIT. It was proven that the application of a voltage stimulation and stimulus current measurement, together with additional buffering of the data acquisition card, made the EIT imaging feasible.

An important issue is ensuring good contact between the tested material and the electrodes. The fact that granulated materials contact with the electrodes only on a finite number of points may lead to a difference between the contact impedance on each electrode. For this reason, it is useful to use electrodes with a relatively large area. However, increasing the electrodes' area limits the number of electrodes able to be installed inside the sensor, which in turn limits its imaging resolution. On the basis of the simulation results, the use of an 8-electrode sensor instead of a 16-electrode sensor might be justified in the considered EIT system.

Identification of the insufficiently dried material can be regarded as satisfactory despite the above-mentioned issues. However, the artefacts appearing in the images make interpretation of the results difficult. However, these artefacts might be significantly reduced through minor modifications of the measurement system. This was shown, in particular, by the comparison between the simulation and experimental results. Since the quality of imaging clearly increases by raising the frequency of the stimulus voltage, future research should be focused on a wider frequency range.

The accuracy achievable with other image reconstruction algorithms has not been studied in this paper, but the results acquired using impedance spectroscopy suggest that it will be possible to improve the imaging quality by taking into account both the real and imaginary parts of complex impedance. Especially at a low moisture content in the material being dried, the measurement of the phase angle seems to be purposeful. Nevertheless, the obtained results indicate the prospective usefulness of the presented methods in the practical, non-intrusive monitoring and control of drying processes.

**Funding:** This research was funded by the Polish Ministry of Science and Higher Education, grant number: S-4/400/2018/DS

**Conflicts of Interest:** The author declares no conflict of interest.

## References


1. Kulling, S.E.; Harshadei, M.R. Chokeberry (Aronia melanocarpa)—A Review on the Characteristic Components and Potential Health Effects. *Planta Med.* **2008**, *74*, 1625–1634. [CrossRef]
2. Holder, D.S. *Electrical Impedance Tomography. Methods, History and Application*; Institute of Physics Publishing: Bristol, UK, 2005.
3. Porzuczek, J. Applications of the electrical capacitance tomography for research on phenomena occurring in the fluidized bed reactors. *Chem. Process Eng.* **2014**, *35*, 397–408. [CrossRef]
4. Tapp, H.S.; Peyton, A.J.; Kemsley, E.K.; Wilson, R.H. Chemical engineering applications of electrical process tomography. *Sens. Actuators B* **2003**, *92*, 17–24. [CrossRef]
5. Wang, M. *Industrial Tomography—Systems and Applications*; Woodhead Publishing: Sawston, UK, 2015.
6. Dauti, D.; Tengattini, A.; Dal Pont, S.; Toropows, N.; Briffaut, M.; Weber, B. Analysis of moisture migration in concrete at high temperature through insitu neutron tomography. *Cem. Concr. Res.* **2018**, *111*, 41–55. [CrossRef]
7. Salchak, Y.; Bulavinov, A.; Pinchuk, R.; Lider, A.; Bolotina, I.; Sednev, D. Dry Storage Casks Monitoring by Means of Ultrasonic Tomography. *Phys. Procedia* **2015**, *70*, 484–487. [CrossRef]
8. Rimpilainen, V.; Heikinnen, L.; Vauhkonen, M. Moisture distribution and hydrodynamics of wet granules during fluidized-bed drying with volumetric electrical capacitance tomography. *Chem. Eng. Sci.* **2012**, *75*, 220–234. [CrossRef]
9. Wang, H.G.; Senior, P.R.; Mann, R.; Yang, W.Q. Online measurement and control of solids moisture in fluidised bed dryers. *Chem. Eng. Sci.* **2009**, *64*, 2893–2902. [CrossRef]
10. Zhang, J.L.; Mao, M.X.; Ye, J.M.; Wang, H.G.; Yang, W.Q. Investigation of wetting and drying process in a gas-solid fluidized bed by electrical capacitance tomography and pressure measurement. *Powder Technol.* **2016**, *301*, 1148–1158. [CrossRef]
11. Barsoukov, E.; Macdonald, J.R. *Impedance Spectroscopy Theory, Experiment and Applications*, 2nd ed.; Wiley-Interscience: Hoboken, NJ, USA, 2005.
12. Rymarczyk, T.; Kłosowski, G.; Kozłowski, E. A Non-Destructive System Based on Electrical Tomography and Machine Learning to Analyze the Moisture of Buildings. *Sensors* **2018**, *18*, 2285. [CrossRef]
13. André, L.; Lamy, E.; Lutz, P.; Pernier, M.; Lespinard, O.; Pauss, A.; Ribeiro, T. Electrical resistivity tomography to quantify in situ liquid content in a full-scale dry anaerobic digestion reactor. *Bioresour. Technol.* **2016**, *201*, 89–96. [CrossRef]
14. Hoyle, B.S. Spectro-tomography. In *Industrial Tomography—Systems and Applications*; Wang, M., Ed.; Woodhead Publishing: Sawston, UK, 2015; pp. 263–284.
15. Oh, T.I.; Koo, H.; Lee, K.H.; Kim, S.M.; Lee, J.; Kim, S.W.; Seo, J.K.; Woo, E.J. Validation of a multi-frequency electrical impedance tomography (mfEIT) system KHU Mark1: Imedance spectroscopy and time-difference imaging. *Physiol. Meas.* **2008**, *29*, 295–307. [CrossRef]
16. Nahvi, M.; Hoyle, B.S. Wideband electrical impedance tomography. *Meas. Sci. Technol.* **2008**, *19*, 1–9. [CrossRef]
17. Hayden, R.I.; Moyse, C.A.; Calder, F.W.; Crawford, D.P.; Fensom, D.S. Electrical Impedance Studies on potato and alfalfa tissue. *J. Exp. Bot.* **1969**, *20*, 177–200. [CrossRef]
18. Toyoda, K.; Kojima, H.; Miyamoto, S.; Takeuchi, R. Measurement and analysis of moisture changes in agricultural products using FFT noise impedance spectroscopy. *Dry Technol.* **1997**, *15*, 2025–2035. [CrossRef]
19. Zhang, M.I.N.; Stout, D.G.; Willson, J.H.M. Plant tissue impedance and cold acclimation: A re-analysis. *J. Exp. Bot.* **1992**, *43*, 263–266. [CrossRef]
20. Bauchot, A.D.; Harker, F.R.; Arnold, W.M. The use of electrical impedance spectroscopy to access the physiological condition of kiwifruit. *Postharvest Biol. Technol.* **2000**, *18*, 9–18. [CrossRef]
21. Calin-Sanchez, A.; Kharaghani, A.; Lech, K.; Figiel, A.; Carbonell-Barrachina, A.; Tsotsas, E. Drying Kinetics and Microstructural and Sensory Properties of Black Chokeberry (Aronia melanocarpa) as Affected by Drying Method. *Food Bioprocess Technol.* **2015**, *8*, 63–74. [CrossRef]
22. Khan, T.A.; Ling, S.H. Review on Electrical Impedance Tomography: Artificial Intelligence Methods and its Applications. *Algorithms* **2019**, *12*, 88. [CrossRef]
23. Liu, S.; Wu, H.; Huang, Y.; Yang, Y.; Jia, J. Accelerated Structure-Aware Sparse Bayesian Learning for 3D Electrical Impedance Tomography. *IEEE Trans. Ind. Inform.* **2019**, in press. [CrossRef]

24. Adler, A.; Lionheart, W. Uses and abuses of EIDORS: An extensible software base for EIT. *Physiol. Meas.* **2006**, *27*, 25–42. [CrossRef]

25. EIDORS Function Documentation. Available online: http://eidors3d.sourceforge.net/doc/index.shtml (accessed on 15 May 2019).

26. Wei, K.; Qiu, C.; Soleimani, M.; Primrose, K. ITS Reconstruction Tool-Suite: An inverse algorithm package for industrial process tomography. *Flow Meas. Instrum.* **2015**, *46*, 292–302. [CrossRef]

27. Netgen Homepage. Available online: https://ngsolve.org/ (accessed on 15 May 2019).

28. Kim, S.; Khambampati, A.K. Mathematical concepts for image reconstruction in tomography. In *Industrial Tomography—Systems and Applications*; Wang, M., Ed.; Woodhead Publishing: Sawston, UK, 2015; pp. 305–346.

29. Zanganeh, M. A new high output impedance wideband AC current source with high current swing authority for electrical impedance tomography. *Int. J. Comput. Netw. Technol.* **2013**, *1*, 205–213. [CrossRef]

30. Baidillah, M.R.; Iman, A.S.; Sue, Y.; Takei, M. Electrical Impedance Spectro-Tomography Based on Dielectric Relaxation Model. *IEEE Sens. J.* **2017**, *17*, 8251–8262. [CrossRef]

31. Wang, M. Electrical impedance tomography. In *Industrial Tomography—Systems and Applications*; Wang, M., Ed.; Woodhead Publishing: Sawston, UK, 2015; pp. 23–60.

32. NI 6351 Device Specification. Available online: http://www.ni.com/pdf/manuals/374591d.pdf (accessed on 18 May 2019).

33. Sumathi, S.; Surekha, P. *LabVIEW Based Advanced Instrumentation Systems*; Springer: Berlin/Heidelberg, Germany, 2007.

34. Cui, Z.; Wang, H.; Xu, Y.; Zhang, L.; Yan, Y. An integrated ECT/ERT dual modality sensor. In Proceedings of the International Instrumentation and Measurement Technology Conference, Singapore, 5–7 May 2009.

*Article*

# Detection of Single Steel Strand Distribution in Grouting Duct Based on Capacitive Sensing Technique [†]

**Nan Li [1,*], Mingchen Cao [2], Hangben Du [1], Cunfu He [3] and Bin Wu [3]**

[1]  School of Automation, Northwestern Polytechnical University, Xi'an 710071, China; xiaobendu@mail.nwpu.edu.cn
[2]  State Key Laboratory for Manufacturing Systems Engineering, Xi'an Jiaotong University, Xi'an 710049, China; mingchen.cao@stu.xjtu.edu.cn
[3]  NDT&E Research Centre, Beijing University of Technology, Beijing 100124, China; hecunfu@bjut.edu.cn (C.H.); wb@bjut.edu.cn (B.W.)
*   Correspondence: nan.li@hotmail.co.uk
†   This paper is an extended version of our paper: Nan Li, Mingchen Cao, Cunfu He and Bin Wu. Detection of steel strand cross section distribution in post-tensioned pre-stressed ducts based on simulation studies. In Proceedings of the 9th World Congress on Industrial Process Tomography, Bath, UK, 2–6 September 2018.

Received: 15 April 2019; Accepted: 30 May 2019; Published: 5 June 2019

**Abstract:** Grouting ducts (containing steel strands) are widely used to increase the structural strengths of infrastructures. The determination of the steel strand's integrity inside of ducts and the grouting quality are important for a strength evaluation of the structure. In this study, a capacitive sensing technique was applied to identify the cross-sectional distribution of the steel strands. The distribution was expressed in polar coordinates in an external post-tensioned pre-stressed duct model. An improved capacitive sensor structure was designed, which consisted of four electrodes, and different electrode-pairs were used to determine various locations' information of the steel strands. Two rounds of measurements were conducted using the designed sensor to detect the angle ($\theta$) and center distance ($r$) of the steel strand in the duct. The simulated and experimental results are presented and analyzed. In general, it is difficult to locate the angle of a steel strand directly from first-round capacitance measurements by analyzing the experimental results. Our method based on Q-factor analysis was presented for the position detection of a steel bar in an external post-tensioned pre-stressed duct. The center distance of the steel bar could be identified by second-round capacitance measurements. The processed results verified the effectiveness of the proposed capacitive sensor structure. Thus, the capacitive sensing technique exhibited potential for steel strand cross-section distribution detection in external post-tensioned pre-stressed ducts.

**Keywords:** grouting duct; capacitive sensor; cross section distribution detecting; measurement

---

## 1. Introduction

External post-tensioned pre-stressed technology has been extensively used in the construction of bridges and tunnels. This technology introduces many advantages, including lower component weights, reductions in the vertical shear forces and main tensile stresses of concrete beams, simple structures, and component installation ease. Martin et al. briefly introduced the structures of post-tensioned pre-stressed ducts [1]. In general, pre-stressed steel strands are placed in a HDPE (high-density polyethylene) tube, and high-pressure cement is filled for grouting purposes to prevent corrosion of the steel strand and increase the strength of the structure. The typical method of production for external post-tensioned pre-stressed concrete is illustrated in Figure 1. In the grouting process, an appropriate amount of water-reducing agent should be added to promote cement coagulation.

However, in practical grouting operations, the proportion of the water-reducing agent and the grouting methods may not be correct. As a result, the pre-stressed duct may not be densely filled and may exhibit stratification. Consequently, the liquid exuded from the cement may promote corrosion of the steel strand, and voids in the pre-stressed duct may lead to a reduction in the service performance and durability of the structure. The safety risk increases significantly because of these types of damage and fracture of pre-stressed ducts during service [2,3]. Therefore, there are two main aspects that must be considered during duct service: The grouting quality of the pre-stressed duct and the state of the steel strand in the duct.

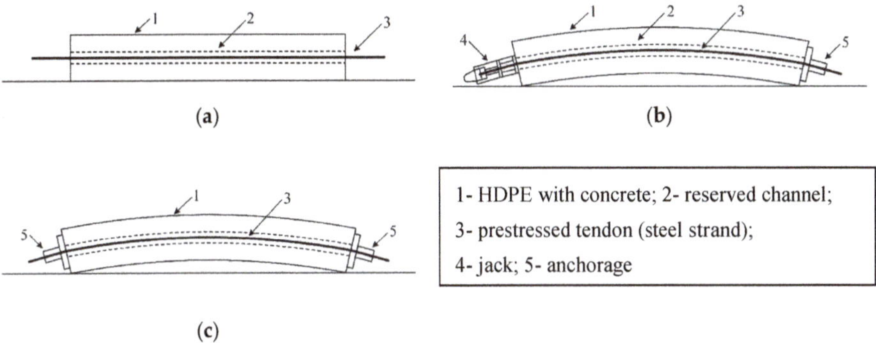



**Figure 1.** Method of production for external post-tensioned pre-stressed concrete. (**a**) Modeling concrete component; (**b**) pulling steel bar; (**c**) anchoring and grouting.

The current research on pre-stressed ducts mainly focuses on the detection of the concrete filling, such as the detection of voids in grouted tendon ducts by the impact-echo method [4,5], and evaluation of the electromagnetic properties of an exuded cement product using capacitive sensors [6,7].

In our previous work on grouting quality evaluation, a proximity capacitive sensor was designed and used to evaluate the grouting quality of the duct. Detection principle is derived from industrial processing tomography (IPT) technique [8]. A new method based on the Q-factor was presented to identify the boundary position of the three-phase layer [9]. Previous studies used simplified conditions, and the presence of steel strands was not considered. However, state detection of steel strands is another important factor for structure safety evaluation.

For the damage detection of steel bars, Moustafa et al. used fractal analysis of guided ultrasonic waves to monitor the corrosion of post-tensioned concrete [10]. Sun et al. re-analyzed the principle of the magnetic flux leakage (MFL) technique, and the 'escape' phenomenon was explained from a physical point of view [11]. Tan et al. presented a fiber Bragg grating method to monitor rebar corrosion, and the corrosion could be detected by observing the Bragg wavelength shift [12]. Deng et al. analyzed the multi-source effect of a magnetization-based eddy current testing (MB-ECT) method using an equivalent source method, and the influence of the magnetizing current and the probe lift-off were investigated by finite element analysis (FEA). The experimental results revealed a multi-source effect in the mechanism of the MB-ECT sensors and provided the basic theory for precision crack evaluation [13]. Liu et al. designed a tunnel magnetoresistive-based MFL sensor array to detect the flaws of steel wire rope. However, the cement layer was not considered properly [14]. Liu and Xiao et al. designed a new type of sensor to measure the biased pulse magnetic response in a large-diameter steel stay cable. A finite element simulation was conducted to optimize the implementation plan of the new sensor, a prototype of the proposed sensor and a biased pulse current supply were developed, and the experimental results suggested the effectiveness of the new sensor for the assessment of surface and internal flaws [15]. The current research on steel bars in external post-tensioned pre-stressed ducts mainly focuses on the cracks and corrosion of steel bars, and scholars worldwide have achieved valuable results [16–20]. However, the steel strand distribution is also important, and little research has been conducted on this aspect.

The cross-sectional distribution of the steel bars is significant because it influences the load capacity of the duct and the bearing capacities and crack resistance of the structural components. According to the Chinese National Standard GB50204-2015, the positional deviation of pre-stressed tendons should be less than ±5 mm for structural components smaller than 300 mm [21]. Furthermore, the Chinese National Standard GB/T 30827-2014 requires that the steel strands should be as close to the center of the duct as possible [22]. However, in the real construction of pre-stressed tendons, the steel bars are not always placed in the designed positions. The buoyancy of the concrete during construction causes the pre-stressed tendons to deviate from the original design position. This deviation causes the structural components to be subjected to additional radial forces. Based on statistics of the bridges, where box girder cracking occurs, the positioning deviation of the steel bars can reach 20 to 30 mm [23]. Under the action of additional radial forces, when the steel bar floats by 20 mm, the maximum vertical stress on the box girder cross-section reaches 1.0 MPa; when the steel bar floats by 30 mm, the maximum vertical stress on the box girder cross-section reaches 1.3 MPa. Therefore, the positioning error of the pre-stressed tendons will greatly increase the local vertical tensile stress of the box girder. Eventually, this leads to cracking of the weak section of the box girder and endangers the safety of the bridge.

Locating the steel bars not only contributes to the assessment of the quality of the construction, but it also helps to prevent bridge accidents during service. In this paper, a capacitive sensing technique-based method is proposed to locate the cross-sectional distribution of a steel bar and provide useful information for follow-up work on the damaged locations of the steel bars. In Section 2, the working principle of the capacitive sensing technology and the detection methodology are briefly introduced and explained. Models with different positions of the steel strands in external post-tensioned pre-stressed ducts were constructed. Simulations and experiments are presented in Section 3, and the results are compared and analyzed simultaneously. A new method is proposed to improve the detection results of the experiment stage. The conclusions are given in Section 4.

## 2. Working Principle and Methodology

### 2.1. Working Principle of Capacitive Sensor

A typical capacitive sensor is made of two traces: The exciting electrode and the receiving electrode. The working principle of the coplanar capacitive sensor is called the fringe effect of capacitance [24]. The electric field is formed between the two traces with an AC (Alternating current) excitation. When an unknown substance is placed inside the sensing area, the electric field and the permittivity between the two electrodes will change according to the material characteristics of the substance. By measuring the capacitance of the sensor, the existence and properties of the substance can be identified [25,26]. A diagram of the working principle is shown in Figure 2.

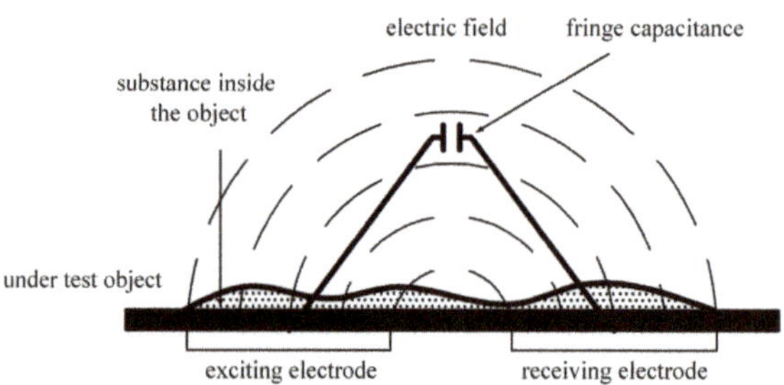


**Figure 2.** The working principle of the coplanar capacitive sensor.

The penetration depth of the capacitive sensor indicates the sensing range of the sensor. The greater the penetration depth is, the larger the sensing range becomes. In our experiment, the test block (made of high-density polyethylene (HDPE)) was placed in the sensing area above the capacitive sensor, and the fringing capacitance of the capacitive sensor was measured with the test block at different lift distances, as shown in Figure 3a. In Figure 3a, $a$, $b$, and $d$ are the width, length, and interval of the traces, respectively. The vertical distance between the test block and the surface of the capacitive sensor is the lift distance, $l$, and the effective penetration depth of the capacitive sensor is $\gamma$. The difference between the capacitance value, C ($l = \gamma$), at the position of the effective penetration depth and the minimum capacitance value, C ($l = \infty$), is equal to 3% of the difference between the maximum capacitance value, C ($l = 0$), and the minimum capacitance value, C ($l = \infty$), as shown in Equation (1) [27]:

$$C(l = \gamma) - C(l = \infty) = (C(l = 0) - C(l = \infty)) \cdot 3\%, \tag{1}$$

where $l = 0$ indicates that the test block is in full contact with the upper surface of the capacitive sensor. Furthermore, $l = \infty$ indicates that the test block is far from the upper surface of the capacitive sensor. In this situation, the change in $l$ does not affect the capacitance between the electrodes. The trend of the capacitance with different lift distances is shown in Figure 3b.

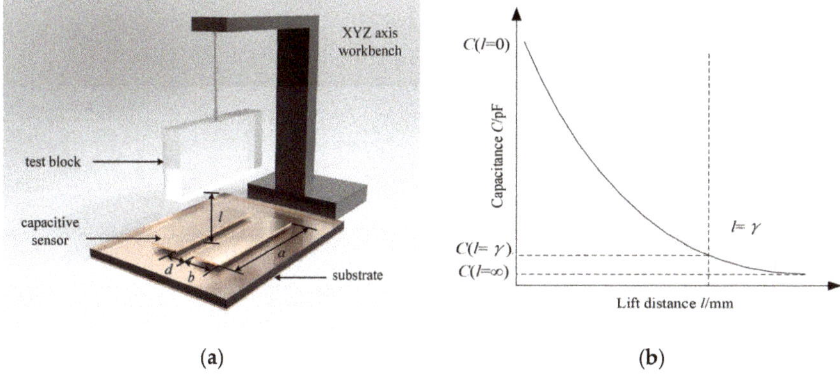


**Figure 3.** Penetration depth of the coplanar capacitive sensor: (**a**) Setup of the experiment; (**b**) the trend of the capacitance with different lift distances.

## 2.2. Methodology

The cross-section of the sensor and pre-stressed duct is shown in Figure 4a, and the capacitive sensor structure consisted of four electrodes, named E1, E2, E3, and E4. Electrodes E1 to E4 were placed against on the external surface of the duct. The widths of electrodes E1, E2, E3, and E4 corresponded to the different central angles of 10°, 20°, 20°, and 10°, as shown in Figure 4b. The intervals between the proximity electrodes all corresponded to 10°. One quarter of the circumference was covered by the sensor. Therefore, the center of the sensor was at 45°. With the support of a flexible plastic belt, the traces could be easily fixed on the tube and rotated around the pipe. The shielding layer was placed on the outermost layer to resist the outside interference.

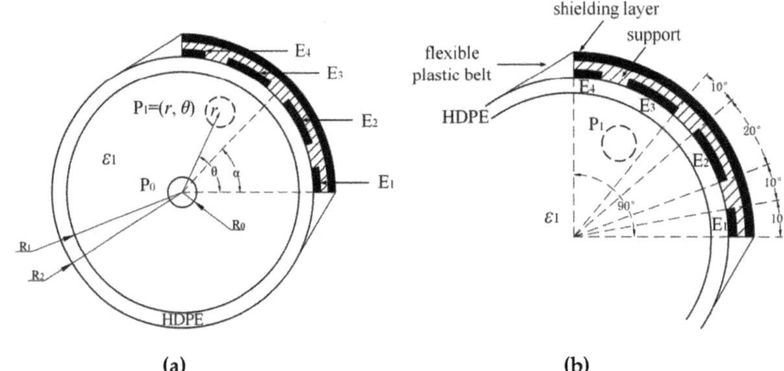


**Figure 4.** Capacitive sensor for the detection of the steel strand cross-section's distribution. (**a**) Sensor placed on the external surface of duct and (**b**) sensor structure with 1/2 duct.

Polar coordinates were used to represent the positions of the steel bar and sensor. The coordinate system origin was the center of the duct. The angles were measured from the horizontal direction and rotated in an anti-clockwise direction. Three parameters, $\alpha$, $\theta$, and $r$, were defined for the following simulations and experiments, as shown in Figure 4a. $\alpha$ is the angle position of the sensor, i.e., the angle between the horizontal direction and the line joining the center of the duct and the midpoint of the sensor. $\theta$ refers to the angle position of the steel bar, i.e., the angle between the horizontal direction and the line joining the center of the duct and the center of the steel bar. $r$ is the distance between the center of the duct and the center of the steel bar. To complete the positioning, the four electrodes were divided into two pairs, E1, E4 and E2, E3. The sensor rotated around the pipe for scanning purposes. At the first stage, E1 was set as an excitation plate and E4 was a receiver. The capacitance between E1 and E4 were measured with rotation intervals of 45° ($\alpha = 0°, 45°, 90°, \ldots, 360°$) to locate the angle ($\theta$) position of the steel strand. In the second round of measurements (second stage), E2 was set as an excitation plate, while E3 was a receiving electrode after $\theta$ was located. The capacitance between the two traces was measured to determine the center distance ($r$) of the steel strand.

To make the comparison between the simulated and measured values more intuitive, normalization was necessary, defined as follows:

$$Nor_{ij} = \frac{C - C_{min}}{C_{max} - C_{min}} \tag{2}$$

where $Nor_{ij}$ is the normalized simulated value between electrode Ei and Ej, $C$ is the simulated or measured value, $C_{min}$ is the minimum of simulated or measured value, and $C_{max}$ is the maximum of the simulated or measured value.

A single capacitance value cannot be used to determine $r$. When $\theta$ was determined by the first round of measurements, the capacitance between E2 and E3 was measured at $\theta$ ($C_1$). In addition, the capacitance at $\theta + 180°$ ($C_2$) should be measured to provide a reference of the capacitance proportion. In this way, the $r$ position of the steel strand could be calculated by the capacitance proportion, $P_c$, as follows:

$$P_C = C_1/(C_1 + C_2) \tag{3}$$

## 3. Simulations and Experiments

### 3.1. Phantom and Setup Conditions

Simulations are needed for the sensor structure design and verification of the proposed two-round measurement. The simulations were a crucial component of the research discussed below.

The simulations were performed using COMSOL Multiphysics, and the relevant parameters of the simulation were set as follows. Three-dimensional (3D) models were used for all the simulations. The dimensions, materials, and the relative permittivity values of the pre-stressed ducts, steel strand, and sensor in the experiment were the same as those in the simulations. The internal radius, $R_1$, and external radius, $R_2$, of the duct were 37.5 and 42.5 mm, respectively. A single steel bar was placed in the duct, which was filled with cement ($\varepsilon_{cement}$ = 4). The radius, $R_0$, of the steel bar was 7.5 mm. The widths of the electrodes were 7.4 mm for E1 and E4 and 14.8 mm for E2 and E3. The lengths and intervals of the electrodes were 150 and 7.4 mm, respectively, for all traces. The ducts and the support for the sensor were made of HDPE, and the entire capacitive sensor and shielding layer were made of copper foil. A schematic diagram of the simulations and experiments is shown in Figure 5a. The cement for grouting was standard Portland cement, and the water:cement ratio was 1:0.65 [28].

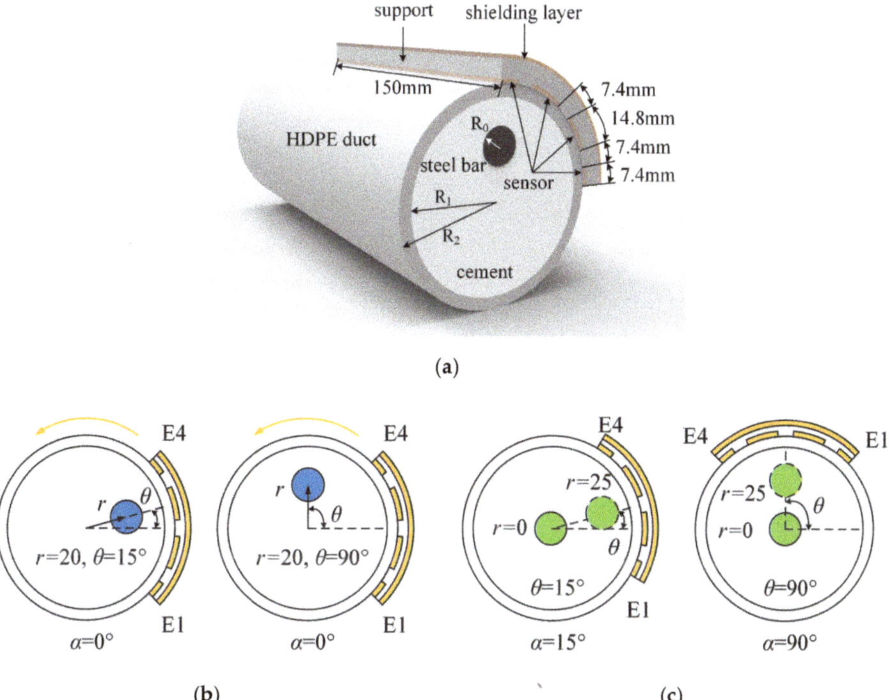


**Figure 5.** The phantoms for the simulations and experiments. (a) Schematic diagram of simulations and experiments; (b) the phantoms for $\theta$ detection in the first-round measurements; (c) the phantoms for $r$ detection in the second-round measurements.

In the simulation stage, the permittivity of cement, HDPE, and water were set to $\varepsilon_{cement}$ = 4, $\varepsilon_{HDPE}$ = 2.3, and $\varepsilon_{warer}$ = 80, respectively. The permittivity depends on various experimental parameters, such as the cement preparation process (water and cement) and chemical reaction process (steel and water). The AC excitation was set to 5 V for the simulations and experiments. The excitation frequency for the experiment was 1 MHz. However, this could not be set in the simulation. The steel bar and shielding layer were grounded in the simulations and experiments.

The location process can be divided into two stages. As mentioned in Section 2, the sensor scanning interval was 45° in the first round of measurements, i.e., $\alpha$ = 0°, 45°, 90°, ... Two positions of the steel bar were considered. $\theta$ was set to 15° and 90° with the same $r$ = 20 mm. Therefore, the position of the steel strand can be represented as $P(r, \theta)$, as shown in Figure 5b. In the second stage, for the detection of $r$, due to its symmetrical structure, two sets of models were set, as shown in Figure 5a: (1) $\theta = 15°$, $\alpha = 15°$, where $r$ varied from the center of the duct to 25 mm with intervals of 5 mm, and (2) $\theta = 90°$, $\alpha = 90°$, where $r$ varied from the center of the duct to 25 mm with intervals of 5 mm. The sensor must also be placed at $\theta + 180°$ to further identify the exact segment containing the steel strand. The phantoms for the simulations and experiments are shown in Figure 5.

Since the diameter of the steel bar was 15 mm in this study, and the inner radius of the duct was 37.5 mm, we divided the entire cross-sectional area into 16 segments with 45° for each section, as shown in Figure 6. Such segmentation is sufficient for engineering applications. Note that the internal circle's radius, $R_3$, was set to 22.5 mm (corresponding to the steel bar at $r = 15$ mm). Thus, we only needed to identify in which area the steel strand is located instead of determining its position accurately.

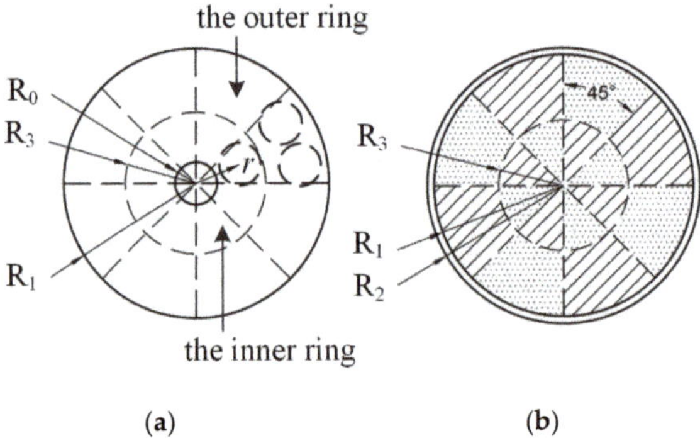


**Figure 6.** Cross-sectional area division of the duct model. (a) Placement of the steel bar in the duct; (b) the 16 segments.

*3.2. Simulation Results and Analysis*

### 3.2.1. A. Detection of the Angle Position, θ, of the Steel Strand

The capacitance between E1 and E4 was obtained by the simulations, and the normalized simulated values were calculated using Equation (2). The simulated capacitance and normalized values between E1 and E4 are presented in Table 1 and Figure 7. The special positions for $\theta$ detection are also presented, and part of the area containing special points is magnified five times in Figure 7.

**Table 1.** The simulated capacitance (pF) and normalized values between E1 and E4.

| $\alpha/°$ | $\Theta = 15°, r = 20$ mm | | $\Theta = 90°, r = 20$ mm | |
|---|---|---|---|---|
| | $C_{14}/\text{pF}$ | Normalized $C_n{}^{14}$ | $C_{14}/\text{pF}$ | Normalized $C_n{}^{14}$ |
| 0° | 0.015 | 0.014 | 0.057 | 0.594 |
| 45° | 0.014 | 0.000 | 0.017 | 0.014 |
| 90° | 0.044 | 0.429 | 0.016 | 0.000 |
| 135° | 0.073 | 0.843 | 0.017 | 0.014 |
| 180° | 0.084 | 1.000 | 0.057 | 0.594 |
| 225° | 0.080 | 0.943 | 0.077 | 0.884 |
| 270° | 0.067 | 0.757 | 0.085 | 1.000 |
| 315° | 0.028 | 0.200 | 0.077 | 0.884 |
| 360° | 0.015 | 0.014 | 0.057 | 0.594 |

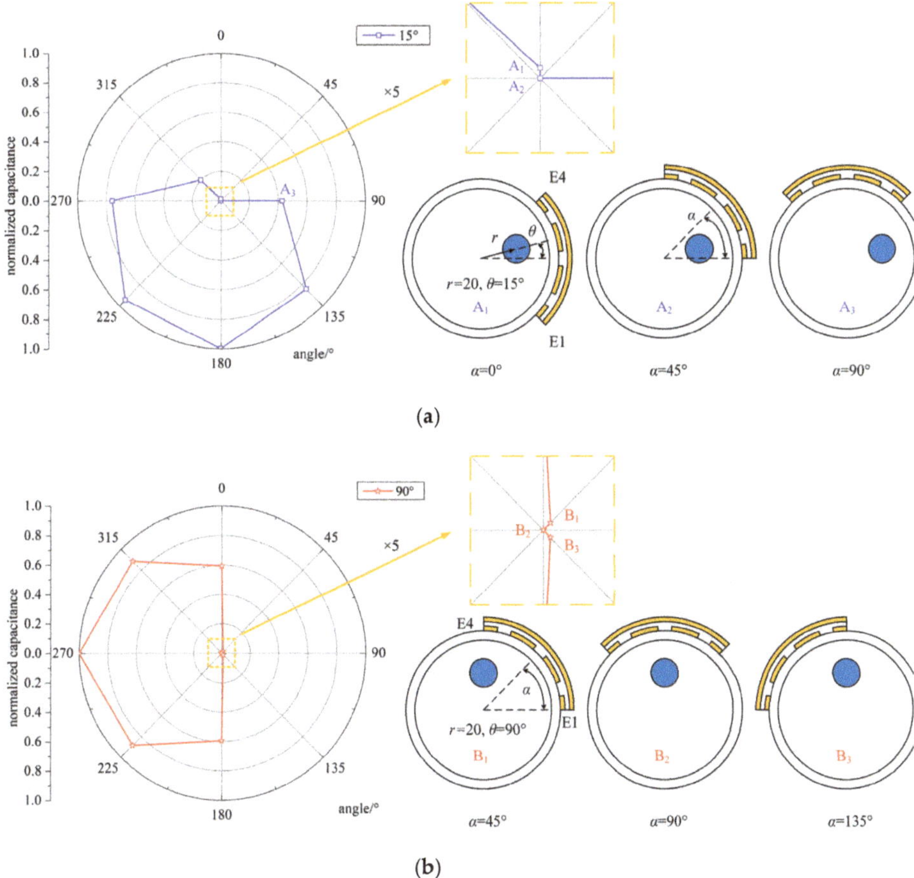

**Figure 7.** The normalized simulated capacitance between E1 and E4 and the special positions for $\theta$ detection. (**a**) $\theta = 15°$, $r = 20$ mm, $\alpha = 0°, 45°, 90°$; (**b**) $\theta = 90°$, $r = 20$ mm, $\alpha = 45°, 90°, 135°$.

For $\theta = 15°$ and $r = 20$ mm, three characteristic points were selected, named $A_1$, $A_2$, and $A_3$. $A_1$, $A_2$, and $A_3$ corresponded to the normalized capacitance value, $C_n^{14}$, when the position of the sensor, $\alpha$, was 0°, 45°, and 90°, respectively. We can derive from Figure 7a that (i) $A_2$ was the lowest point among the normalized values. If the steel bar was in the sensing area of the capacitive sensor, the capacitance between the two electrodes would decrease. Therefore, the steel bar was in the area of the sensor electrode at 45°, which was in the 0° to 90° range. (ii) Note that $A_1$ was basically equal to 0, as was $A_2$, which was different from the other points. Thus, the steel strand was also in the area of the sensor electrode at 0°, corresponding to the −45° to 45° range. (iii) Based on the above analysis, the steel bar was placed in the 0° to 45° segment, which was the overlapped range.

For $\theta = 90°$ and $r = 20$ mm, three characteristic points were marked, named $B_1$, $B_2$, and $B_3$. $B_1$, $B_2$, and $B_3$ corresponded to the normalized capacitance value, $C_n^{14}$, when the position of the sensor, $\alpha$, was 45°, 90°, and 135°, respectively. We can derive from Figure 7b that (i) $B_2$ was the lowest point among the normalized simulated values. If the steel bar was in the sensing area of the capacitive sensor, the capacitance between the two electrodes would decrease. Thus, the steel bar was in the area of the sensor electrode at 90°, which was in the 45° to 135° range. (ii) Note that $B_1$ and $B_3$ were basically equal to 0, as was $B_2$, which meant that the steel bar was also in the area of the sensor electrode at 45° and 135°, corresponding to the 0° to 90° and 90° to 180° ranges, respectively. (iii) Combining the information above, the steel bar was placed in the 45° to 135° segment. In addition, $B_1$ was equal to $B_3$, which meant that the steel bar was in a symmetric position when the sensor electrode was at 45° and 135°. Therefore, the steel bar was in the middle of the 45° to 135° range, and its position angle was 90°.

By the method above, we located the steel bar at least in one 45° segment, and for the special angles (90°), the proposed method can even determine the exact position of the steel bar.

### 3.2.2. B. Detection of the Center Distance, r, of the Steel Strand

The capacitance between E2 and E3 ($C_1^{23}$) was acquired by simulations. The capacitance at $\theta + 180°$ ($C_2^{23}$) was also simulated to obtain the area of the steel bar. To compare the capacitance trend more conveniently, the relative proportion of capacitance was calculated using Equation (3), as demonstrated in Table 2 and Figure 8. When the angle of the sensor was fixed at the angle of the steel bar, the sensor and duct had symmetrical structures. Thus, there was no difference between $\theta = 15°$ and $\theta = 90°$. We only plotted the relative proportion of the capacitance for one phantom in Figure 8. Accordingly, for $\theta = 90$ and $\alpha = 90°$, $r$ ranged from 0 to 25 mm with increments of 5 mm.

**Table 2.** The simulated capacitance (pF) and relative proportion of capacitance between E2 and E3.

| r/mm | r/30 | 15°($C_1^{23}$) | 15° + 180°($C_2^{23}$) | 15° $P_c$ | 90°($C_1^{23}$) | 90° + 180°($C_2^{23}$) | 90° $P_c$ |
|---|---|---|---|---|---|---|---|
| 0 | 0.000 | 1.218 | 1.212 | 0.501 | 1.205 | 1.209 | 0.499 |
| 5 | 0.167 | 1.168 | 1.246 | 0.484 | 1.155 | 1.244 | 0.482 |
| 10 | 0.333 | 1.095 | 1.272 | 0.463 | 1.083 | 1.269 | 0.460 |
| 15 | 0.500 | 0.986 | 1.291 | 0.433 | 0.974 | 1.288 | 0.431 |
| 20 | 0.667 | 0.823 | 1.306 | 0.386 | 0.812 | 1.303 | 0.384 |
| 25 | 0.833 | 0.599 | 1.319 | 0.312 | 0.589 | 1.316 | 0.309 |

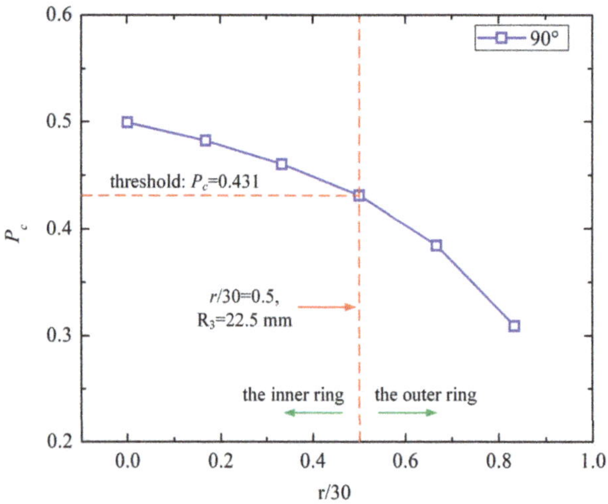


**Figure 8.** The trend of the relative proportion of capacitance and center distance percentage.

From Table 2 and Figure 8, some conclusions can be derived:

(i) As the center distance, $r$, of the steel bar increased, the capacitance between E2 and E3 ($C_1^{23}$) decreased rapidly and the capacitance at $\theta + 180°$ ($C_1^{23}$) increased gradually, which contributed to the fast decline of $C_1^{23}/(C_1^2 + C_2^{23})$.

(ii) When the relative position between the sensor and steel bar was fixed, it was symmetrical for every $\theta$ in the one-phase model. Thus, the two sets of simulations shared the same trend. Therefore, we set $P_c = C_1^{23}/(C_1^{23} + C_2^{23}) = 0.431$ as the threshold of $r = 15$ mm ($r$/30 = 0.5, $R_3 = 22.5$ mm).

We identified the steel bar location as either in the inner circle (0 mm $\leq r \leq$ 15 mm) or the outer ring (15 mm $\leq r \leq$ 30 mm). With the two rounds of measurements, the steel bar could be located in 1 of 16 segments. The improved sensor structure could locate the steel bar positions based on the simulated results.

### 3.3. Experimental Results and Analysis

#### 3.3.1. A. Detection of the Angular Position, θ, of the Steel Strand

The parameters of the experiments were identical to the simulations. The capacitance was measured using an impedance analyzer (Agilent 4294A, Beijing, China), and a diagram of the experimental device is shown in Figure 9.

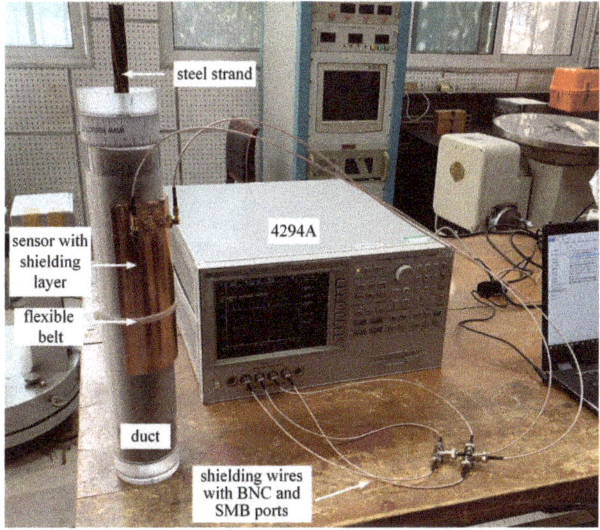


**Figure 9.** Diagram of the experimental device.

In the first round of measurement for the angular position, the experimental capacitance values between E1 and E4 were negative, as shown in Table 3, which were invalid measurements. It is the stray capacitance that influenced the experimental results. When the steel bar appeared in the sector that was covered by the sensing area of the capacitive sensor, four capacitors were formed. For example, when E1 was used for excitation, the original capacitance, C, was formed by E1 and E4. The stray capacitance, $C_{s1}$, $C_{s2}$, and $C_{s3}$, were formed by E1 and the steel bar, E1 and the shielding layer, and E1 and the cement, respectively. As a result, the measured capacitance between E1 and E4 dropped below zero. Therefore, the quality factor (Q-factor) was introduced for further experiments of the angular position's identification.

**Table 3.** The measured capacitance (fF) and normalized values between E1 and E4.

| $\alpha/°$ | $\theta = 15°, r = 20$ mm | | $\theta = 90°, r = 20$ mm | |
| --- | --- | --- | --- | --- |
| | $C_{14}$/pF | Normalized $C_n$[14] | $C_{14}$/pF | Normalized $C_n$[14] |
| 0° | −19 | 1.000 | −28 | 0.583 |
| 45° | −19 | 1.000 | −20 | 0.917 |
| 90° | −27 | 0.652 | −18 | 1.000 |
| 135° | −37 | 0.217 | −20 | 0.917 |
| 180° | −42 | 0.000 | −30 | 0.500 |
| 225° | −40 | 0.087 | −38 | 0.167 |
| 270° | −31 | 0.478 | −42 | 0.000 |
| 315° | −21 | 0.913 | −37 | 0.208 |
| 360° | −19 | 1.000 | −28 | 0.583 |

In our previous research, the boundary detecting method for post-tensioned pre-stressed ducts based on Q-factor analysis was introduced, which could effectively identify the boundary position of the three-phase duct model. Generally, $Q$ is calculated for a capacitor as follows:

$$Q = \frac{X_C}{R_C} = \frac{1}{\omega_0 C R_C} \tag{4}$$

where $\omega_0$ is the resonance frequency (radians per second), $C$ is the capacitance, $X_C$ is the capacitive reactance, and $R_C$ is the series resistance of the capacitor.

The frequency sweep was performed using an Agilent 4294A. The frequency range was from 40 Hz to 100 MHz. The frequency sweep was conducted for each $\alpha$ (0°, 45°, 90° ... ). The maximum value of the Q-factor was defined as MAX, and the corresponding frequency was $f_{max}$. Since MAX and $f_{max}$ were unstable within the frequency range, trace bandwidth analysis was performed. The cutoff point was obtained by dividing the MAX by 2. The cutoff points were searched for toward both sides of the measurement parameter axis, using the current position of the MAX as the center. The bandwidth (distance between the two cutoff points), center value (midpoint of the two cutoff points), and corresponding frequency, $f_c$, were obtained. The steps for calculating the center frequency, $f_c$, were introduced in detail in our previous work [9].

The center frequency, $f_c$, was selected for angle position detection, and the measured data are listed in Table 4. Normalization was conducted for better identification of the steel strand angle position. The sensor positions for $\theta = 15°, r = 20$ mm and $\theta = 90°, r = 20$ mm are also shown in Figure 10.

**Table 4.** The measured $f_c$ (MHz) and normalized values between E1 and E4.

| $\alpha/°$ | $\theta = 15°, r = 20$ mm | | $\theta = 90°, r = 20$ mm | |
| --- | --- | --- | --- | --- |
| | $f_c$/MHz | Normalized $f_c$ | $f_c$/MHz | Normalized $f_c$ |
| 0° | 0.538 | 0.000 | 1.079 | 0.943 |
| 45° | 1.126 | 1.000 | 1.112 | 1.000 |
| 90° | 1.099 | 0.955 | 0.534 | 0.000 |
| 135° | 1.066 | 0.898 | 1.102 | 0.982 |
| 180° | 0.743 | 0.350 | 1.081 | 0.946 |
| 225° | 0.726 | 0.320 | 0.748 | 0.370 |
| 270° | 1.065 | 0.897 | 0.707 | 0.299 |
| 315° | 1.112 | 0.976 | 0.727 | 0.334 |
| 360° | 0.538 | 0.000 | 1.079 | 0.943 |

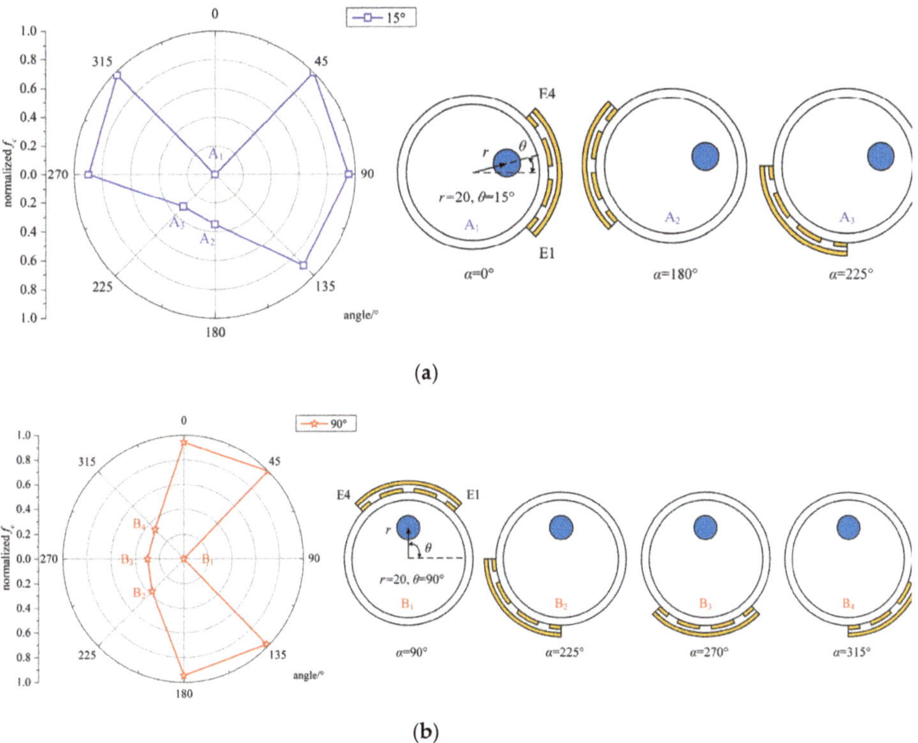

**Figure 10.** The normalized simulated capacitance between E1 and E4 and the special positions for $\theta$ detection. (**a**) $\theta = 15°$, $r = 20$ mm, $\alpha = 0°, 180°, 225°$; (**b**) $\theta = 90°$, $r = 20$ mm, $\alpha = 90°, 225°, 270°, 315°$.

For $\theta = 15°$ and $r = 20$ mm, three characteristic points were marked, named $A_1$, $A_2$, and $A_3$. $A_1$, $A_2$, and $A_3$ corresponded to the normalized capacitance value, $C_n{}^{14}$, when the position of the sensor, $\alpha$, was $0°$, $180°$, and $225°$, respectively. (i) $A_1$ was the lowest point among the normalized values. Therefore, the steel bar was in the area of the sensor electrode at $0°$, which was in the $-45°$ to $45°$ range. (ii) $A_2$ and $A_3$ were the lowest points other than $A_1$, and their normalized values were around 0.3, which means that the steel strand was in the opposite area of $\alpha = 180°$ and $\alpha = 225°$, corresponding to the $-45°$ to $90°$ range, respectively. (iii) Note that $A_2$ was approximately equal to $A_3$, so the steel bar was in the middle region of $-45°$ to $90°$, that is $0°$ to $45°$. Therefore, the steel bar could be located in the $0°$ to $45°$ segment.

For $\theta = 90°$ and $r = 20$ mm, four characteristic points were marked, named $B_1$, $B_2$, $B_3$ and $B_4$. $B_1$, $B_2$, $B_3$, and $B_4$ corresponded to the normalized capacitance value, $C_n{}^{14}$, when the position of the sensor, $\alpha$, was $90°$, $225°$, $270°$, and $315°$, respectively. (i) $B_1$ was the lowest point among the normalized values. Thus, the steel bar was in the area of the sensor electrode at $90°$, which was in the $45°$ to $135°$ range. (ii) $B_2$, $B_3$, and $B_4$ were the lowest points other than $B_1$, and their normalized values were around 0.3, which means the steel strand was in the opposite area of $\alpha = 225°, 270°, 315°$, i.e., the range from $0°$ to $180°$. (iii) It is remarkable that the steel bar was in the symmetrical position when the sensor electrode was at $45°$ and $135°$, $0°$ and $180°$, and $225°$ and $315°$, since their values were approximately the same. Therefore, the steel bar was in the middle of the $45°$ to $135°$ range, and its position angle was $90°$.

By the method above, we could locate the steel bar at least in one $45°$ segment, and for the special angles ($90°$), the proposed method could even determine the exact position of the steel bar. Furthermore, the accuracy of the positioning could be improved by reducing the scanning interval of the sensor. For instance, we could set $\alpha = 30°$ or $15°$, and the steel bar could be located in one segment of $30°$ or $15°$.

### 3.3.2. B. Detection of the Center Distance, r, of the Steel Strand

In the second-round of measurements for the center distance, only one configuration was studied: $\theta = 90°$ and $\alpha = 90°$, where $r$ ranged from 0 to 25 mm with increments of 5 mm, due to its symmetrical structure. The capacitance between E2 and E3 ($C_1{}^{23}$), and the capacitance of $\theta + 180°$ ($C_2{}^{23}$) were measured using an Agilent 4294A, as described above. The excitation frequency was 1 MHz. The measured capacitance and relative proportion of the capacitance between E2 and E3 ($P_c$) are demonstrated in Table 5 and Figure 11. In addition, normalization was conducted for a better comparison between the simulation and measurement, as illustrated in Table 6 and Figure 12.

**Table 5.** The measured capacitance (pF) and relative proportion of capacitance between E2 and E3 ($P_c$).

| r/mm | r/30 | $90°(C_1{}^{23})$ | $90°+180°(C_2{}^{23})$ | $90°\ P_c$ |
|---|---|---|---|---|
| 0 | 0.000 | 0.258 | 0.253 | 0.505 |
| 5 | 0.167 | 0.273 | 0.272 | 0.501 |
| 10 | 0.333 | 0.255 | 0.257 | 0.498 |
| 15 | 0.500 | 0.243 | 0.257 | 0.486 |
| 20 | 0.667 | 0.245 | 0.258 | 0.487 |
| 25 | 0.833 | 0.239 | 0.275 | 0.465 |

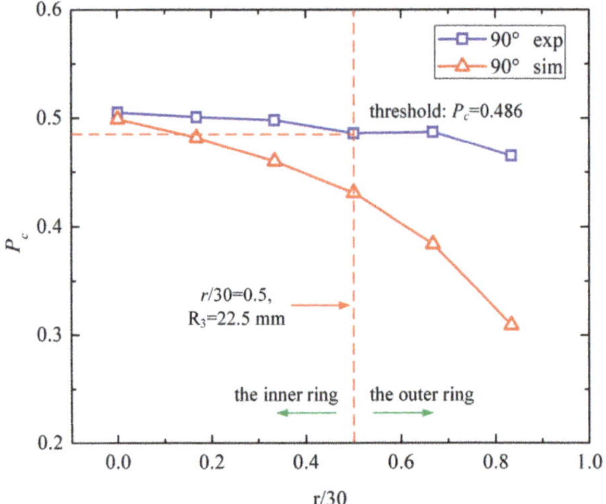


**Figure 11.** The measured and simulated relative proportion of the capacitance between E2 and E3 ($P_c$).

**Table 6.** The normalized $P_c$ of the experiment and simulation.

| r/mm | r/30 | 90° Experiment | | 90° Simulation | |
|---|---|---|---|---|---|
| | | $P_c$ | Normalized $P_c$ | $P_c$ | Normalized $P_c$ |
| 0 | 0.000 | 0.505 | 1.000 | 0.499 | 1.000 |
| 5 | 0.167 | 0.501 | 0.900 | 0.482 | 0.906 |
| 10 | 0.333 | 0.498 | 0.828 | 0.460 | 0.796 |
| 15 | 0.500 | 0.486 | 0.527 | 0.431 | 0.639 |
| 20 | 0.667 | 0.487 | 0.554 | 0.384 | 0.393 |
| 25 | 0.833 | 0.465 | 0.000 | 0.309 | 0.000 |

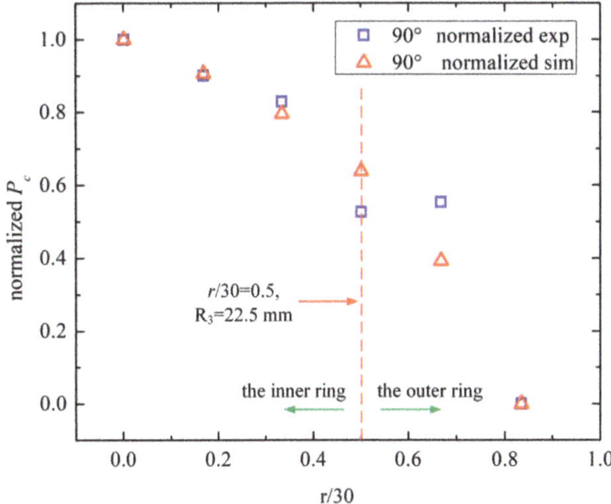



**Figure 12.** The normalized $P_c$ of the experiment and simulation.

From Tables 5 and 6 and Figures 11 and 12, some conclusions can be drawn:

(i) As the center distance, $r$, of the steel bar increases, the capacitance between E2 and E3 ($C_1^{23}$) decreased gradually, and the capacitance of $\theta + 180°$ ($C_2^{23}$) increased slowly, which contributed to the slight decrease in $C_1^{23}/(C_1^{23} + C_2^{23})$.

(ii) In Figure 11, the experimental and simulated results ($P_c$) followed the same decreasing trend as $r$ increased, indicating the validity of the experiments. The measured $P_c$ decreased more slowly than the simulated one because the stray capacitance weakened the influence of the center distance, $r$, on $P_c$.

(iii) For Figure 12, the normalized $P_c$ of the experiments and simulations shared the same decreasing tendency. There were gaps between the three measured values and the simulation, that is $r/30 = 0.333$, 0.5, and 0.667. On the one hand, due to the small variation of $P_c$ in the measurements, the normalization amplified the variable quantity. On the other hand, the position error of the steel bar caused by the twisting of the steel strand and the setting of the cement contributed to the differences between the experiments and simulations.

Therefore, we set $P_c = C_1^{23}/(C_1^{23} + C_2^{23}) = 0.486$ as the threshold of $r = 15$ mm ($r/30 = 0.5$, $R_3 = 22.5$ mm) to determine whether the steel bar was located in the inner circle (0 mm $\leq r \leq 15$ mm) or the outer ring (15 mm $\leq r \leq 30$ mm). With the two rounds of measurements, the steel bar could be located in 1 of 16 segments. In the first-round of measurements, its angle ($\theta$) was inspected in the 45° range. In the second round of measurements, its center distance ($r$) could be identified in one of the two rings. Thus, the improved sensor structure could locate the steel bar positions.

## 4. Conclusions

In this research project, the position of a single steel strand in a grouting duct was successfully identified by simulations and experiments based on capacitive sensing technology. From the simulated and experimental results, we reached the following conclusions:

(1) The proposed sensor structure with four electrodes demonstrated the effectiveness of detecting the distribution of a single steel bar in an external post-tensioned pre-stressed duct using simulations and experiments. Two measurement steps were presented for the detection of the angular position and center distance of the steel strand.

(2) For the detection of the angular position, $\theta$, of the steel bar, the capacitance between E1 and E4 was measured, and it was effective for the simulations. However, it was invalid for practical

experiments. A method based on the Q-factor was presented, and it successfully located the steel strand position in the 45° area.

(3) For the detection of the center distance, *r*, of the steel bar, the capacitance between E2 and E4 was measured, and it was valid for the simulations. The detection differences between the simulations and experiments at center distance, *r*, were analyzed, and the results were acceptable. After two steps of measurements in the simulations and experiments, the steel cross-section distribution could be identified in 1 of 16 segments.

(4) Future research shall place an emphasis on: (a) How the stray capacitance quantitatively affects the experimental results; (b) determining the dependence of the Q-factor on the steel and center frequency; and (c) sensor structure optimization.

**Author Contributions:** Conceptualization, N.L. and M.C.; methodology, N.L. and M.C.; software, N.L. and M.C.; formal analysis, N.L. and H.D.; investigation, N.L., M.C., and H.D.; writing—original draft preparation, N.L. and M.C.; writing—review and editing, N.L., C.H., and B.W.

**Funding:** This research was funded by the National Natural Science Foundation of China (No. 51875477, No. 51475013). The research was partly supported by the Top International University Visiting Program for Outstanding Young scholars of Northwestern Polytechnical University. The research works was partly supported by the China Scholarship Council (No. 201806295037)

**Conflicts of Interest:** The authors declare that there is no conflict of interest regarding the publication of this paper.

# References


1. Martin, J.; Broughton, K.J.; Giannopolous, A.; Hardy, M.S.A.; Forde, M.C. Ultrasonic tomography of grouted duct post-tensioned reinforced concrete bridge beams. *NDT E Int.* **2001**, *34*, 107–113. [CrossRef]

2. Jiang, T.; Kong, Q.; Wang, W.; Huo, L.; Song, G. Monitoring of grouting compactness in a post-tensioning tendon duct using piezoceramic transducers. *Sensors* **2016**, *16*, 1343. [CrossRef] [PubMed]

3. Jiang, T.; Zheng, J.; Huo, L.; Song, G. Finite element analysis of grouting compactness monitoring in a post-tensioning tendon duct using piezoceramic transducers. *Sensors* **2017**, *17*, 2239. [CrossRef] [PubMed]

4. Jaeger, B.J.; Sansalone, M.J.; Poston, R.W. Detecting voids in grouted tendon ducts of post-tensioned concrete structures using the impact-echo method. *Struct. J.* **1996**, *93*, 462–473.

5. Im, S.B.; Hurlebaus, S. Non-destructive testing methods to identify voids in external post-tensioned tendons. *KSCE J. Civ. Eng.* **2012**, *16*, 388–397. [CrossRef]

6. Bore, T.; Placko, D.; Taillade, F.; Himbert, M. Capacitive Sensor for Measuring the Filled of Post-Tensioned Ducts. Experimental Setup, Modeling and Signal Processing. *IEEE Sens. J.* **2013**, *13*, 457–465. [CrossRef]

7. Bore, T.; Placko, D.; Taillade, F.; Sabouroux, P. Electromagnetic characterization of grouting materials of bridge post tensioned ducts for NDT using capacitive probe. *NDT E Int.* **2013**, *60*, 110–120. [CrossRef]

8. Yu, X.; Liu, X.; Chen, S.; Luo, Y.; Wang, X.; Liu, L. High-resolution extended source optical coherence tomography. *Opt. Express* **2015**, *23*, 26399–26413. [CrossRef]

9. Li, N.; Cao, M.; Liu, K.; He, C.; Wu, B. A boundary detecting method for post-tensioned pre-stressed ducts based on Q-factor analysis. *Sens. Actuators A-Phys.* **2016**, *248*, 88–93. [CrossRef]

10. Moustafa, A.; Niri, E.D.; Farhidzadeh, A.; Salamone, S. Corrosion monitoring of post-tensioned concrete structures using fractal analysis of guided ultrasonic waves. *Struct. Control Health Monit.* **2014**, *21*, 438–448. [CrossRef]

11. Sun, Y.; Kang, Y. Magnetic mechanisms of magnetic flux leakage nondestructive testing. *Appl. Phys. Lett.* **2013**, *103*, 184104. [CrossRef]

12. Tan, C.H.; Shee, Y.G.; Yap, B.K.; Adikan, F.M. Fiber Bragg grating based sensing system: Early corrosion detection for structural health monitoring. *Sens. Actuators A-Phys.* **2016**, *246*, 123–128. [CrossRef]

13. Deng, Z.; Kang, Y.; Zhang, J.; Song, K. Multi-source effect in magnetizing-based eddy current testing sensor for surface crack in ferromagnetic materials. *Sens. Actuators A-Phys.* **2018**, *271*, 24–36. [CrossRef]

14. Liu, X.; Wang, Y.; Wu, B.; Gao, Z.; He, C. Design of tunnel magnetoresistive-based circular MFL sensor array for the detection of flaws in steel wire rope. *J. Sens.* **2016**, *2016*. [CrossRef]

15. Liu, X.; Xiao, J.; Wu, B.; He, C. A novel sensor to measure the biased pulse magnetic response in steel stay cable for the detection of surface and internal flaws. *Sens. Actuators A-Phys.* **2018**, *269*, 218–226. [CrossRef]

16. Zhang, X.; Zhang, L.; Liu, L.; Huo, L. Prestress monitoring of a steel strand in an anchorage connection using piezoceramic transducers and time reversal method. *Sensors* **2018**, *18*, 4018. [CrossRef] [PubMed]

17. Hong, X.; Zhou, J.; He, Y. Damage detection of anchored region on the messenger cable based on matching pursuit algorithm. *Mech. Syst. Signal Process.* **2019**, *130*, 221–247. [CrossRef]

18. Xia, R.; Zhou, J.; Zhang, H.; Zhou, D.; Zhang, Z. Experimental Study on Corrosion of Unstressed Steel Strand based on Metal Magnetic Memory. *KSCE J. Civ. Eng.* **2019**, *23*, 1320–1329. [CrossRef]

19. Dang, N.L.; Huynh, T.C.; Kim, J.T. Local Strand-Breakage Detection in Multi-Strand Anchorage System Using an Impedance-Based Stress Monitoring Method—Feasibility Study. *Sensors* **2019**, *19*, 1054. [CrossRef]

20. Zhang, Q.; Xin, R. The defect-length effect in corrosion detection with magnetic method for bridge cables. *Front. Struct. Civ. Eng.* **2018**, *12*, 662–671. [CrossRef]

21. Ministry of Housing and Urban-Rural Development of PRC. *The Quality Acceptance Specification for Concrete Structure Engineering Construction GB50204-2015*; China Architecture & Building Press: Beijing, China, 2015.

22. Ministry of Housing and Urban-Rural Development of PRC. *External Prestressed Cable Technical Conditions GB/T 30827-2014*; China Architecture & Building Press: Beijing, China, 2014.

23. Lou, L. Bursting Crack Mechanism of Closure Segment Bottom Slab for the Certain Prestressed Variable Depth Box-section Bridge. *Sci. Technol. Rev.* **2012**, *30*, 52–56.

24. Li, N.; Zhu, H.; Wang, W.; Gong, Y. Parallel double-plate capacitive proximity sensor modelling based on effective theory. *AIP Adv.* **2014**, *4*, 027119. [CrossRef]

25. Li, N.; Yang, X.; Gong, Y.; Wang, P. Enhancing electrical capacitance tomographic sensor design using fuzzy theory based quantifiers. *Meas. Sci. Technol.* **2014**, *25*, 125401. [CrossRef]

26. Li, N.; Cao, M.; He, C.; Wu, B.; Jiao, J.; Yang, X. Multi-parametric indicator design for ECT sensor optimization used in oil transmission. *IEEE Sens. J.* **2017**, *17*, 2074–2088. [CrossRef]

27. Li, X.B.; Larson, S.D.; Zyuzin, A.S.; Mamishev, A.V. Design principles for multichannel fringing electric field sensors. *IEEE Sens. J.* **2006**, *6*, 434–440. [CrossRef]

28. Ministry of Housing and Urban-Rural Development of PRC. *Code for Application Technique of Cementitious Grout GB/T50448-2015*; China Architecture & Building Press: Beijing, China, 2015.

*Article*

# On the Design of a New Simulated Inductor Using a Contactless Electrical Tomography System as an Example

Xin Ye, Yuxin Wang, Xiao-Yu Tang, Haifeng Ji, Baoliang Wang and Zhiyao Huang *

State Key Laboratory of Industrial Control Technology, College of Control Science and Engineering, Zhejiang University, Hangzhou 310027, China; yexin0601@zju.edu.cn (X.Y.); wangyx2015@zju.edu.cn (Y.W.); xytang@zju.edu.cn (X.-Y.T.); hfji@zju.edu.cn (H.J.); wangbl@zju.edu.cn (B.W.)
* Correspondence: zy_huang@zju.edu.cn; Tel.: +86-571-87952145

Received: 7 May 2019; Accepted: 24 May 2019; Published: 29 May 2019

**Abstract:** This work reports a new simulated inductor which is suitable for a Contactless Electrical Tomography (CET) system and can effectively overcome the unfavorable influence of coupling capacitance on the measurement results. By detailed analysis and comparison, it is found that the grounded simulated inductor has a simple circuit construction but its output current is not equal to its input current, while the floating simulated inductor can be used as an independent inductor module but its circuit structure is relatively complex. A new simulated inductor is designed by compensating the currents from the common node of an introduced independent power source to the main circuit. The new simulated inductor combines the advantages of the grounded simulated inductor and the floating simulated inductor. It has the simple construction similar to that of the grounded simulated inductor and its input current is equal to the output current, which means it can be used as an independent module. The impedance measurement and practical image reconstruction experiments were carried out to verify the effectiveness of the new simulated inductor. The experimental results show that the design of the new simulated inductor is successful, and the performance of the impedance measurement is satisfactory. The signal-to-noise ratio of the CET system is improved. Meanwhile, the research work also indicates that in the case when the independent power source is not available, the new simulated inductor is also an effective alternative method. But the phase difference between input signal and output signal is approximately 90° when the elimination principle is realized.

**Keywords:** simulated inductor technique; process tomography; contactless electrical tomography

## 1. Introduction

As basic electromagnetic devices, inductors are widely used in circuit design and signal processing [1–3]. Inductance is the circuit parameter used to describe an inductor, which relates the induced voltage to the current. However, the conventional practical inductors are usually made up of coils and magnetic cores. It is always difficult to implement a practical inductor with large inductance value and small physical size. Besides, the inductance value of a practical inductor can't be adjusted easily. Even for the practical adjustable inductor, its adjustment range is usually less than 15%.

The simulated inductor technique was developed and studied to replace a practical inductor effectively in the research field of integrated circuits [3–14]. A simulated inductor is an active circuit for generating an equivalent inductive reactance, which is implemented with active and passive components (such as resistors, capacitors and operational a mplifiers) [6]. Compared with a practical inductor, a simulated inductor has comparable function performance. It also has the advantages of easily implementing a larger inductance value and acquiring a wider adjustable range of inductance.

Meanwhile, because the miniaturization of the components (resistors, capacitors and operational amplifiers) used in simulated inductors is easy to realize at the current technique level, it is easy to achieve miniaturization of simulated inductors.

It is worth mentioning that the simulated inductor has high skill requirements for the circuit design, although it has significant advantages over practical inductors [4]. The simulated inductors can be roughly divided into two types: the grounded simulated inductor and the floating simulated inductor [10–15]. The grounded simulated inductor has the advantage of a simple circuit structure, but one of its terminals needs to be grounded directly. The floating simulated inductor does not have to be grounded directly, and it can be regarded as an independent module and be connected to the required position of an application circuit, but it has the disadvantages of a complicated circuit structure and high component matching requirements. These design requirements of the two simulated inductors more or less limit the practical applications of the simulated inductor technique. Up to date, the simulated inductor technique has mainly been studied and used as part of the active filters in the field of electronic communication. Few research works on the application of the simulated inductor technique in other research fields are reported. Therefore, it will be of great significance if we can seek an effective approach to develop a new kind of simulated inductor which combines the advantages of the two conventional simulated inductors mentioned above and overcomes the existing disadvantages of each type. That would extend the application fields of the simulated inductor technique and satisfy the wide requirements of an inductor (or inductor module) with large inductance value, wide adjustable range of inductance value and small size.

Currently, contactless electrical tomography (CET) has received increasing attention in the field of process tomography [16,17]. The latest research progresses have displayed the potential and broad perspective of the contactless electrical tomography (CET) techniques [18–27]. However, the current CET systems still exist a drawback, i.e., the existence of the coupling capacitances (formed by the two electrodes, insulating pipe, and the measured fluid) are adverse to the detection of the useful measurement signal (the details will be discussed in Section 5). Research works have verified that the impedance elimination principle can provide an effective approach [28–32], i.e., by introducing an inductor module, using the inductive reactance of the inductor module to eliminate the capacitive reactance of the coupling capacitances. However, that needs large-value and adjustable independent inductors. The emergence of simulated inductor technique provides an effective approach to solve this problem. Unfortunately, at the current stage, our knowledge and experience on applying simulated inductor technique to the field of process tomography is limited. More research work needs to be undertaken.

This work aims to report a new simulated inductor which is suitable for CET systems. The new simulated inductor, which is designed on the basis of the classic Riordan circuit, has a simple circuit structure and can be regarded as an independent module. The research work mainly includes the following four parts:

(1)    The analysis and discussions on the characteristics of the Riordan simulated inductors.
(2)    The requirements of the inductor module in CET system.
(3)    The design of the new simulated inductor module.
(4)    The experimental results with the new simulated inductor.

## 2. Riordan Simulated Inductor

Simulated inductors based on the Riordan circuit are widely accepted and studied by many scholars [3,4]. The Riordan simulated inductors can also be divided into two types, the Riordan grounded simulated inductor and the Riordan floating simulated inductor. Figure 1a,b show the typical circuits of a Riordan grounded simulated inductor and a Riordan floating simulated inductor, respectively. For both Riordan simulated inductors, the circuits input impedance $Z_{in}$ both behave as an equivalent inductance $L_{eq}$, i.e.,:

$$Z_{in} = \frac{u_{in}}{i_{in}} = j2\pi f L_{eq} \tag{1}$$

where $f$ is the frequency of the excitation signal and $j$ is the imaginary unit.

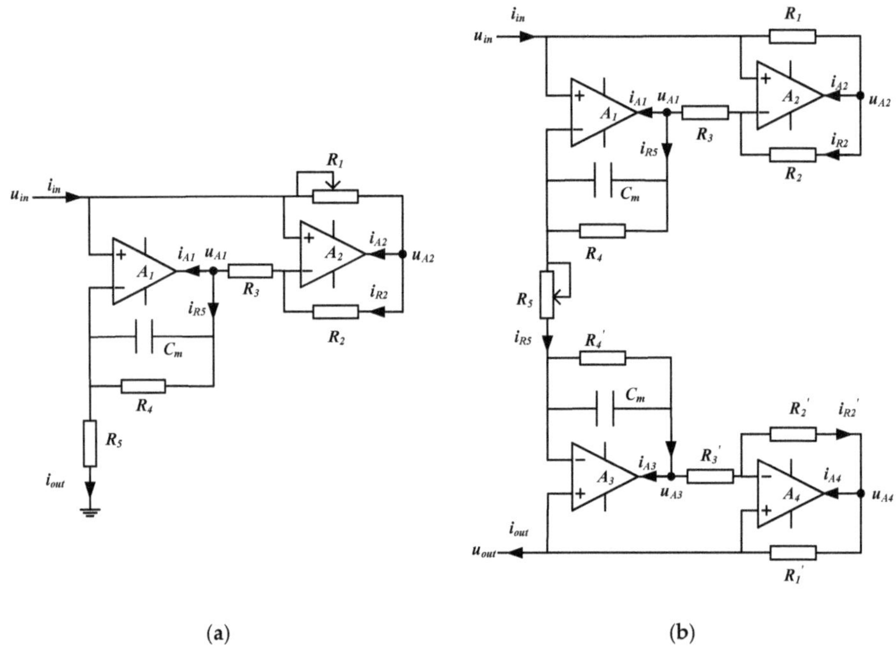

**Figure 1.** The application circuits of two types Riordan simulated inductors: (**a**) the Riordan grounded simulated inductor; (**b**) the Riordan floating simulated inductor.

The Riordan grounded simulated inductor has a simple circuit structure, as shown in Figure 1a. One of its terminals must be grounded directly. The Riordan floating simulated inductor uses two Riordan grounded simulated inductors which are connected in cascade (back to back), as shown in Figure 1b. It has a symmetrical circuit structure. Its terminals need not to be grounded directly. It can be regarded as an independent module (like a practical inductor) and be connected to the required position of an electrical circuit freely.

The detailed discussions and analysis on the characteristics of the Riordan grounded simulated inductor and the Riordan floating simulated inductor are given in the following sections.

### 2.1. The Characteristics of the Riordan Grounded Simulated Inductor

Let $u_{in}$ be the input voltage, $i_{in}$ be the input current, $i_{out}$ be the output current, $i_{A1}$ and $i_{A2}$ be the current flowing into the operational amplifier $A_1$ and $A_2$ respectively. The following equations can be obtained:

$$i_{in} = i_{A2} + i_{R2} = i_{A2} + i_{A1} + i_{out} \tag{2}$$

$$u_{A2} = u_{in} - i_{in}R_1 \tag{3}$$

$$i_{R2} = \frac{u_{A2} - u_{in}}{R_2} = -\frac{R_1}{R_2}i_{in} \tag{4}$$

$$i_{A2} = i_{in} - i_{R2} = \frac{R_1 + R_2}{R_2}i_{in} \tag{5}$$

$$u_{A1} = u_{in} - i_{R2}R_3 = u_{in} + \frac{R_1 R_3}{R_2} i_{in} \tag{6}$$

$$i_{out} = \frac{u_{A1} - u_{in}}{\frac{1}{1/R_4 + j2\pi f C_m}} = \frac{\frac{R_1 R_3}{R_2} i_{in}}{\frac{1}{1/R_4 + j2\pi f C_m}} = \frac{(1/R_4 + j2\pi f C_m)R_1 R_3}{R_2} i_{in} \tag{7}$$

$$i_{A1} = i_{R2} - i_{out} = -\frac{R_1 + (1/R_4 + j2\pi f C_m)R_1 R_3}{R_2} i_{in} \tag{8}$$

Meanwhile, the output current $i_{out}$ can also be described as:

$$i_{out} = i_{R5} = \frac{u_{in}}{R_5} \tag{9}$$

According to Equations (7) and (9), input impedance of the Riordan grounded simulated inductor $Z_{in}$ is:

$$Z_{in} = \frac{u_{in}}{i_{in}} = \frac{(1/R_4 + j2\pi f C_m)R_1 R_3 R_5}{R_2} = \frac{R_1 R_3 R_5}{R_2 R_4} + j2\pi f \frac{C_m R_1 R_3 R_5}{R_2} \tag{10}$$

Thus, the equivalent inductance $L_{eq}$ of the Riordan grounded simulated inductor is:

$$L_{eq} = \frac{C_m R_3 R_5}{R_2} R_1 \tag{11}$$

The equivalent internal resistance $r_{eq}$ of the Riordan grounded simulated inductor is:

$$r_{eq} = \frac{R_3 R_5}{R_2 R_4} R_1 \tag{12}$$

According to Equations (11) and (12), the equivalent inductance $L_{eq}$ is determined by the resistances $R_1$ $R_2$, $R_3$, $R_5$ and the capacitance $C_m$, and the equivalent internal resistance $r_{eq}$ is determined by the resistances $R_1$ $R_2$, $R_3$, $R_4$ and $R_5$. Usually, in practical applications, in order to adjust the value of $L_{eq}$ and $r_{eq}$ conveniently and independently, the values of $R_2$, $R_3$, $R_5$ and $C_m$ are fixed. The value of $L_{eq}$ is mainly adjusted by changing the value of the high-precision adjustable resistor $R_1$, while the value of $r_{eq}$. is mainly determined by the value of $R_4$. Because the equivalent internal resistance of an inductor module should be as small as possible, the value of $R_4$ is usually much larger than those of $R_1, R_2, R_3$ and $R_5$, respectively.

It is necessary to indicate that the value of the $L_{eq}$ is affected by $R_5$. In practical applications, the value of $L_{eq}$ is expected as a constant, which means the value of $R_5$ should also be a constant. That is the reason why one terminal of the circuit is grounded. Otherwise, it is difficult to guarantee the value of $L_{eq}$ to be a constant and the value of $L_{eq}$ will vary with the impedance of the succeeding circuit. In other words, $R_5$ could be regarded as the load of the circuit. If the output terminal is not grounded, the load of the circuit will change, and then $L_{eq}$ can not be fixed.

Meanwhile, Equation (7) also shows that the input current $i_{in}$ is not equal to the output current $i_{out}$, (i.e., $i_{in} \neq i_{out}$) and there exists a phase difference (approximately 90°) between $i_{in}$ and $i_{out}$. From Equations (2), (5) and (8), it can be found that there are currents flowing through the output of the operational amplifiers ($A_1$ and $A_2$), leading to the difference between $i_{in}$ and $i_{out}$.

Based on the discussions above, it can be found that although the Riordan grounded simulated inductor has a simple circuit construction, it has the disadvantage that it can't constitute an independent module like the practical inductor yet. The reason is that the currents through the two terminals of the inductor module are not equivalent to each other.

### 2.2. The Characteristics of the Riordan Floating Simulated Inductor

As shown in Figure 1b, the Riordan floating simulated inductor has a symmetrical circuit structure. It is a combination of two identical Riordan grounded simulated inductors, i.e., the values of the resistances and capacitances of both sides are equal, i.e., $R_1 = R_1'$, $R_2 = R_2'$, $R_3 = R_3'$, $R_4 = R_4'$, $C_m = C_m'$. Based on Equation (7), the current flow through $R_5$ is:

$$i_{R5} = \frac{(1/R_4 + j2\pi f C_m)R_1 R_3}{R_2} i_{in} \tag{13}$$

Further:

$$u_{out} = u_{in} - i_{R5} R_5 \tag{14}$$

$$u_{A3} = u_{out} - i_{R5}\frac{1}{1/R_4' + j2\pi f C_m} = \frac{1/R_4 + j2\pi f C_m}{1/R_4' + j2\pi f C_m}\frac{R_1 R_3}{R_2} i_{in} \tag{15}$$

$$i_{R2}' = \frac{u_{A3} - u_{out}}{R_3'} \tag{16}$$

$$u_{A4} = u_{out} - i_{R2}' R_2' \tag{17}$$

$$i_{out} = \frac{u_{A4} - u_{out}}{R_1'} = -\frac{R_2'}{R_1'}i_{R2}' \tag{18}$$

According to Equations (13) and (14), the input impedance $Z_{in}$ of the Riordan floating simulated inductor can be described as:

$$Z_{in} = \frac{u_{in} - u_{out}}{i_{in}} = \frac{i_{R5} R_5}{i_{in}} = \frac{R_1 R_3 R_5}{R_2 R_4} + j2\pi f\frac{C_m R_1 R_3 R_5}{R_2} \tag{19}$$

Thus, the equivalent inductance and the internal resistance $r_{eq}$ of the Riordan floating simulated inductor can be expressed as:

$$L_{eq} = \frac{C_m R_1 R_3}{R_2} R_5 \tag{20}$$

$$r_{eq} = \frac{R_1 R_3}{R_2 R_4} R_5. \tag{21}$$

By comparison of Equations (11)–(12) and (20)–(21), respectively, we can see that the Riordan grounded simulated inductor and the Riordan floating simulated inductor have the same equivalent inductance and the same equivalent internal resistance. However, the way to adjust the value of $L_{eq}$ of the Riordan floating simulated inductor is different from the Riordan grounded simulated inductor. In practical applications, the value of $L_{eq}$ of the Riordan floating simulated inductor is mainly adjusted by changing the value of the adjustable resistor $R_5$. The value of $r_{eq}$ of the Riordan floating simulated inductor is mainly determined by the value of $R_4$.

Meanwhile, according to Equations (15) to (19), the output impedance $Z_{out}$ of the Riordan floating simulated inductor can be described as:

$$Z_{out} = \frac{u_{in} - u_{out}}{i_{out}} = \frac{i_{R5} R_5}{-\frac{R_2'}{R_1'}\left(\dfrac{-i_{R5}\frac{1}{1/R_4' + j2\pi f C_m}}{R_3'}\right)} = \frac{R_1' R_3' R_5}{R_2' R_4'} + j2\pi f\frac{C_m' R_1' R_3' R_5}{R_2'} \tag{22}$$

Because the Riordan floating simulated inductor is formed by two mirrorly-connected identical Riordan grounded simulated inductor. According to Equations (19) and (22), the input impedance $Z_{in}$ of the Riordan floating simulated inductor is equal to the output impedance $Z_{out}$, i.e.,

$$Z_{in} = Z_{out} \tag{23}$$

Further, from Equations (15), (16) and (18):

$$
\begin{aligned}
i_{out} &= -\frac{R_2'}{R_1'} i'_{R2} \\
&= -\frac{R_2'}{R_1'}\left(\frac{u_{A3}-u_{out}}{R_3'}\right) \\
&= -\frac{R_2'}{R_1'}\left(-\frac{1/R_4+j2\pi f C_m}{1/R_4'+j2\pi f C_m}\frac{R_1 R_3}{R_2} i_{in}\right)\frac{1}{R_3'} \\
&= \frac{R_2' R_1 R_3}{R_2 R_1' R_3'} i_{in} \\
&= i_{in}
\end{aligned}
\tag{24}
$$

i.e., the input current $i_{in}$ is equal to the output current $i_{out}$ and there is no phase difference between $i_{in}$ and $i_{out}$. Besides, the relationship between the currents flowing into the operational amplifiers $A_3$ and $A_4$ can be described respectively as:

$$
i_{A3} = i_{R5} - i_{R2}'
\tag{25}
$$

$$
i_{A4} = i_{R2}' - i_{out}
\tag{26}
$$

According to Equations (5), (8), (14), (22)–(24), it can be found that the sum of the total current flowing into the output terminals of four operational amplifiers is zero, which means that the current flowing into the operational amplifiers $A_1$ and $A_2$ can be compensated by the current flowing into the operational amplifiers $A_3$ and $A_4$, i.e.,

$$
(i_{A1} + i_{A2}) + (i_{A3} + i_{A4}) = 0
\tag{27}
$$

Therefore, based on the discussions above, the Riordan floating simulated inductor can be regarded as an independent module like a practical inductor. That is convenient for practical applications. However, the electronic components in the Riordan floating simulated inductor are required to be strictly equal to their mirror components. The circuit structure of the Riordan floating simulated inductor is relatively complex and it includes four closed-loops, so, its realization needs higher quality components and higher circuit design skill. Meanwhile, the stability of the Riordan floating simulated inductor should be seriously considered, because it is a multi-closed-loop system.

Based on the above analysis and discussions, it can be found that the two simulated inductors both have their advantages and some limitations or preconditions for practical applications. Therefore, up to date, the simulated inductor technique is mainly studied and applied in the field of electronic communications, for either the Riordan grounded simulated inductor or the Riordan floating simulated inductor.

## 3. The Requirement of the Inductor Module in CET System

In the field of process tomography, different kinds of CET systems have been studied, including the capacitively coupled electrical resistance tomography (CCERT) system, the capacitively coupled electrical impedance tomography (CCEIT) system and the electrical capacitance tomography (ECT) system [18–20]. Figure 2a shows a sketch of a 12-electrode CET sensor. The electrodes of the CET sensor are mounted symmetrically around the outer surface of the insulating pipe. Figure 2b shows the equivalent circuit of an electrode pair in the CET sensor. It is simplified as two coupling capacitances $C_1$ and $C_2$ (formed by the two electrodes, the insulating pipe, and the measured fluid), and an impedance $Z_x$ of the measured fluid. For the CCERT system, the measured fluid is equivalent to a resistance $R_x$, i.e., $Z_x = R_x$. For the ECT system, the measured fluid is equivalent to a capacitance $C_x$, i.e., $Z_x = C_x$. For the CCEIT system, the measured fluid is equivalent to an impedance $Z_x$.

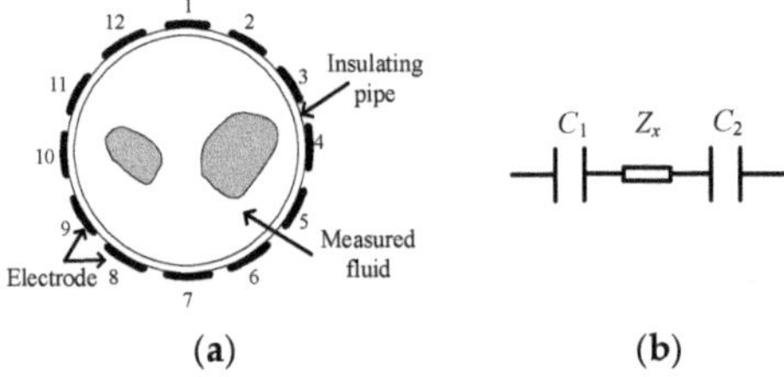

**Figure 2.** Measurement principle of the CET sensor. (**a**) The sketch of a 12-electrode CET sensor; (**b**) The equivalent circuit of a measurement electrode pair in the CET sensor.

It is worth mentioning that the existence of the two coupling capacitances $C_1$ and $C_2$ makes the contactless measurement possible. However, from the viewpoint of electrical impedance measurement, only the $Z_x$ is the useful signal. The capacitive reactance of $C_1$ and $C_2$ is the background signal which will limit the signal-to-noise ratio (SNR) of impedance measurement and should be overcome.

Research works have verified that the impedance elimination principle can provide an effective approach to overcome the unfavorable influences of the coupling capacitances, i.e., by introducing an inductor module and using the inductive reactance of the inductor module to eliminate the capacitive reactance of the coupling capacitances.

Figure 3 shows the flowchart of the impedance elimination principle for the measurement of an electrode pair in the CET sensor where $f$ is the frequency of the excitation signal and $j$ is the imaginary unit. The overall impedance $Z$ of the measurement path is:

$$Z = \frac{1}{j2\pi f C_1} + Z_x + \frac{1}{j2\pi f C_2} + j2\pi f L \tag{28}$$

With the application of the impedance elimination principle, the imaginary part of the $Z$ should be zero, i.e.:

$$\frac{1}{j2\pi f C_1} + \frac{1}{j2\pi f C_2} + j2\pi f L = 0 \tag{29}$$

Thus, from Equation (29), the excitation frequency $f$. is determined by:

$$f = \frac{1}{2\pi} \sqrt{\frac{C_1 + C_2}{L C_1 C_2}} \tag{30}$$

Equations (28) to (30) show that if the excitation frequency is set by Equation (30), the impedance of the measurement path only consists of the impedance of measured fluid $Z_x$, i.e.:

$$Z = Z_x \tag{31}$$

Therefore, the capacitive reactance of the coupling capacitances can be successfully eliminated by the inductive reactance of the introduced inductor module. Thus, the negative influence of the background signal (coupling capacitances $C_1$ and $C_2$) can be effectively overcome. The SNR of the CET system can be improved.

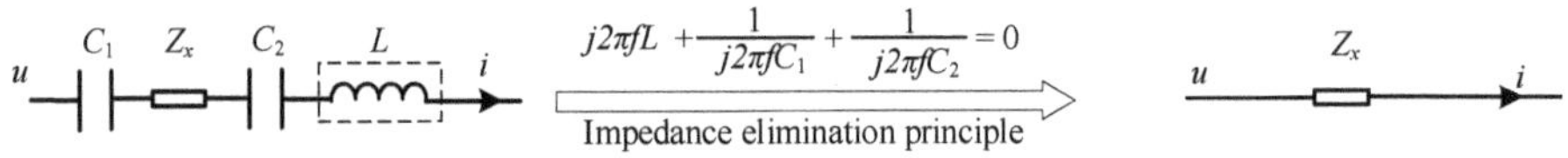

**Figure 3.** The flowchart of the impedance elimination principle in the CET systems.

According to Equation (30), it can be found that the system excitation frequency $f$ is related to the inductance value $L$ of the introduced inductor module and the coupling capacitances $C_1$ and $C_2$.

In practical application, the value of the excitation frequency $f$ shouldn't be set too large in order to avoid imposing extra high requirements on the circuit design and system complexity. Meanwhile, the values of the coupling capacitances are usually small (about 30~50 pF), so the introduced inductor module is expected to have a large inductance value.

Further, the materials, the wall thickness and the diameters of the insulating pipes as well as the electrode angle and the length of the electrode, etc., will also affect the values of the coupling capacitances, so the inductance value $L$ of the inductor module should be adjustable and the adjustment range is desired to be as wide as possible. That means the inductor module should have good compatibility and adaptability.

Besides, the CET system is required to be compacted for miniaturization and integration. So the introduced inductor module should also have the capability to be minimized and can be easily integrated in the CET system.

Based on the above considerations, the problems have clearly indicated that the practical inductor still cannot satisfy the requirements of the practical application in the CET system, even if it is a simple way to use a practical inductor to form the inductor module.

Obviously, the simulated inductor technique provides an attractive approach to develop the inductor module and hence effectively overcome the unfavorable influences of the coupling capacitances. However, in the research field of CET, the applications and research of the simulated inductor are not sufficient. More research work should be undertaken.

## 4. New Simulated Inductor Module

### 4.1. The Design of the New Simulated Inductor Module

This work aims to develop a new simulated inductor which is suitable for the CET systems. Based on the discussions in Sections 2 and 3, the new simulated inductor module should have the following features:

(1)    The new simulated inductor module should have a simple circuit construction.
(2)    The new simulated inductor module should be regarded as an independent module and can be connected into the circuit flexibly.
(3)    The input current of the new simulated inductor should be equal to the output current, because the CET system implements the impedance measurements by measuring the current flowing through the detection path, as shown in Figure 3.

Figure 4 illustrates the circuit of the new simulated inductor module.

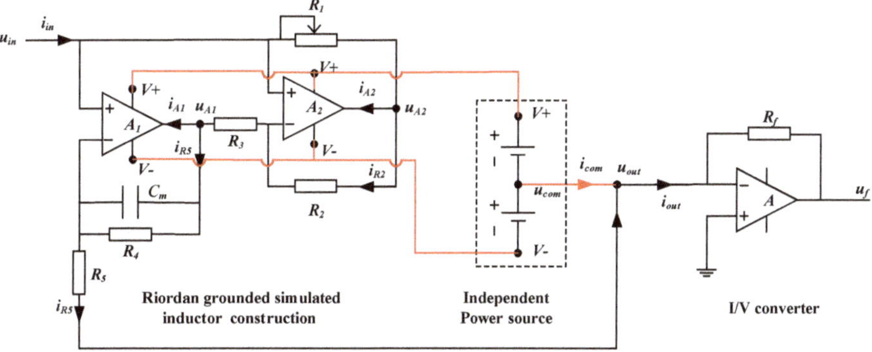


**Figure 4.** The circuit of the new simulated inductor module.

In Figure 4, the new simulated inductor consists of a Riordan grounded simulated inductor with an independent power source and an I/V (current to voltage) convertor. Compared with the typical Riordan grounded simulated inductor in Figure 1a, the new simulated inductor has two key improvements (or differences):

(1)  The output terminal (marked as $u_{out}$) is not connected to the ground directly. Instead, its output terminal is connected to the inverting input terminal of the operational amplifier of the I/V converter.

(2)  The operational amplifiers in the Riordan grounded simulated inductor are supplied by an independent power source, while other operational amplifiers are powered by the system power source. Meanwhile, the common node ($u_{com}$) of the independent power is connected to the output terminal.

From Section 2, the output terminal of the Riordan grounded simulated inductor needs to be grounded, to ensure that the simulated inductor module has a fixed equivalent inductance value $L_{eq}$ (as discussed in Section 2.1, the pre-condition to obtain a fixed value of $L_{eq}$ is that the load of the circuit is fixed). However, in practical applications, changes of the succeeding circuit should not affect the inductance value $L_{eq}$ of an independent inductor module. Therefore, as shown in Figure 4, an equivalent grounded method is introduced to meet the application pre-condition of the Riordan grounded simulated inductor, i.e., connecting the output terminal to the inverting input terminal of an operational amplifier, and the noninverting terminal of the operational amplifier is directly connected to the ground. Meanwhile, as shown in Figure 4, the detection current signal $i_{out}$ which flows through the new simulated inductor is transmitted to a output voltage signal $u_f$ by the I/V convertor.

Additionally, according to the discussion in Section 2.1, since the currents flow into the operational amplifiers ($A_1$ and $A_2$) in Figure 1a, the output current of the Riordan grounded simulated inductor is different from the input current, so the Riordan grounded simulated inductor can't be used as an independent inductor. To solve this problem, we should seek an effective way to make the input current equal to the output current. Our method is to compensate the currents, which flow into the operational amplifiers, to main circuit.

Figure 5 shows the operation characteristics of a typical operational amplifier [1,33,34]. In Figure 5, $u_{com}$ is the common node of the independent power source. $i_n$ is the current into the inverting input terminal. $i_p$ is the current into the noninverting input terminal. $i_o$ is the current into the output terminal. $i_{c+}$ is the current flowing from the positive power supply terminals to common node. $i_{c-}$ is the current flowing from the negative power supply terminal to common node. $i_c$ is the current obtained from the common node.

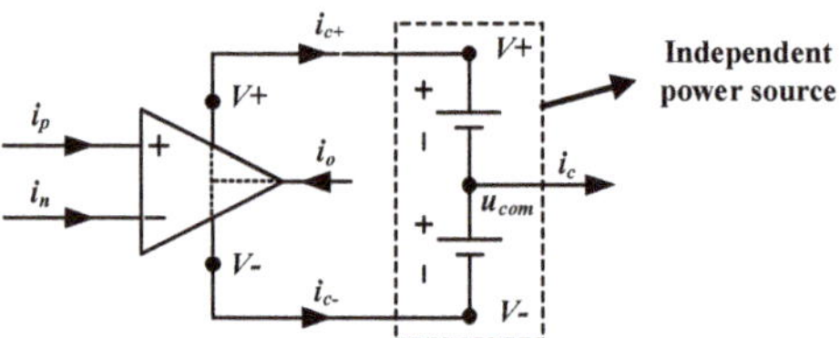


**Figure 5.** The operation characteristics of a typical operational amplifier.

The sum of the currents entering the operational amplifier should be zero:

$$i_p + i_n + i_o - (i_{c+} + i_{c-}) = 0. \tag{32}$$

Meanwhile, there is no current flowing into either input terminal (i.e., $i_p = i_n = 0$). The current $i_o$ flowing into the output terminal equals to the sum of the currents $i_{c+}$ and $i_{c-}$. The current obtained from the common node $u_{com}$ is also the sum of the currents $i_{c+}$ and $i_{c-}$. Thus, the following equation can be obtained:

$$i_o = i_{c+} + i_{c-} = i_c \tag{33}$$

Equation (33) indicates that the current which flows into the output terminal of operational amplifier equals to the current which is obtained from the common node of the independent power source. Obviously, it is independent on the power sources of other operational amplifiers.

In the case of this work, two operational amplifiers $A_1$ and $A_2$ are used in the grounded simulated inductor as shown in Figure 5. $i_{A1}$ is the current into the output terminal of $A_1$. $i_{A2}$ is the current into the output terminal of $A_2$. As mentioned in Section 2.1, the input current of a grounded simulated inductor is not equal to the output current. There exist some differences and the differences between the input current and the output current are $i_{A1}$ and $i_{A2}$ (as shown in Figure 1). In another respect, based on the above discussion in this section, the current which flows into the output terminal of an operational amplifier can be obtained from the common node of the independent power source. Therefore, making the input current equal to the output current can be realized by introducing the current from the common node of the independent power source into the main circuit.

The new simulated inductor module has two currents ($i_{A1}$ and $i_{A2}$) which should be compensated. Simply, we can use two independent power sources and introduce the currents from the two common nodes of the two independent power sources. It is a direct treatment. But, it needs two independent power sources. In fact, in this work, it is not necessary to use two independent power sources, only one independent power source can also meet the requirement, because the currents ($i_{A1}$ and $i_{A2}$) can only flow from the common node of the independent power source. There is no other way. So, in this work, the two operational amplifiers share one independent power source which is independent of the power sources of other operational amplifiers. Thus, our method to make the input current equal to the output current is implemented by introducing the current from the common node into the main circuit. As shown in Figure 4, $i_{A1}$ and $i_{A2}$ can be obtained from one common node $u_{com}$:

$$i_{com} = i_{A1} + i_{A2} \tag{34}$$

Further, the current $i_{out}$ is:

$$i_{out} = i_{R5} + i_{com} = i_{R5} + (i_{A1} + i_{A2}) = i_{in} \tag{35}$$

From the above discussions, it can be found that:

(1) By connecting the output terminal to the inverting input terminal of the operational amplifier, equivalent grounded is realized, which meets the requirement of pre-condition of the grounded simulated inductor. The equivalent inductance $L_{eq}$ of the new simulated inductor module is guaranteed to be fixed. It is not affected by the load of the succeeding circuit.

(2) By connecting the common node of the independent power source to the main circuit, the currents which flow into the output terminals of the operational amplifiers $A_1$ and $A_2$ are compensated. The input current of the new simulated inductor module equals to the output current ($i_{in} = i_{out}$). Further, the input impedance $Z_{in}$ of the new simulated inductor equals to its output impedance $Z_{out}$.

Therefore, the new simulated inductor not only has the simple construction and good stability similar to that of the grounded simulated inductor, but also can be used conveniently in practical applications like the floating simulated inductor. It reduces the requirements for components and can be used as an independent inductor module to effectively replace the practical inductor. Correspondingly, the equivalent inductance. $L_{eq}$ and the internal resistance $r_{eq}$ of the new simulated inductor module are:

$$L_{eq} = \frac{C_m R_3 R_5}{R_2} R_1 \tag{36}$$

$$r_{eq} = \frac{R_3 R_5}{R_2 R_4} R_1 \tag{37}$$

Comparing Equations (36) and (37) with Equations (11) and (12), it can be found that the $L_{eq}$ and the $r_{eq}$ of the new simulated inductor are the same as those of the grounded simulated inductor.

### 4.2. Impedance Measurement Principle by Using the New Simulated Inductor Module

Figure 6 shows the simplified measurement principle circuit of the CET system by using the new simulated inductor module, where the impedance of the measured fluid is $Z_x$.

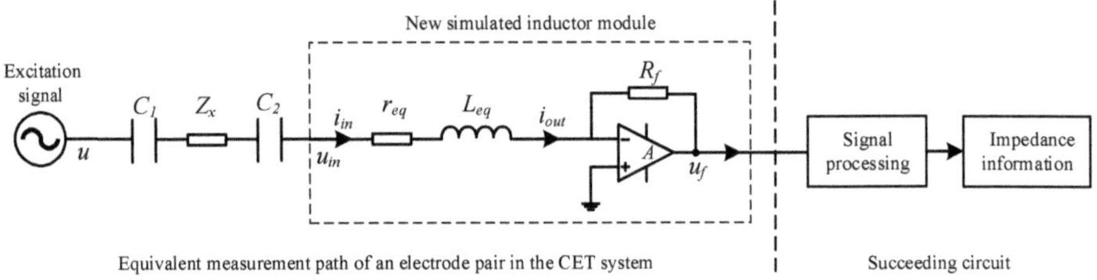


**Figure 6.** The simplified measurement principle circuit of the CET system.

The overall impedance $Z$ of the measurement path is:

$$Z = \frac{1}{j2\pi f C_1} + Z_x + \frac{1}{j2\pi f C_2} + r_{eq} + j2\pi f L_{eq} \tag{38}$$

The excitation frequency $f$ can be determined by the following equations:

$$\frac{1}{j2\pi f C_1} + \frac{1}{j2\pi f C_2} + j2\pi f L_{eq} = 0 \tag{39}$$

$$f = \frac{1}{2\pi} \sqrt{\frac{C_1 + C_2}{L_{eq} C_1 C_2}} \tag{40}$$

At the excitation frequency, the capacitance reactance of $C_1$ and $C_2$ is eliminated by the inductive reactance of $L_{eq}$. The unfavorable influences of the coupling capacitances are overcome by the impedance elimination principle. As a result, the overall impedance of the measurement path $Z$ consists of the impedance of measured fluid $Z_x$ and the internal resistance $r_{eq}$, i.e.:

$$Z = Z_x + r_{eq} \tag{41}$$

After the operation of the I/V convertor, the output voltage signal $u_f$ is:

$$u_f = -i_{out} R_f = -\frac{u}{Z_x + r_{eq}} R_f \tag{42}$$

From Equation (37), it can be seen that if $R_4$ is selected as a large-valued resistor, the value of $r_{eq}$ can be very small compared to the impedance of the measured fluid $Z_x$. Thus, Equation (40) can be further simplified as:

$$u_f \approx -\frac{u}{Z_x} R_f \tag{43}$$

Equation (43) indicates that the phase difference between $u$ and $u_f$ becomes 180° when the impedance elimination principle is realized. In practical measurement process, it is not necessary to know the exact value of $C_1$ or $C_2$. The excitation frequency can be predetermined by an oscilloscope.

In the case when the independent power source is not available, the combination of the Riordan grounded simulated inductor and the I/V convertor can also construct a simulated inductor module as shown in Figure 7. The simulated inductor module can also be workable. It has the same equivalent inductance $L_{eq}$ and internal resistance $r_{eq}$ as those of the new simulated inductor module, but there are two problems:

(1)   The input current is not equal to the output current ($i_{in} \neq i_{out}$).
(2)   When the impedance elimination principle is realized, the phase difference between the input current $i_{in}$ and the output current $i_{out}$ is approximately 90° rather than 180°.

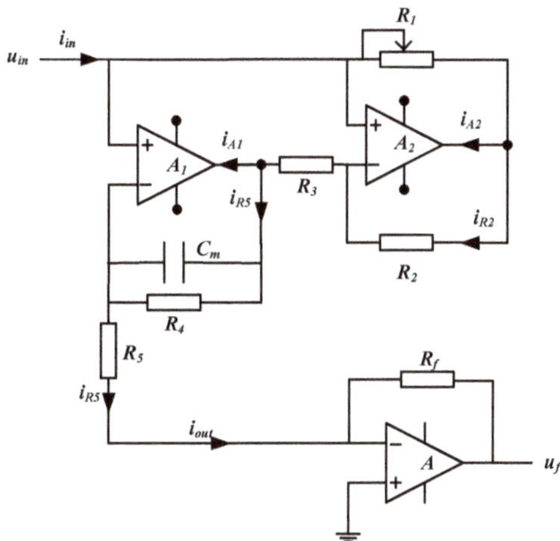

**Figure 7.** The simulated inductor module based on a Riordan grounded simulated inductor and a I/V convertor.

In this case, as discussion in Section 2.1, the input current $i_{in}$ is not equal to the output current $i_{out}$. The output voltage signal $u_f$ is:

$$u_f = -i_{out}R_f \tag{44}$$

Meanwhile, according to Figure 7 and the discussions in Section 2.1, the following equations can be obtained:

$$i_{in} = \frac{u - u_{in}}{\frac{1}{j2\pi f C_1} + Z_x + \frac{1}{j2\pi f C_2}} \tag{45}$$

$$i_{out} = \frac{(1/R_4 + j2\pi f C_m)R_1 R_3}{R_2} i_{in} = \frac{u_{in}}{R_5} \tag{46}$$

$$i_{out} = \frac{(1/R_4 + j2\pi f C_m)R_1 R_3}{R_2} \cdot \frac{u}{\frac{1}{j2\pi f C_1} + Z_x + \frac{1}{j2\pi f C_2} + \frac{(1/R_4 + j2\pi f C_m)R_1 R_3}{R_2} \cdot R_5} \tag{47}$$

$$i_{out} = \frac{r_{eq} + j2\pi f L_{eq}}{R_5} \cdot \frac{u}{\frac{1}{j2\pi f C_1} + Z_x + \frac{1}{j2\pi f C_2} + r_{eq} + j2\pi f L_{eq}} \tag{48}$$

When the impedance elimination principle is realized, Equation (48) is simplified as:

$$i_{out} = \frac{r_{eq} + j2\pi f L_{eq}}{R_5} \cdot \frac{u}{Z_x + r_{eq}} \tag{49}$$

Thus:

$$u_f = -\frac{r_{eq} + j2\pi f L_{eq}}{R_5} \cdot \frac{u}{Z_x + r_{eq}} R_f \tag{50}$$

Further, when the value of $r_{eq}$ is relatively small to the impedance of measured fluid $Z_x$, Equation (50) can be further simplified as:

$$u_f \approx -j\frac{2\pi f L_{eq}}{R_5} \cdot \frac{u}{Z_x} R_f \tag{51}$$

Therefore in this case, from Equation (51), the phase difference between the excitation signal $u$ and the output voltage signal $u_f$ is approximately 90° when the impedance elimination principle is realized.

Based on the above discussion, the new simulated inductor module is also workable in the case when an independent power source is not available. It is also an effective alternative method. However, it is necessary to indicate that, in this case, the input current of the simulated inductor module is not equal to the output current ($i_{in} \neq i_{out}$) and the phase difference between input signal and output signal is approximately 90°. That will more or less have influence on the measurement results.

## 5. Experimental Results and Discussion

To test the effectiveness and evaluate the performance of the new simulated inductor with independent power source in the CET system, two sets of experiments were carried out: impedance measurement experiments simulating the characteristics of the measured fluid (hereinafter referred to as simulation experiments), and image reconstruction experiments based on a CET system using the classic linear back projection (LBP) algorithm. More detailed information on the construction of the CET system can be found in [32] ([32] focuses on the hardware improvement of the CCERT system which used the simulated inductor technique. But the detailed information of the simulated inductor module are not discussed and provided in [32]).

The components information of the new simulated inductors (Figure 4) was: the value of the adjustable resistor $R_1$ ranged from 0 to 10.0 kΩ, $R_2 = R_3 = 3.30$ kΩ, $R_4 = 1.00$ MΩ, $R_5 = 5.10$ kΩ, and $C_m = 2.20$ nF. The operational amplifiers ($A_1$ and $A_2$) were AD817 (Analog Devices, Inc., Norwood, MA, USA). The operational amplifier ($A_3$) of the I/V converter was LM6172 (Texas Instruments, Inc., Dallas, TX, USA). The feedback resistor $R_f$ was 200 Ω.

### 5.1. The Simulation Experiments

In the simulation experiments, the equivalent coupling capacitances of the CET system were represented by two capacitors $C_{1s}$ and $C_{2s}$ with suitable fixed values. The practical measured fluid was represented by RC series combinations $Z_x$ with different impedance values.

The simulation experimental setup is shown in Figure 8.

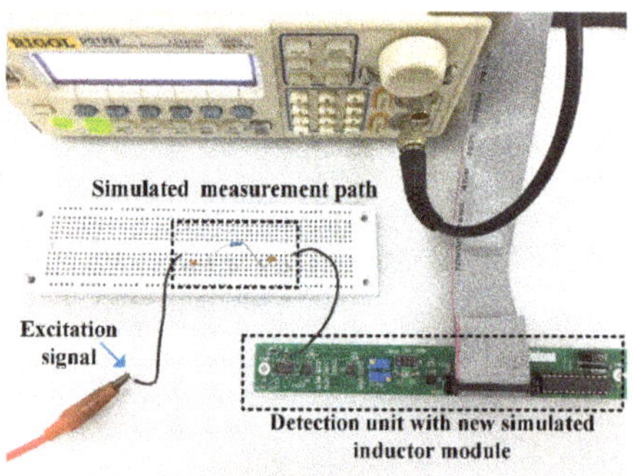


**Figure 8.** The simulation experimental setup.

In the simulated measurement path, the true values of the capacitor $C_{1s}$ and $C_{2s}$ were 9.97 pF and 9.99 pF, respectively. Correspondingly, the equivalent inductance value of new simulated inductor module was adjusted to 81.23 mH. Three groups of resistors, capacitors, and their combinations were measured:

(1) Group 1 used some different resistors to simulate the ERT system, the true values of the resistors were 10.06 k$\Omega$, 20.16 k$\Omega$, 30.42 k$\Omega$, 39.64 k$\Omega$, 47.58 k$\Omega$, 57.73 k$\Omega$, 76.65 k$\Omega$, 83.30 k$\Omega$ and 100.36 k$\Omega$.
(2) Group 2 used some different capacitors to simulate ECT system, the true values of the capacitors were 2.12 pF, 5.52 pF, 9.97 pF, 15.32 pF and 33.87 pF.
(3) Group 3 used some RC series combinations to simulate the EIT system, six kinds of RC series combinations were formed by resistors of 20.16 k$\Omega$, 47.58 k$\Omega$, 76.65 k$\Omega$ and capacitors of 5.52 pF, 15.32 pF.

The true values of the resistors and the capacitors were calibrated by a commercial impedance analyzer (Keysight 4294A, Santa Rosa, CA, USA, 40 Hz to 110 MHz) at 250 kHz. The AC excitation signal was generated by a digital signal generator. The AC excitation signal had the amplitude value of 1.52 V and its frequency is 250 kHz. The total impedance information was measured by the signal processing module.

To test the performance of the measurement, four performance indexes are selected, including $e_R$, $e_C$, $\sigma_R$ and $\sigma_C$. The $e_R$ and $e_C$ are the relative errors (%) of the resistance measurement and the capacitance measurement. $\sigma_R$ and $\sigma_C$ are the standard deviations of the resistance measurement and the capacitance measurement, respectively. The four performance indexes are defined as follows:

$$e_R = \frac{R}{R_r} - 1 \tag{52}$$

$$e_C = \frac{C}{C_r} - 1 \tag{53}$$

$$\sigma_R = \sqrt{\frac{1}{K-1} \sum_{k=1}^{K} \left(R_k - \frac{1}{K} \sum_{k=1}^{K} R_k\right)^2} \tag{54}$$

$$\sigma_C = \sqrt{\frac{1}{K-1} \sum_{k=1}^{K} \left(C_k - \frac{1}{K} \sum_{k=1}^{K} C_k\right)^2} \tag{55}$$

where $R_m$ and $C_m$ are the mean values of the measurement results of the resistance and the capacitance of the impedance, $R_m = \frac{1}{K} \sum_{k=1}^{K} R_k$ and $C_m = \frac{1}{K} \sum_{k=1}^{K} C_k$. $R_r$ and $C_r$ are the true values of the resistance and the capacitance. $R_k$ and $C_k$ are the $k$th measurement results of the resistance and the capacitance. $K$ is the total number of repeated measurements (in this work, $K = 100$).

Figure 9a,b shows the measurement results of Group 1 and Group 2, respectively. The maximum relative error of the resistance measurement $e_R$ was 2.15%. The maximum relative error of the capacitance measurement $e_C$ was 2.99%. The maximum standard deviation of the resistance measurement was 0.50 k$\Omega$. The standard deviation of the capacitance measurement was 0.05 pF.

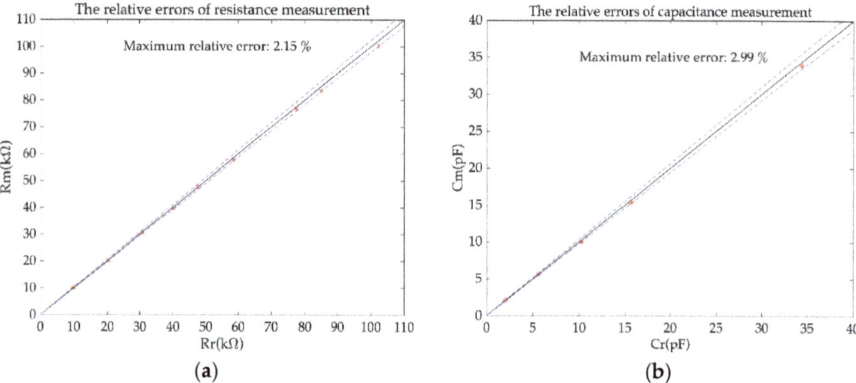

**Figure 9.** The measurement results of the simulation experiments. (**a**) the relative errors of resistance measurement; (**b**) the relative errors of capacitance measurement.

Table 1 shows the experimental results of the impedance measurement experiments (Group 3). From Table 1, it can be seen that the CET system based on the new simulated inductor module can effectively obtain the total impedance information. The maximum relative error of the impedance measurement of RC series combination was 4.77%. The maximum standard deviation of the resistance measurement was 0.27 kΩ. The maximum standard deviation of the capacitance measurement was 0.07 pF.

**Table 1.** Impedance measurement results of the CET system with independent power source (250 kHz).

| $R_r$ (kΩ) | $C_r$ (pF) | $R$ (kΩ) | $C$ (pF) | $e_R$ (%) | $e_C$ (%) | $\sigma_R$ (kΩ) | $\sigma_C$ (pF) |
|---|---|---|---|---|---|---|---|
| 20.16 | 5.52 | 21.11 | 5.63 | 4.74 | 2.01 | 0.44 | 0.01 |
| 20.16 | 15.32 | 20.33 | 15.70 | 0.89 | 2.51 | 0.03 | 0.03 |
| 47.58 | 5.52 | 47.55 | 5.64 | −0.04 | 2.20 | 0.20 | 0.02 |
| 47.58 | 15.32 | 46.98 | 15.74 | −1.25 | 2.78 | 0.09 | 0.06 |
| 76.65 | 5.52 | 74.35 | 5.39 | −2.99 | −2.21 | 0.27 | 0.02 |
| 76.65 | 15.32 | 75.81 | 14.59 | −1.09 | −4.77 | 0.14 | 0.07 |

Correspondingly, in the case when the independent power source of the new simulated inductor module is not available, the impedance measurement experiments were also carried out. The measurement results of Group 1, Group 2 and Group 3 are shown in Figure 10a, Figure 10b and Table 2, respectively. In this case, the maximum relative error of the resistance measurement $e_R$ was 2.66%. The maximum relative error of the capacitance measurement $e_C$ was 3.13%. The maximum relative error of the impedance measurement of RC series combination was 4.84%. The maximum standard deviation of the resistance measurement was 0.43 kΩ. The maximum standard deviation of the capacitance measurement was 0.08 pF.

**Table 2.** Impedance measurement results of the CET system without independent power source (250 kHz).

| $R_r$ (kΩ) | $C_r$ (pF) | $R$ (kΩ) | $C$ (pF) | $e_R$ (%) | $e_C$ (%) | $\sigma_R$ (kΩ) | $\sigma_C$ (pF) |
|---|---|---|---|---|---|---|---|
| 20.16 | 5.52 | 20.81 | 5.60 | 3.26 | 1.47 | 0.43 | 0.01 |
| 20.16 | 15.32 | 20.83 | 15.84 | 3.37 | 3.45 | 0.03 | 0.03 |
| 47.58 | 5.52 | 47.43 | 5.61 | −0.31 | 1.66 | 0.21 | 0.02 |
| 47.58 | 15.32 | 47.18 | 15.78 | −0.83 | 3.04 | 0.08 | 0.07 |
| 76.65 | 5.52 | 74.47 | 5.45 | −2.84 | −1.12 | 0.30 | 0.02 |
| 76.65 | 15.32 | 75.52 | 14.57 | −1.47 | −4.84 | 0.14 | 0.08 |

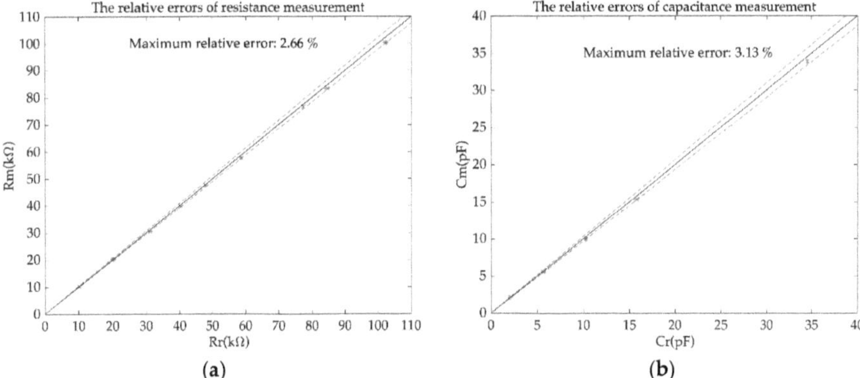

**Figure 10.** The measurement results of the simulation experiments without independent power source. (**a**) the relative errors of resistance measurement; (**b**) the relative errors of capacitance measurement.

The simulated experimental results have demonstrated that the design of the new simulated inductor module is effective, and it can replace the practical inductor to achieve impedance elimination in the measurement path of the CET system, and its performance was satisfactory. The unfavorable influences of coupling capacitances can be successfully overcome by using the new simulated inductor modules, and the SNR of the CET system is improved.

Meanwhile, it is also found that the performance of the new simulated inductor without the independent power source is comparable to that of the new simulated inductor with the independent power source. There only exists slight difference between the experimental results of these two cases. So in the case when the independent power source is not available, the new simulated inductor is also an effective alternative method, but the phase difference between input signal and output signal is approximately 90° when the elimination principle is realized.

## 5.2. The Practical Image Reconstruction Experiments

To test the image reconstruction quality of the CET system using new simulated inductors modules, the practical image reconstruction experiments were carried out. The material of insulating pipe of the CET sensor was polyvinyl chloride (PVC). The outer diameter and the thickness of the PVC pipe were 110 mm and 2 mm, respectively. The length of electrode and the electrode angel were 150 mm and 26°, respectively. Figure 11 is a photo of the CET system prototype.

Tap water with the conductivity of 160 μS/cm was used as the continuous phase. Non-conductive plastic (polyethylene) rods with different diameters (20 mm and 35 mm) were used as the discrete phase. During the experiments, the plastic rods were put into different positions in the pipe which was full of tap water. Then, the conductivity changes in the pipe can be detected. The classic LBP algorithm was used to implement the image reconstruction.

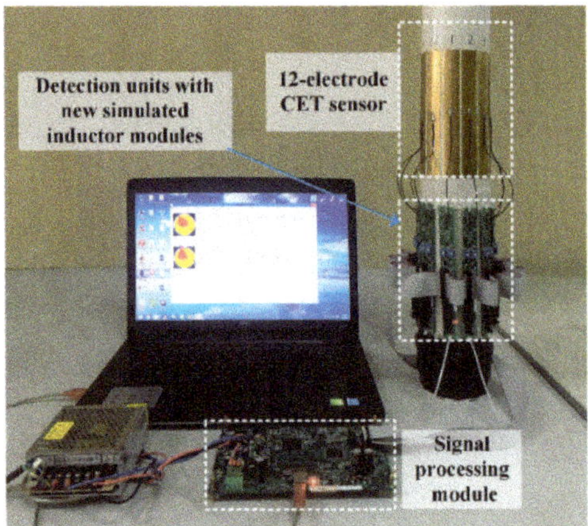


**Figure 11.** The photo of the CET system prototype.

In order to compare the image reproduction effect, three image quality indexes were used [21]: mean squared error (MSE), relative image error (RIE) and image correlation coefficient (ICC). MSE is evaluated according to Equation (56). RIE is evaluated according to Equation (57). ICC is evaluated according to Equation (58):

$$MSE = \frac{1}{N}\sum_{n=1}^{N}(\hat{g} - g)^2 \tag{56}$$

$$RIE = \frac{\|\hat{g} - g\|}{\|g\|} \tag{57}$$

$$ICC = \frac{\sum_{n=1}^{N}(\hat{g}_n - \overline{\hat{g}})(g_n - \overline{g})}{\sqrt{\sum_{n=1}^{N}(\hat{g}_n - \overline{\hat{g}})^2 \sum_{n=1}^{N}(g_n - \overline{g})^2}} \tag{58}$$

where, $\hat{g}_n$ is the gray level of the $n$th pixel of the reconstructed images and $\hat{g} = [\hat{g}_1, \hat{g}_2 \ldots \hat{g}_n \ldots \hat{g}_N]^T$ is the vector of the gray levels of the reconstructed images. $g_n$ is the practical gray level of the $n$th pixel of the conductivity distributions and $g = [g_1, g_2 \ldots g_n \ldots g_N]^T$ is the vector of the gray levels of the practical conductivity distributions. $\overline{\hat{g}}$ and $\overline{g}$ are the mean value of $\hat{g}$ and $g$ respectively. $N$ is the number of the pixel of the reconstructed images, in this work, $N = 856$.

Figure 12 is the experimental results of image reconstruction. Table 3 lists the information of the three image quality indexes.

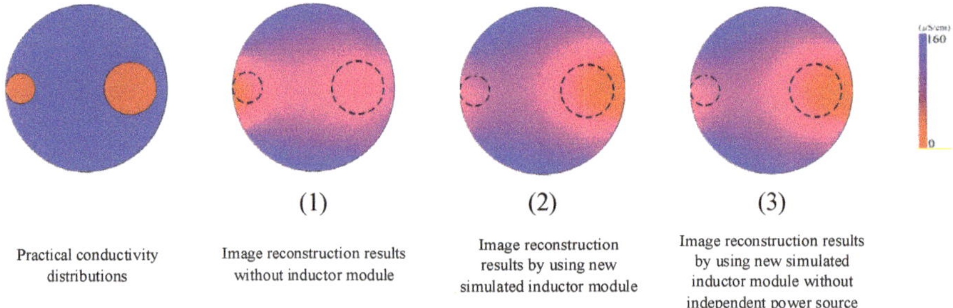


**Figure 12.** Practical conductivity distributions and image reconstruction results of CET systems.

**Table 3.** Three image quality indexes of image reconstruction results.

| Image Reconstruction Results | (1) | (2) | (3) |
|---|---|---|---|
| MSE | 0.0142 | 0.0138 | 0.0138 |
| RIE | 0.4805 | 0.4376 | 0.4408 |
| ICC | 0.5389 | 0.5514 | 0.5484 |

The practical experimental image reconstruction results indicate that the quality of the image reconstruction of the CET system prototype using new simulated inductors modules is satisfactory. Compared with the CET system without inductor module, the MSE and RIE of the CET system using new simulated inductor module are smaller. The quality of the image reconstruction of CET system is improved by introducing the new simulated inductor module. The application of simulated inductor technique to the research field of the process tomography is effective.

## 6. Conclusions

This work analyzes the characteristics of the simulated inductor based on the Riordan circuit in detail. Through comparison and analysis, it is found that the grounded simulated inductor has a simple circuit construction but one terminal should be grounded. The floating simulated inductor can be used as an independent inductor module while its circuit structure is relatively complex. Meanwhile, it is also pointed out that, the reason why the grounded simulated inductor can't be used as an independent module is that the output current of the grounded simulated inductor is different from the input current ($i_{in} \neq i_{out}$).

A new simulated inductor module which is suitable for CET system is reported in this work. By compensating the currents from the common node of the introduced independent power source to the main circuit, a new simulated inductor is designed. The new simulated inductor not only has a simple construction similar to the grounded simulated inductor, but also can be used as an independent module like the floating simulated inductor. It reduces the requirements for components and can be used conveniently in practical applications like a practical inductor. The new simulated inductor module guarantees the equivalence between the input current and the output current. These two advantages make it an attractive approach to apply the simulated inductor technique to the research fields of process tomography.

To test the effectiveness and evaluate the performance of the new designed simulated inductor, two sets of experiments were carried out. The experimental results show that the design of the new simulated inductor is successful and the performance of the new simulated inductor is satisfactory. The maximum relative error of the resistances measurement is 2.15%, the maximum relative error of the capacitances measurement is 2.99%, and the maximum relative error of the impedance measurement is 4.77%. The practical image reconstruction of the CET system results indicate that the quality of the image reconstruction is satisfactory and the application of simulated inductor technique to the research field of the process tomography is effective. The SNR of the CET system can be improved.

Besides, the experimental results also indicate that, in the case when the independent power source is not available, the performance of the new simulated inductor is comparable to that of the new simulated inductor with the independent power source. So in this case, the new simulated inductor is also an effective alternative method, but the phase difference between input signal and output signal is approximately 90° when the elimination principle is realized (the input current is not equal to the output current). That may more or less have influence on measurement results and practical application.

This work provides the detailed information on simulated inductors based on Riordan circuit, including the detailed measurement circuits and related analysis, the technical parameters of the components of the simulated inductor, the obtained experience/knowledge and the latest progresses, as well as the design of a new simulated inductor. According to the technical content in the paper, readers can duplicate the measurement circuit of the new simulated inductor. We hope that this paper can provide useful reference/experience for other researchers' work and extend the application fields of the simulated inductor technique.

**Author Contributions:** Conceptualization, Z.H.; methodology, Z.H., B.W. and H.J.; software, X.Y. and Y.W.; validation, X.Y., Y.W. and X.-Y.T.; formal analysis, X.Y., Y.W. and X.-Y.T.; investigation, X.Y., Y.W. and X.-Y.T.; resources, Z.H.; data curation, X.Y., Y.W. and X.-Y.T.; writing—original draft preparation, X.Y. and Y.W.; writing—review and editing, B.W., H.J. and X.-Y.T.; visualization, X.Y. and Y.W.; supervision, Z.H.; project administration, H.J.; funding acquisition, H.J.

**Funding:** This work is supported by the National Nature Science Foundation of China No. 61573312 and the project of State Key Laboratory of Industrial Control Technology (Zhejiang University) ICT1909.

**Conflicts of Interest:** The authors declare no conflict of interest.

## References

1.  James, W.; Susan, A. *Electric Circuits*, 9th ed.; Pearson Education: Upper Saddle River, NJ, USA, 2011.
2.  William, H.; Jack, E.; Steven, M. *Engineering Circuit Analysis*, 7th ed.; McGraw Hill Companies: New York, NY, USA, 2007.
3.  Christiansen, D.; Alexander, C.; Jurgen, R. *Standard Handbook of Electronic Engineering*, 5th ed.; McGraw Hill Professional: Boston, MA, USA, 2004.
4.  Kumar, U.; Shukla, S.K. Analytical study of inductor simulation circuits. *Act. Passive Electron. Compon.* **1989**, *13*, 211–227. [CrossRef]
5.  Riordan, R.H.S. Simulated inductors using differential amplifiers. *Electron. Lett.* **1967**, *3*, 50–51. [CrossRef]
6.  Antoniou, A. Realisation of gyrators using operational amplifiers, and their use in RC-active-network synthesis. *Proc. Inst. Electr. Eng.* **1969**, *116*, 1838–1850. [CrossRef]
7.  Dutta Roy, S.C. A circuit for floating inductance simulation. *Proc. IEEE* **1974**, *62*, 521–523. [CrossRef]
8.  Susan, D.; Jayalalitha, S. Analog filters using simulated inductor. In Proceedings of the 2010 International Conference on Mechanical and Electrical Technology, Singapore, 10–12 September 2010.
9.  Fakhfakh, M.; Pierzchała, M.; Rodanski, B. On the design of active inductors with current-controlled voltage sources. *Analog Digit. Signal Process.* **2012**, *73*, 89–98. [CrossRef]
10. Abuelma'atti, M.T. New grounded immittance function simulators using single current feedback operational amplifier. *Analog Integr. Circuits Signal Process.* **2012**, *71*, 95–100. [CrossRef]
11. Jayalalitha, D.S.; Susan, D. Grounded simulated inductor-a review. *Middle East J. Sci. Res.* **2013**, *15*, 278–286.
12. Yuce, E.; Minaei, S.; Cicekoglu, O. A novel grounded inductor realization using a minimum number of active and passive components. *Etri J.* **2005**, *27*, 427–432. [CrossRef]
13. Soliman, A.M. On the realization of floating inductors. *Nat. Sci.* **2010**, *8*, 167–180.
14. Reddy, M.A. Some new operational-amplifier circuits for the realization of the losses floating inductance. *IEEE Trans. Circuits Syst.* **1976**, *23*, 171–173. [CrossRef]
15. Singh, V. On floating impedance simulation. *IEEE Trans. Circuits Syst.* **1989**, *36*, 161–162. [CrossRef]
16. Crowe, C.T. *Multiphase Flow Handbook*; CRC Press: Boca Raton, FL, USA, 2006.
17. Falcone, G.; Hewitt, G.F.; Alimonti, G.F. *Multiphase Flow Metering*; Elsevier Press: Oxford, UK, 2010.
18. Wang, B.L.; Zhang, W.B.; Huang, Z.Y.; Ji, H.F.; Li, H.Q. Modeling and optimal design of sensor for capacitively coupled electrical resistance tomography system. *Flow Meas. Instrum.* **2013**, *31*, 3–9. [CrossRef]
19. Wahab, Y.A.; Rahim, R.A.; Rahiman, M.H.F. Non-invasive process tomography in chemical mixtures—A review. *Sens. Actuators B* **2015**, *210*, 602–617. [CrossRef]
20. Holder, D.S. *Electracal Impedance Tomography: Methods, History and Applications*; Institute of Physics Publishing: Bristol, UK, 2005.
21. Rymarczyk, T.; Kłosowski, G.; Kozłowski, E.; Tchórzewski, P. Comparison of Selected Machine Learning Algorithms for Industrial Electrical Tomography. *Sensors* **2019**, *19*, 1521. [CrossRef] [PubMed]
22. Rymarczyk, T.; Kłosowski, G.; Kozłowski, E. A Non-Destructive System Based on Electrical Tomography and Machine Learning to Analyze the Moisture of Buildings. *Sensors* **2018**, *18*, 2285. [CrossRef]
23. Stefani, F.; Gundrum, T.; Gerbeth, G. Contactless inductive flow tomography. *Phys. Rev. E* **2004**, *70*, 056306. [CrossRef] [PubMed]
24. Stefani, F.; Rüdiger, G.; Szklarski, J.; Gundrum, T.; Gerbeth, G.; Hollerbach, R. Experiments on the magnetorotational instability in helical magnetic fields. *New J. Phys.* **2007**, *9*, 10949. [CrossRef]

25. Ziolkowski, M.; Gratkowski, S.; Żywica, A.R. Analytical and numerical models of the magnetoacoustic tomography with magnetic induction. *COMPEL Int. J. Comput. Math. Electr.* **2018**, *37*, 538–548. [CrossRef]
26. Stawicki, K.; Gratkowski, S.; Komorowski, M.; Pietrusewicz, T. A new transducer for magnetic induction tomography. *IEEE Trans. Magn.* **2009**, *45*, 1832–1835. [CrossRef]
27. Xie, C.G.; Huang, S.M.; Hoyle, B.S. Electrical capacitance tomography for flow imaging: System model for development of image reconstruction algorithms and design of primary sensors. *IEE Proc. G* **1992**, *139*, 89–98. [CrossRef]
28. Huang, Z.Y.; Jiang, W.W.; Zhou, X.M.; Wang, B.L.; Ji, H.F.; Li, H.Q. A new method of capacitively coupled contactless conductivity detection based on series resonance. *Sens. Actuators B Chem.* **2009**, *143*, 239–245. [CrossRef]
29. Lyu, Y.C.; Huang, J.C.; Huang, Z.Y.; Ji, H.F.; Wang, B.L.; Li, H.Q. Study on the application of simulated inductor technique to the design of C4D sensor. *Sens. Actuators A* **2017**, *264*, 195–204. [CrossRef]
30. Lyu, Y.C.; Ji, H.F.; Yang, S.J.; Huang, Z.Y.; Wang, B.L.; Li, H.Q. New C4D Sensor with a Simulated Inductor. *Sensors* **2016**, *16*, 165. [CrossRef] [PubMed]
31. Wang, Y.X.; Ji, H.F.; Huang, Z.Y.; Wang, B.L.; Li, H.Q. Online measurement of conductivity/permittivity of fluid by a new contactless impedance sensor. *Rev. Sci. Instrum.* **2017**, *88*, 055111. [CrossRef]
32. Wang, Y.X.; Wang, B.L.; Huang, Z.Y.; Ji, H.F.; Li, H.Q. New capacitively coupled electrical resistance tomography (CCERT) system. *Meas. Sci. Technol.* **2018**, *29*, 104007. [CrossRef]
33. Huijsing, J.H. Design and applications of the operational floating amplifier (OFA): The most universal operational amplifier. *Analog Integr. Circuits Signal Process.* **1993**, *4*, 115–129. [CrossRef]
34. Huijsing, J.H.; Veelenturf, C.J. Monolithic class AB operational mirrored amplifier. *Electron. Lett.* **2007**, *17*, 119–120. [CrossRef]

*Article*

# Focusing Sensor Design for Open Electrical Impedance Tomography Based on Shape Conformal Transformation [†]

Yu Wang, Shangjie Ren * and Feng Dong

Tianjin Key Laboratory of Process Measurement and Control, School of Electrical and Information Engineering, Tianjin University, Tianjin 300072, China; tju_wangyu@tju.edu.cn (Y.W.); fdong@tju.edu.cn (F.D.)
* Correspondence: rensjie@tju.edu.cn
† This paper is an extended version of our paper Optimized Stimulation Patterns for Miniscopic Electrical Impedance Tomography with Planar Electrodes Array, published in Proceedings of the 9th World Congress on Industrial Process Tomography, Bath, UK, 2–6 September 2018.

Received: 30 March 2019; Accepted: 30 April 2019; Published: 2 May 2019

**Abstract:** Electrical Impedance Tomography (EIT) is a non-invasive detection method to image the conductivity changes inside an observation region by using the electrical measurements at the boundary of this region. In some applications of EIT, the observation domain is infinite and is only accessible from one side, which leads to the so-called open EIT (OEIT) problem. Compared with conventional EIT problems, the observation region in OEIT can only be measured from limited projection directions, which makes high resolution imaging much more challenging. To improve the imaging quality of OEIT, a focusing sensor design strategy is proposed based on shape conformal theory. The conformal bijection is used to map a standard EIT sensor defined at a unit circle to a focusing OEIT sensor defined at an upper half plane. A series of numerical and experimental testes are conducted. Compared with the traditional sensor structure, the proposed focusing sensor has higher spatial resolution at the near-electrode region and is good at distinguishing multi-inclusions which are close to each other.

**Keywords:** open electrical impedance tomography; sensor design; conformal transformation; focusing sensor; open domain imaging

## 1. Introduction

Electrical Impedance Tomography (EIT) is a noninvasive imaging method, which reconstructs the conductivity distribution of the imaging field [1]. It applies an electrical excitation signal to the target field through electrodes placed at the boundary of an observation area, and then obtains the electrical responses reflecting the conductivity distribution within the observed field. This technology has the advantages of portability, low cost and high time resolution. EIT has been widely applied in the monitoring of many industrial and biomedical processes [2,3]. Conventional EIT aims at reconstructing the conductivity distribution within a closed domain [4–6]. However, in practical applications, the observed domain is not always closed or is big enough to be approximated to infinity compared with the size of the EIT sensors. This leads to the open EIT (OEIT) problem.

OEIT usually takes an open area with an unclosed boundary as observation field for measurements. OEIT was used in surface geophysics for surface geological exploration in an earlier time. Then it was also used in structural damage detection, landmine detection and biomedical tissue imaging. Youssef et al. [7] used OEIT to detect sinkholes in Saudi Arabia, and the results showed that the OEIT was effective in detecting and mapping a known sinkhole in the study area. Church et al. [8] constructed a prototype confirmatory landmine detector based on OEIT to image buried landmines. The results demonstrated that it is possible to reliably reconstruct conductivity perturbations of a shallow buried antitank mine or similar object in a variety of soils. Baltopoulos et al. [9] using OEIT to map damage in carbon fiber reinforced polymer plates. The methodology is validated based on experimental damage scenarios, successfully identifying the induced damage. Allen et al. [10] used an electrical impedance tomography-based sensing skin for quantitative imaging of damage in concrete. In the meantime, OEIT has been extensively studied in clinical medicine. Mueller et al. [11] developed an OEIT sensor with $4 \times 4$ electrodes for detecting conductivity changes during ventilation and perfusion. Time traces of the reconstructed conductivity distribution demonstrated the detected changes in conductivity were due to ventilation and perfusion. Borsic et al. [12] used a cylindrical probe based on EIT with an array of electrodes on the front surface to detect prostate disease. They also developed a novel reconstruction algorithm used for conductivity estimation. The simulation results demonstrated the feasibility of imaging moderately contrasting inclusions at distances of three times the probe radius from the probe surface. Cherepenin et al. [13] developed a 48-electrodes OEIT system combined with a vaginal probe for early detection of cervical neoplasia. They checked their system on a saline solution tank containing different small objects. The results showed the system could distinguish the size and location of single targets and the type for different targets. Aristovich et al. [14] used a 30-electrodes OEIT array to reconstruct images of fast neural-evoked activity in the rat cerebral cortex and developed a novel noise-based image post-processing technique. The results indicated that the developed methods may be expected to be reliably applied for imaging neural activity with planar arrays. Murphy et al. [15] used an end-fired microendoscopic EIT probe for surgical margin assessment and developed a novel regularization technique using the dual-mesh method. The results showed the feasibility of surgical margin detection using microendoscopic EIT.

OEIT thus has broad application prospects, however, reconstructing images in the open-domain geometry poses additional challenges compared with conventional EIT systems. In OEIT, the electrodes are usually arranged on one surface of the observation field, thus the current density decreases rapidly with the distance from the electrode surface. In the meantime, the incomplete boundary conditions also increase the difficulty of solving the OEIT problem, making the imaging quality of OEIT relatively low and sensitive to noise. Thus, there is still a large scope for the improvement for OEIT. Improved imaging algorithms can improve the reconstruction quality of OEIT to a certain extent, but algorithm improvement cannot fundamentally improve the spatial and temporal resolution of OEIT. This is because the electric field distribution in the observation field is fixed under the premise of not changing the electrode configuration and excitation strategy. The improvement of the imaging algorithm can only increase the acquisition of measurement information, but cannot improve the electric field distribution. However, different electrode configurations and data collection patterns will affect the penetration depth of the electric field, thus affecting the spatial resolution and imaging depth of OEIT. Perez et al. [16] designed a novel rectangular array of 20 active electrodes to inject the external currents, and 16 passive electrodes to measure the induced voltages. The results showed that this kind of electrode arrangement can avoid problems due to the unknown contact impedance. The reconstructions have good spatial resolution in the $xy$-plane and are quite stable with respect to the noise level in the data. Liu et al. [17] used a scanning linear electrode for the imaging of open domains and proposed a novel measurement and stimulation pattern. This research is dedicated to increasing the number of independent voltage measurements to improve the imaging quality of OEIT. Aiming at the problem of poor image reconstruction quality and low reconstruction accuracy of OEIT, Wang et al. [18] proposed a novel image reconstruction method based on conformal transformation. By this method the imaging quality and accuracy of OEIT are improved, especially for the inclusions located far from the electrode area.

For two-dimensional OEIT, the most used sensor is the uniform sensor that uses electrodes of the same size and places the electrodes evenly at the boundary of the open field [17]. This simple electrode configuration allows the observation of conductivity changes in the open field. However, previous studies showed that the reconstruction quality of uniform sensors for inclusions near the electrode region and those far from the electrode was relatively poor [18]. When the sensor array and the excitation strategy are fixed, the electric field distribution in the observation field is uniquely determined. Therefore, the most fundamental solution to improve the poor quality of OEIT imaging is to optimize the electrode configuration, as well as the excitation strategy.

In this research, a novel focusing sensor is designed based on shape conformal theory to improve the imaging quality of OEIT in the near-electrode region. The conformal bijection is used to map a standard EIT sensor defined at a unit circle to a focusing OEIT sensor defined at an upper half plane. To evaluate the proposed focusing sensor, a series of simulation and experiment analyses are carried out. The results from the novel OEIT sensor are compared with those of a conventional OEIT sensor.

## 2. Materials and Methods

### 2.1. Principle of OEIT

Figure 1 gives an illustration of an OEIT system with the conventional sensor structure discussed in this paper. The EIT system includes an electrode array, data acquisition system and image reconstruction unit (usually in a PC). The electrode array is uniformly fixed at the boundary $\Gamma$ of the open domain $\Omega$. The excitation strategy of OEIT system is the conventional adjacent current stimulation and adjacent voltages measurement [19]. At each measuring timepoint, one of the adjacent electrode pairs is excited with currents, and voltage measurements are made between the other adjacent pairs not involving the driven electrodes. The process will continue until all the electrodes have been excited. For a system consisting of $L$ electrodes, $L \times (L-3)$ voltage measurements are collected. Based on the reciprocal theory, half of them are independent. Aiming at the sensor with 16 electrodes in this research, there are 208 measurement datapoints in all, and the independent data is the half of this number. When low-frequency currents are applied to the active electrodes, the electric potential $\varphi(x, y)$ at $\Omega$ satisfies the Gaussian equation:

$$\nabla \cdot [\sigma(x, y)\nabla\phi(x, y)] = 0, \left\{(x, y)\middle| x \in (-\infty, +\infty), y \in (0, +\infty)\right\}, \tag{1}$$

where $\sigma(x, y)$ is the conductivity distribution. Following the complete electrode model (CEM) [20], the boundary condition of the OEIT forward problem can be written as follows:

$$\begin{cases} \sigma(x, y)\frac{\partial\phi(x,y)}{\partial y}\big|_{y=0} = 0, & x \in \Gamma \backslash \overset{L}{\underset{l=1}{\cup}} e_l \\ \phi(x, y) + \rho_l\sigma(x, y)\frac{\partial\phi(x,y)}{\partial y}\big|_{y=0} = U_l, & x \in e_l,\, l = 1, 2, \ldots, L \\ \int_{e_l} \sigma(x, y)\frac{\partial\phi(x,y)}{\partial y}\big|_{y=0}ds = I_l, & l = 1, 2, \ldots, L \\ \overset{L}{\underset{l=1}{\sum}} I_l = 0 \\ \overset{L}{\underset{l=1}{\sum}} U_l = 0 \end{cases} \tag{2}$$

where $\Gamma \in (-\infty, +\infty)$ is the actual boundary of open domain, $e_l$ denotes the electrode $l$, $\rho_l$ is the contact impedance between electrode $l$ and internal medium, $U_l$ is the voltage on the electrode $l$, $I_l$ is current injected into field $\Omega$ from electrode $l$.

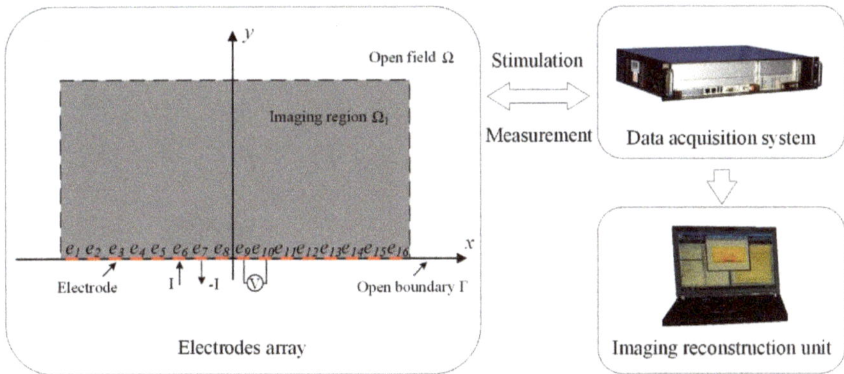


**Figure 1.** Illustration of the OEIT system with conventional sensor structure.

The conventional method to calculate the OEIT problem is to simplify the infinite observation domain $\Omega$ to a finite domain $\Omega_1$ [17]. However, this method will lead to truncation errors not only in the forward problem solution but also the inverse problem solution. In our previous research, a transformation domain method was proposed to directly solve the OEIT problem in the infinite observation domain $\Omega$. According to the Riemann mapping theorem, the infinite upper half plane can be transferred into a simple closed circular domain by bijective holomorphic mapping [21]. After conformal mapping, the voltage potential function and the conductivity distribution function still satisfy the Laplace equation and are related to the original equation before transformation. The boundary condition can be calculated by CEM and it has the same form as (2). According to the conformal transformation, the electrical field distribution of the infinite upper half plane can be evaluated by calculating the electrical field distribution in the conformal circular domain. Thus, the process of the imaging method based on conformal transformation has the follow steps:

(1)   The boundary voltage measurement data is collected by the electrodes placed on the boundary of the open field.
(2)   The reconstructed image is obtained in the conformal circular domain according to the boundary voltage measurement.
(3)   The conductivity distribution image of the open field is mapped from the conformal circular domain by the mapping relation [18].

*2.2. Focusing Sensor Design*

In previous studies, for two-dimensional OEIT, the most commonly used sensor is the uniform sensor. As shown in Figure 1, electrodes of the same size are fixed at the boundary of the observation domain. This simple electrode configuration allows the observation of conductivity changes in the open field. However, in previous studies, we found that the reconstruction quality of uniform sensor for inclusions near the electrode region and those far from the electrode was relatively poor [18]. Figure 2 shows the mapping of inclusions and electrode positions after conformal transformation based on the conventional uniform sensor structure. Five inclusions of the same size are located at different locations in the observation field. As the inclusions move away from the electrode, their mapping moves to the right semicircular domain, and the relative size increases first and then decreases rapidly. When the uniform sensor in the open field is mapped to the conformal circular domain, the electrode position will be offset, resulting in a sparse electrode distribution on the boundary of the left semicircle domain. When the inclusion near the electrode area is mapped to the conformal circular domain, it is located in the left semicircular domain and the size is relatively enlarged. In this situation, the uniform sensors do not provide a good reconstruction of the inclusions. Therefore, we focus on the high resolution reconstruction in the near electrode region and a focusing sensor optimization scheme based on conformal transformation is proposed.

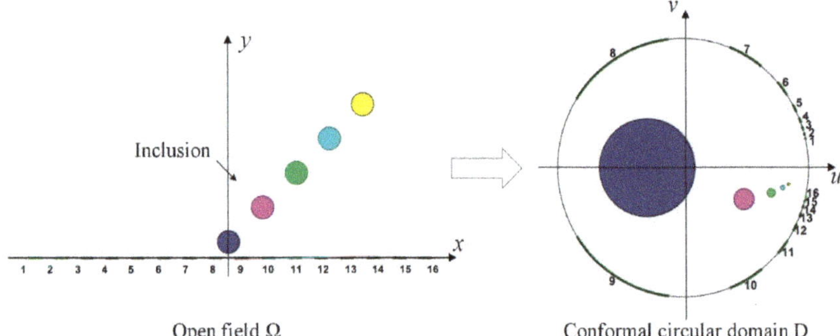


**Figure 2.** Mapping of inclusions and electrode positions after conformal transformation based on conventional sensor structure.

According to the conformal transformation method, the point $(x, y)$ of the infinite upper half plane $\Omega$ can be mapped to the point $(u, v)$ in unit circular domain D one by one, the transformation mapping is [22]:

$$w = \frac{(1+z)}{(1-z)}i \tag{3}$$

where $z = x + yi$ and $w = u + vi$. In the meantime, the point $(u, v)$ in unit circular domain can be mapped to point $(x, y)$ in the infinite upper half plane by:

$$z = \frac{w - i}{w + i} \tag{4}$$

The realization of the proposed focusing sensor is to map the uniform sensor distribution in the circular domain to the infinite upper half plane through the above mapping relations. Figure 3 shows the process of conformal transformation from the unit circular circle domain to upper half plane. After conformal transformation, the electrode arrangement of focusing sensor expands from the center to both sides, and the electrode spacing also expands.

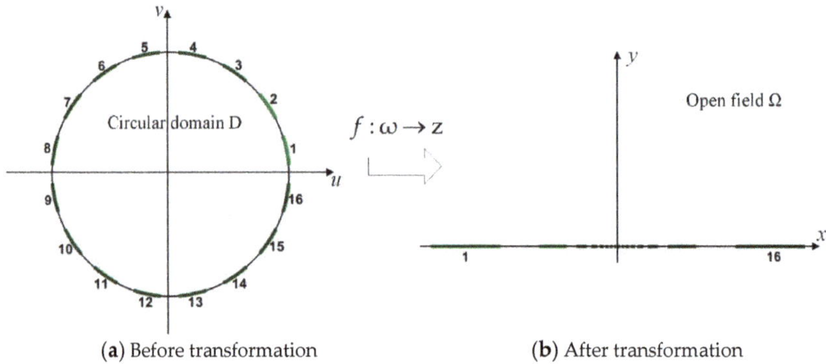


**Figure 3.** Conformal transformation from unit circular domain to upper half plane.

Since the sensor with uniform distribution in the circular region also has a higher sensitivity distribution in the left semicircle, the proposed focusing sensor in the open region can improve the sensitivity to the change of conductivity in the near electrode region, which is conducive to the realization of high-resolution reconstruction in the near electrode region.

### 2.3. Forward Problem Solutions

For the solution of EIT, the forward problem is to calculate the boundary voltages from a given conductivity distribution. The forward model is established in EIDORS [23] and solved by the BEM method [24,25]. For the uniform sensor, $16 \times 6$ mm wide electrodes are evenly placed at the range of $[-10, 10]$ of $x$-axis and the region of imaging is set to be $[-10, 10]$ of $x$-axis and $[0, 10]$ of $y$-axis. For the proposed focusing sensor, according to the above mapping relation, different electrode widths are calculated, and the electrodes are arranged at the center boundary of the open boundary. The region of imaging is set to be the same as the uniform sensor.

### 2.4. Inverse Problem Solutions

The inverse problem is solved by FEM method [26]. As shown in Figure 4, the mesh of the open image region is transformed from the uniform mesh in circular field according to the conformal mapping relation. The imaging region is a part of open field, thus the mesh is also a part of the circular field mesh. The color in the figure indicates the area of different grids. The total number of mesh is 850 grids. In this way, the mesh density in the near-electrode region can be greatly improved, which is helpful to realize high-precision image reconstruction.

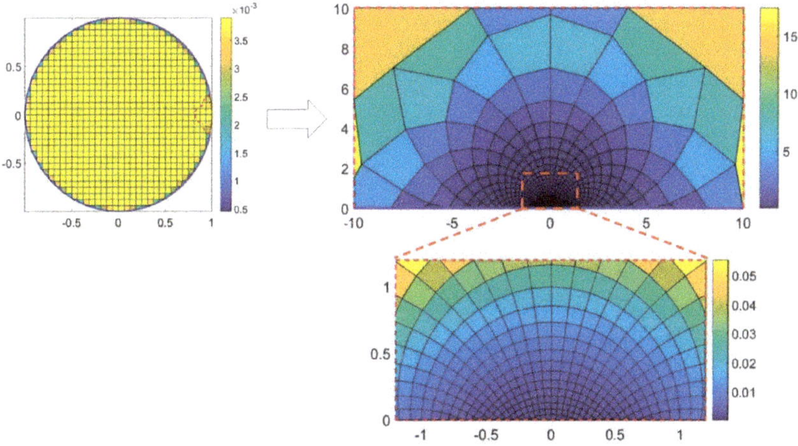

**Figure 4.** Inverse problem mesh based on conformal transformation.

The inverse problem is to estimate the conductivity distribution from boundary voltage measurements. In EIT, absolute imaging and difference imaging are commonly used for image reconstruction. In this research, the difference image approach is used, which helps to suppress the effect of contact impedance. The reconstruction problem is not well-posed as there is not a single valid solution and small changes in input mean big changes in the output. Also the physics of electric field means that the inverse problem is also ill-conditioned [27]. In this condition, regularization methods have been proved to work well towards these problems which usually have a formal minimization objective function as:

$$\|Jh - b\|_2^2 + \lambda G(h) \tag{5}$$

where $J$ is the sensitivity matrix, $b$ is the difference of the boundary voltage measurement vector between the reference field and the object field, $h$ is the conductivity distribution to be solved, $\lambda$ is the regularization factor. Tikhonov regularization algorithm [28] is a widely used non-iterative regularization algorithm. The penalty term of Tikhonov regularization is $G(h) = \|h\|_2^2$. According to the Gaussian Newton method [29], the conductivity image $h$ is estimated by the following one-step reconstruction:

$$h = \left(J^T J + \lambda I\right)^{-1} J^T b \tag{6}$$

where $I$ is an identity matrix. The image reconstruction quality depends on the selection of the regularization factor $\lambda$. The value is used to affect how much the penalty term controls the final solution. Although there exists the theoretical optimal solution of $\lambda$, they are usually computational intensive and cannot be used for the real-time imaging. Here, the selection of the regularization factor $\lambda$ is mainly based on the empirical method.

## 3. Results

The proposed focusing sensor is evaluated by both numerical simulation and experimental study. Simulation is carried out using Matlab with EIDORS on a PC equipped with a 3.2 GHz Intel Core i5 processor.

### 3.1. Numerical Simulation

#### 3.1.1. Boundary Measurement Consistency in Transformation Domain

In order to simulate the open region, the forward model should be very large to satisfy the potential distribution of an open area. Due to the limitation of the computational cost and the actual application, the forward model can't be infinite. Considered the simulation precision and calculation cost, according to the previous research in [18], for the forward problem, the modeling region is set to be $[-40, 40]$ on the $x$-axis and $[0, 40]$ on the $y$-axis. Figure 5 shows the boundary potential measurements of the reference field based on the open modeling region and its conformal circular domain. The blue line with circle dot is the boundary potential of reference field based on the open modeling field, while the red dots are the boundary potential based on the conformal circular domain. It can be seen that the boundary measurements of open field are almost consistent with the conformal circular domain. It can be proved that the focusing sensor has same physical properties as its conformal circular domain. At the same time, because the focusing sensor is mapped by the uniform sensor in the conformal circular domain, its boundary potential measurements of each channel is consistent.

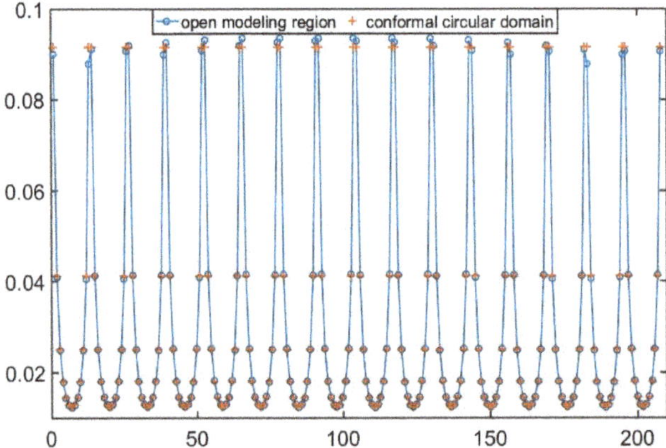


**Figure 5.** Boundary potential of reference field based on the open modeling region and its conformal circular domain.

The distribution of sensitivity fields based on the focusing sensor and the traditional uniform sensor are illustrated in Figure 6. The sensitivity distribution indicates the sensitivity of the boundary voltage response to the variation of conductivity at every pixel in the imaging region. The sensitivity at a certain imaging pixel is defined as the Euclidean norm of the Jacobian values at this pixel [30]. The large values represent the change of conductivity at this pixel will cause a large change in boundary voltage response, while the small values represent a low sensitivity to the conductivity variation. As shown in Figure 6, the sensitivity distribution is mapped from the conformal circular domain. It can be seen that the sensitivity distribution of the focusing sensor has higher sensitivity in the central region near the electrodes, and the sensitivity is decreasing rapidly with respect to the direction away from the electrodes. As for the sensitivity distribution of the uniform sensor, the area near the boundary on both sides of the electrodes has high sensitivity. However, the sensitivity distribution in the region near the central electrode is relatively low. According to the comparative result, the focusing sensor is much more sensitive in the central region compared with the uniform sensor. That means the focusing sensors make it easier to detect small changes in electrical conductivity near the central electrode region. This has certain significance for improving the resolution of OEIT detection in the central region near the electrodes.

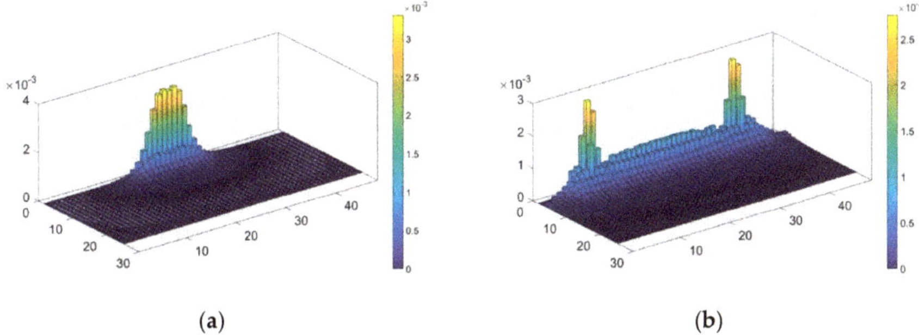

**Figure 6.** The distribution of sensitivity fields based on (**a**) focusing sensor and (**b**) uniform sensor.

### 3.1.2. Quantitative Index

To characterize the performance of the focusing sensor compared with the uniform sensor, a series of simulation test are carried out. Then a series of metrics were used to assess the quality of the reconstructed images. The descriptions of these metrics are as follows:

- Resolution (RES) measures the ratio of the number of pixels in inclusions to the total number of pixels, as shown in (7). In the reconstructed images, the reconstructed inclusions are segmented from the reconstructed images by using a threshold of 75% maximum pixel amplitudes [31]. The total number of pixels represents the area of the region of interest for imaging. In this article, RES is used to measure the relative resolution of the focusing sensor and conventional uniform sensor. A smaller RES means the reconstructed image has high resolution for a similar inclusion:

$$\text{RES} = \sqrt{\frac{Aq}{An}},\tag{7}$$

where $Aq$ is the number of pixels in inclusions, $An$ is the total number of pixels of the image region.
- Resolution ability (Ra) is defined as the measurement ability of distinguish between multiple inclusions. Figure 7 illustrates the schematic diagram of Ra with two inclusions as an example. The bottom row shows the reconstructed image, while the top row plots the amplitude across a row through the center of the multiple simulation inclusions. Ra compares the relationship between the amplitude at the midpoint of the centers of the two inclusions and the 75% maximum

pixel amplitudes to measure whether the two inclusions can be clearly distinguished in the reconstructed image. The definition of Ra is shown in (8):

$$Ra = \frac{h_m - 0.75h_{max}}{h_{min} - 0.75h_{max}},\tag{8}$$

where $h_m$ is the amplitude value at the midpoint of the centers of the two inclusions, $h_{max}$ and $h_{min}$ is the maximum and minimum value of the amplitude across the row through the center of the multiple simulation inclusions. For a certain reconstructed image, a Ra greater than 0 indicates that multiple inclusions can be distinguished; a Ra closer to 1 indicates a stronger ability to distinguish inclusions; while a Ra less than 0 indicates that multiple inclusions cannot be distinguished.

- Average gradient (Ag) is defined to measure the steepness of the multiple reconstructed inclusions boundary. As shown in Figure 7, Ag calculates the mean of the absolute value of gradient of the reconstructed pixels between the centers of the multiple simulation inclusions. The definition of Ag is shown in (9):

$$Ag = ave\left(\sum_{s=1}^{t}\left|\frac{h(c_1 + s\Delta c) - h(c_1 + (s-1)\Delta c)}{\Delta c}\right|\right),\tag{9}$$

where $t = \frac{c_2 - c_1}{\Delta c}$. $\Delta c$ is the length between adjacent pixels. A large Ag means a large relative steepness of the reconstructed inclusion boundary.

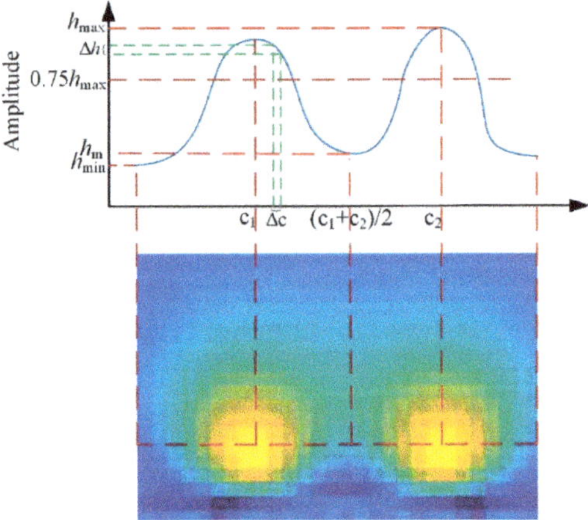


**Figure 7.** The schematic diagram of Ra and Ag with two inclusions.

In the meanwhile, the relative image error (RE) and the correlation coefficient (CC) between the true conductivity distributions and reconstructed images are also determined. The definitions of the RE and CC can be seen in (10) and (11), respectively [32]:

$$RE = \frac{\left\|h_q - h\right\|}{\|h\|} \times 100\%,\tag{10}$$

$$CC = \frac{\sum\limits_{i=1}^{P}(h_i - \bar{h})(h_{qi} - \bar{h}_q)}{\sqrt{\sum\limits_{i=1}^{P}(h_i - \bar{h})^2 \sum\limits_{i=1}^{P}(h_{qi} - \bar{h}_q)^2}},\tag{11}$$

where $h_q$ is the normalized conductivity vector of reconstructed distribution, $h_{qi}$ is the *i*-th element in the vector $h_q$; and $h$ is the normalized conductivity vector of a true distribution, $h_i$ is the *i*-th element in the vector $h$; P is the length of the conductivity vector. For a good reconstruction performance, RE should be low and CC should be relatively high.

### 3.1.3. Reconstruction Analyses of Multiple Positions

During the simulation tests of a single target, highly conductive circular inclusions with diameter of 2–10% of the region width are simulated. The evolutions of the target inclusion with respect to the vertical, horizontal and oblique positions are shown in Figure 8a. In each direction, the inclusions of each size traversed 50 different positions. Direction 1 is the evolution of the horizontal direction near the electrode region and the central coordinate change range is from (0, 1) to (8.5, 1). Direction 2 shows the evolution along the diagonal direction of the field domain and the end of the central coordinate is (−8.5, 1.5). Direction 3 is the evolution keeping away from the center electrode along the vertical direction, and the central coordinates of the inclusions change from (0, 1) to (0, 8.5). During the simulation of double targets, two high conductivity circular inclusions with diameter of 6% of the region width are simulated. The evolution of the target inclusions is along the horizontal direction, and 40 positions are simulated along that direction, as shown in Figure 8b. At each location, the distances between the centers of two inclusions were selected as 0.7, 1.1, and 1.5, respectively, and the degrees of differentiation of the two inclusions measured by different sensors were calculated at each location along with the distance of the inclusions.

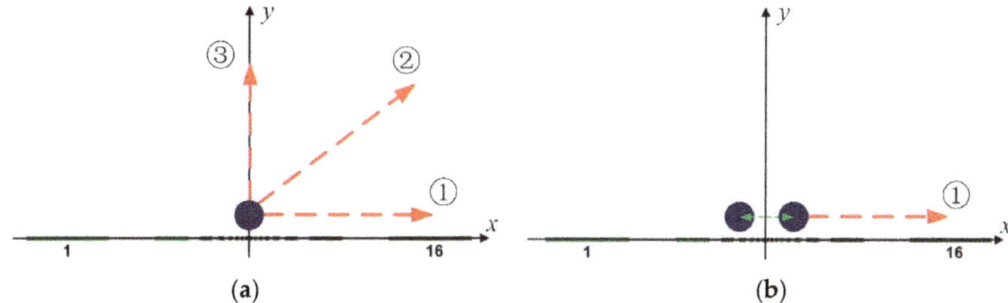

(a)  (b)

**Figure 8.** Traces of inclusions involved in dynamic tests: (**a**) single inclusion; (**b**) double inclusions.

Figure 9 shows the evolutions of minimum RES with respect to the inclusion positions in the single target simulations. The minimum RES means the resolution of the smallest inclusions that can be reconstructed by two sensors for different sized inclusions for the same location. A small RES indicates that the sensor is better at recognizing small sized inclusions. The red lines show the minimum RES performance of the uniform sensor, while the blue lines are the performance of the proposed focusing sensor. The green dotted line indicates the 10% resolution, which is usually considered as the normal resolution for EIT. The solving process uses the Tikhonov regularization algorithm. The regularization parameters are determined by an empirical method and they are invariant during each evolution process. As shown in Figure 9, in Direction 1, along with the position of the target moving to the side boundary of the region, the performance of the focusing sensor has more stable and smaller RES, it is better than the uniform sensor in almost all cases. When the inclusions are located in the central region, the resolution of the uniform sensor for the small inclusions is almost stable at 10%, while the minimum resolution of the focusing sensor is up to 2%. During the traversal process of Direction 1, the minimum RES of the focusing sensor is significantly smaller than that of the uniform sensor, and is less than 10% at most positions. This indicates that the focusing sensor can realize high resolution reconstruction of small size inclusions in the near-electrode region, that is, smaller size inclusions can be reconstructed by the focusing sensor. As the inclusion moves from the center to the boundary and gradually away from the electrode with respect to the oblique direction, as shown in Direction 2, a similar conclusion can be reached. When the inclusions are close to the electrode area, the minimum resolution of the focusing sensor has a significant advantage, which is basically less than 10%. However, as the inclusion moves away from the electrode and is located near the side of the imaging area, the focusing sensor gets worse results than the uniform sensor. That means the focusing sensor has poor resolution for inclusions far from the electrode. During the evolution of Direction 3, it can be seen that the focusing sensor has smaller RES than the uniform sensor when the inclusion is near the electrode, which means the focusing sensor can reconstruct smaller inclusions in this situation. When the inclusion moves away from the central electrode, the difference in RES between the two sensors is decreasing. To sum up, the proposed focusing sensor has significant advantages in the reconstruction of small inclusions near the electrode area compared with the uniform sensor. In the simulation tests, the minimum RES of the focusing sensor can up to 2%, which is a significant improvement over conventional uniform sensors. Meanwhile, as shown in Figure 9, when the position number is lower than 15, it can be seen the minimum resolution of the focusing sensor in different traversal directions is better than the uniform sensor, which roughly determines that the high-resolution imaging area of the focusing sensor is $[-2.5, 2.5]$ on the $x$-axis and $[0, 2.5]$ on the $y$-axis. However, the results also show that the focusing sensor has poor resolution for the inclusions that are far away from the central electrode region and near the boundary.

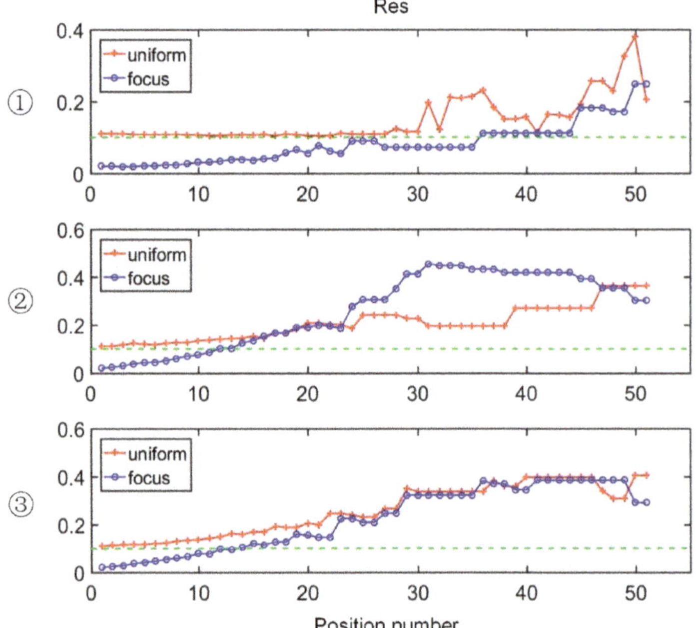

**Figure 9.** The evolutions of minimum RES with respect to the inclusion positions: ①, ② and ③ denotes the evaluations at Direction 1, 2 and 3, respectively.

Figure 10 shows the evolution of Ra with respect to the inclusion positions in the double targets simulations. This simulation compares the ability of two sensors to distinguish multiple inclusions in the near-electrode region. In Figure 10, the top image is the Ra of the focusing sensor, while the bottom image is the Ra of the uniform sensor. The $d$ is the distance between the centers of two targets. The horizontal black dotted line indicates the zero value of Ra. If the value is above 0, the higher the value is, the stronger the ability to distinguish between the two inclusions is. If the value is lower than 0, it cannot distinguish the two inclusions. As it can be seen from Figure 10, when the inclusions are located near the central electrodes region, the focusing sensor can distinguish the two inclusions with a center distance of 0.7, while the uniform sensor can only distinguish the two inclusions with a center distance of 1.5. As the inclusions move away from the central region, the ability of the focusing sensor to distinguish close inclusions gradually weakens. However, the uniform sensor can hardly distinguish two inclusions that are close to each other. When the position number is lower than 15, the Ra of the focusing sensor is better than the uniform sensor, which is consistent with the simulation results of the minimum resolution of a single inclusion. In summary, the proposed focusing sensor, compared with the uniform sensor, has a high resolution in the near center electrode region and a strong ability to distinguish the two inclusions in close proximity. According to the above simulation results, the focusing region of the proposed focusing sensor is about $[-2.5, 2.5]$ on the $x$-axis and $[0, 2.5]$ on the $y$-axis.

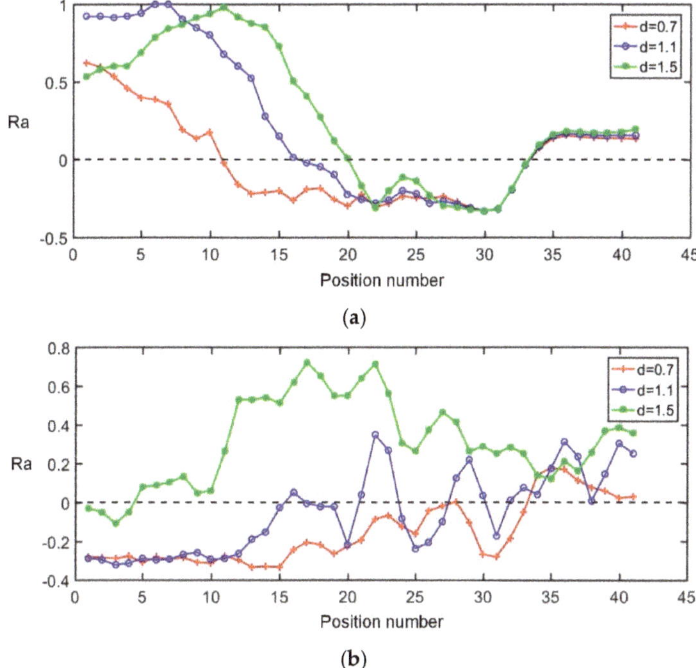

**Figure 10.** The evolutions of **Ra** with respect to the inclusion positions based on (**a**) focusing sensor and (**b**) uniform sensor.

Figure 11 shows the reconstructed results of the single circle target with same size distributed at different positions that are gradually away from the electrode with respect to the oblique direction. In this simulation, the conductivity of the single circle target is 2 S/m, while the background conductivity is 1 S/m. In the meanwhile, Figure 11 also gives the position of the phantom in the conformal circular domain. Compared the reconstructed results of the focusing sensor and the uniform sensor, it can be shown that when the target positions are close to the central electrodes (T1), the focusing sensor leads to better reconstructed results both in location and shape of the circle target and has the highest reconstruction image quality, while the reconstructed image of the uniform sensor has some distortion for the size and shape of the circle target. As the target moves away from the central electrode and close to the boundary, the reconstructed image of focusing sensor becomes bad and gradually worse than that of the uniform sensor. Therefore, the subsequent simulation focused on the comparison of the imaging results of the two sensors in the area near the central electrodes.

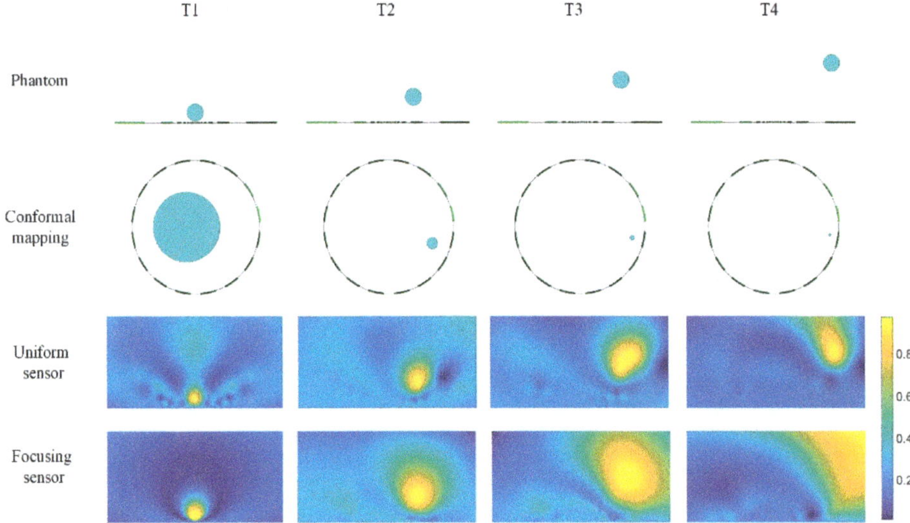

**Figure 11.** Reconstruction results of single target based on uniform and focusing sensors.

Figure 12 shows the reconstructed results of double targets with different size and different distance based on the uniform sensor and the focusing sensor. T5 and T6 are two conductive inclusions with same radius of 0.5 and at different distance from each other. It can be seen that when two inclusions are far apart (T5), both sensors can distinguish the two inclusions. When the two inclusions are close to each other, they cannot be distinguished in the reconstructed image of the uniform sensor. In this case, the reconstructed results of uniform sensor can only reflect the presence of the inclusions, but does not distinguish the number and size of the inclusions. However, from the results of the focusing sensor, he number and size of the inclusions can be distinguished clearly. Then the size and distance of the inclusions are gradually decreased, as shown in T7 and T8. Due to the small size of the inclusions, the distribution of T7 and T8 and the reconstruction results are locally amplified in Figure 13a. T7 shows two conductive inclusions with same radius of 0.2 and relatively close to each other. It can be found that the uniform sensor cannot get a good performance for the two small inclusions and the reconstructed image is similar to the image of T6, which means the uniform sensor cannot distinguish the two small inclusions. However, the focusing sensor has a great performance in this case and the number and size of the inclusions can be clearly distinguished. T8 is two conductive inclusions with the same radius of 0.1 and closer to each other. In this case, the uniform sensor can only reconstruct a large region of conductivity changes and the exact distribution of the inclusions is unknown. On the contrary, the focusing sensor still has great reconstruction quality, and it can represent well not only the size of the inclusions, but also the location of the inclusions. In this case, the focusing sensor has a minimum resolution of 2%.

A quantitative analysis of these reconstruction results is shown in Table 1. As it can be seen from the table, in all four cases, the results of the focusing sensor have better Ra than the uniform sensor. Except for T5, the Ra of the uniform sensor is negative, which means the uniform sensor cannot distinguish small inclusions that are close to each other. Ag measures the steepness of the multiple reconstructed inclusions boundary. When the inclusions are far from each other, the Ag values obtained by the two sensors are similar. When the size and distance of the inclusions are gradually decreased, the reconstruction results of the uniform sensor can no longer distinguish the two inclusions, and then Ag of the focusing sensor becomes much larger than that of the uniform sensor. Therefore, the focusing sensor has strong resolution ability for the inclusions located in the area nearby the central electrodes. Meanwhile, the results based on the focusing sensor have lower RE and higher

CC in almost all simulated situations. This means that the images of the focusing sensor show better image quality and reconstruction accuracy. Moreover, as the inclusion size and spacing decreases, the advantages of focusing sensor become more and more obvious.

**Table 1.** Quantitative analyses of the reconstruction results shown in Figure 12.

|      | T5 | | T6 | | T7 | | T8 | |
|------|---------|----------|---------|----------|---------|----------|---------|----------|
|      | Uniform | Focusing | Uniform | Focusing | Uniform | Focusing | Uniform | Focusing |
| Ra   | 0.642   | 0.813    | −0.132  | 0.611    | −0.149  | 0.545    | −0.217  | 0.498    |
| Ag   | 4.531   | 6.045    | 0.728   | 12.145   | 0.538   | 14.266   | 0.911   | 28.644   |
| RE   | 0.132   | 0.143    | 0.177   | 0.101    | 0.233   | 0.131    | 0.250   | 0.089    |
| CC   | 0.855   | 0.872    | 0.780   | 0.842    | 0.750   | 0.851    | 0.714   | 0.887    |

In order to explore the performance of different sensors in the presence of noise, T7 and T8 are simulated in the case of no noise and 40dB SNR noise, respectively. The result is shown in Figure 13. The uniform sensor results have deformations in the shape of the inclusions and cannot distinguish the two inclusions. On the contrary, the result of the proposed focusing sensor has less deformation and can reflect the actual circumstances of the circle target. In the meantime, when 40 dB SNR random noise is added to the simulation data the uniform sensor leads to bad results in which the reconstructed target has large distortion and image artifacts. However, under the influence of noise, the focusing sensor can reconstruct the true distribution of inclusions, especially for small inclusions close to each other.

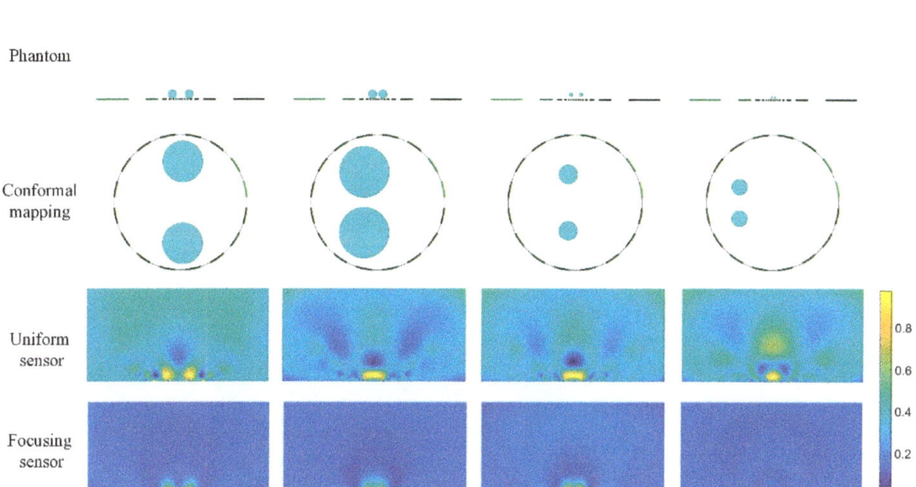


**Figure 12.** Reconstruction results of double targets based on uniform and focusing sensors.

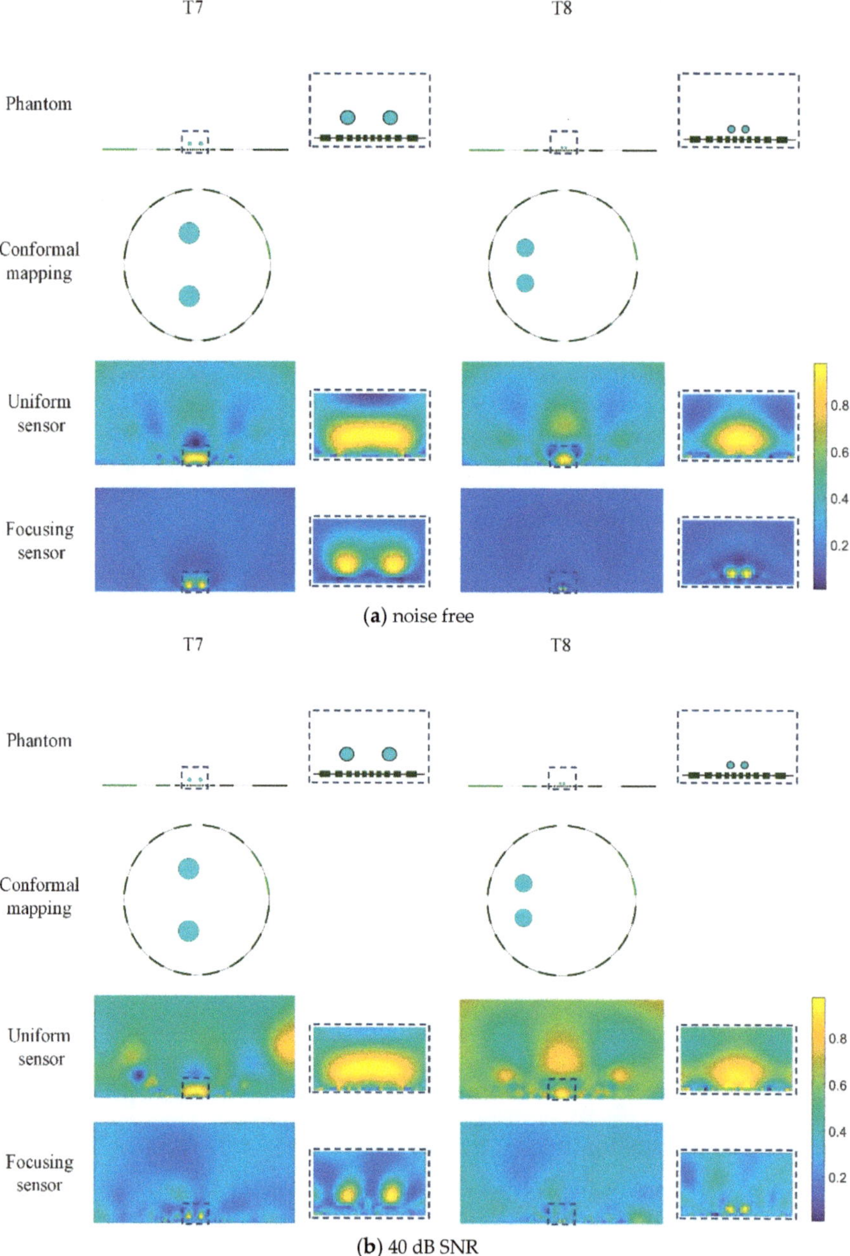

(a) noise free

(b) 40 dB SNR

**Figure 13.** Reconstruction results of double targets based on uniform and focusing sensors with noise free data and 40 dB SNR data.

Table 2 shows the quantitative analyses of these reconstructed results. The results of the proposed focusing sensor have lower RE and higher CC in all of the tested phantoms compared with the uniform sensor results. In these cases, the Ag of the focusing sensor is much better than the uniform sensor and the Ag becomes larger as the size and distance of the inclusions decrease. This indicates that the boundary of the inclusions reconstructed by the focusing sensor is steep, and the steepness increases with the decrease of the distance between inclusions. Meanwhile, the Ra of the proposed focusing sensor are obviously better than the uniform sensor results, which means that under the influence of noise, the focusing sensor can still reconstruct multiple small inclusions located at the area near the central electrodes. Moreover, the reconstructed images of the focusing sensor have less image artifacts and higher reconstruction quality compared with the uniform sensor ones.

**Table 2.** Quantitative analyses of reconstruction results shown in Figure 13.

| | Noise Free | | | | 40 dB SNR | | | |
|---|---|---|---|---|---|---|---|---|
| | T7 | | T8 | | T7 | | T8 | |
| | Uniform | Focusing | Uniform | Focusing | Uniform | Focusing | Uniform | Focusing |
| Ra | −0.149 | 0.545 | −0.217 | 0.498 | −0.161 | 0.441 | −0.117 | 0.312 |
| Ag | 0.538 | 14.266 | 0.911 | 28.644 | 0.507 | 13.889 | 0.912 | 28.627 |
| RE | 0.233 | 0.131 | 0.250 | 0.089 | 0.251 | 0.142 | 0.370 | 0.136 |
| CC | 0.750 | 0.851 | 0.714 | 0.887 | 0.678 | 0.788 | 0.623 | 0.811 |

### 3.1.4. Reconstruction Results Combining the Focusing Sensor and Uniform Sensor

The former discussion shows the focusing sensor can reconstruct smaller targets near the central electrode, but its reconstructed results for targets far away from the electrodes are worse than those of the uniform sensor. Therefore, if a measurement of the same object was taken using each individual sensor, in the same environment and in the same location then it could be assumed that all of the measurements were from one device rather than individual sensors. The reconstruction could then use all of the combined measurements in order to reconstruct the conductivity distribution. There are two ways to fuse the data from two different sensors. The first method (Method 1) is to directly superimpose the reconstructed images of the two sensors. The second method (Method 2) is to combine all of the measurements from two sensors and then the augmented sensitivity matrix is used to solve the same solution vector. This means that each calculation now uses $2 \times 208$ measurements, where 208 is the number of measurements of each sensor. This improves the overall solution as more measurements are used to calculate the same number of answers.

Figure 14 shows reconstructed results of two different phantoms based on each individual sensor as well as the two data fusion methods. When there exists a small inclusion close to the electrode and large inclusions far from the electrode, the reconstructed image of focusing sensor can clearly recognize the two small inclusions that are close to the electrode and close to each other, but the large inclusion far from the electrode can hardly be recognized. On the contrary, the uniform sensor can reconstruct the large inclusion far from the electrode, but cannot recognize the two small inclusions that are close to the electrode. Through the data fusion of two different sensors, from the reconstructed image, it can be found that the combination of two measurement data can effectively improve the reconstruction quality of multiple inclusions. Method 1 and Method 2 can both reconstruct the multiple inclusions compared with the reconstructed results of each individual sensor. These results indicate that the fusion of the measurements from two sensors can not only realize high-resolution reconstruction of multiple inclusions near the electrode, but also realize high-quality reconstruction of distant inclusions.

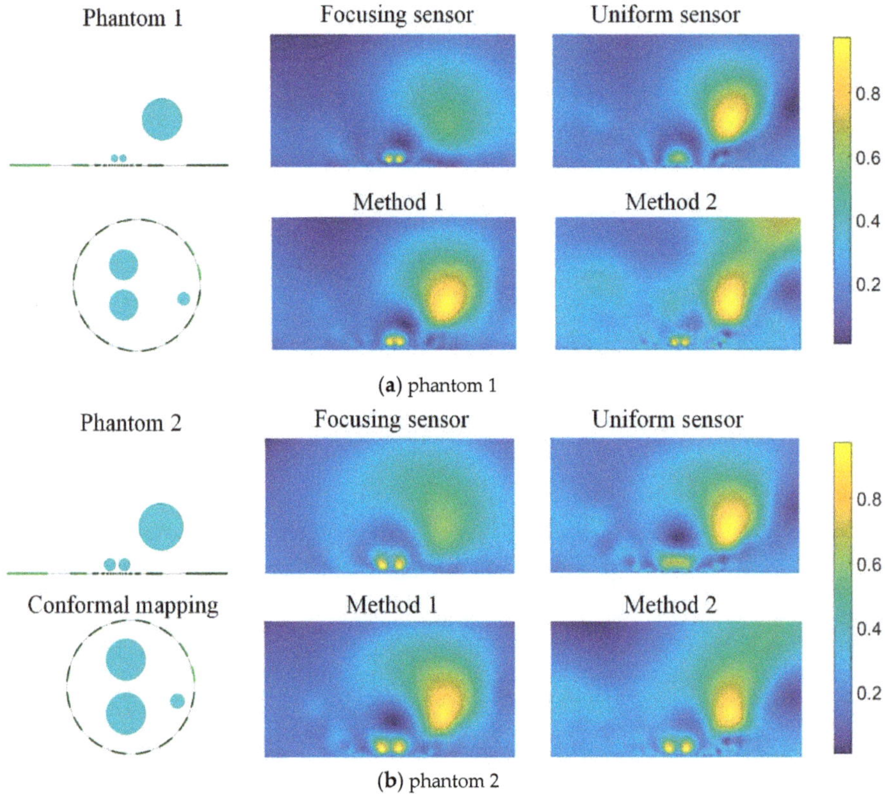

**Figure 14.** Reconstructed results of two different phantoms from focusing sensor, uniform sensor, and two fusion methods.

### 3.2. Experimental Analysis

To further test the performance of the proposed focusing sensor, a set of experimental studies are carried out. The experimental equipment is show in Figure 15. It consists of a rectangular tank and a data acquisition system. The EIT system was developed in Tianjin University, China. It is a sixteen-channel high-speed serial data acquisition (DAQ) system [33]. The rectangular tank is 80 cm in length and 40 cm in width, which is used to simulate the open region. For the proposed focusing sensor, the electrodes are 16 copper plates with different widths which are consistent with the simulation settings. They are placed on the central boundary of the rectangular tank. Table 3 shows the actual dimensions of the focusing sensor electrodes used in experimental studies. For comparison, the electrodes of the uniform sensor are 16 copper plates with 6 mm width and they are uniformly placed on the central boundary of the rectangular tank. According to the simulation results, the imaging region is focused on the shallow region near the electrodes. Thus, the region of interest is set to 20 cm in length and 10 cm in width. Figure 16 shows the schematic diagram of the region of interest. The experiments are conducted by placing different size of plastic (nonconductive) or metallic (conductive) rods placed at different locations. $Na_2SO_4$ solution with a conductivity of $6.22 \times 10^{-2}$ S/m is selected as the background medium.

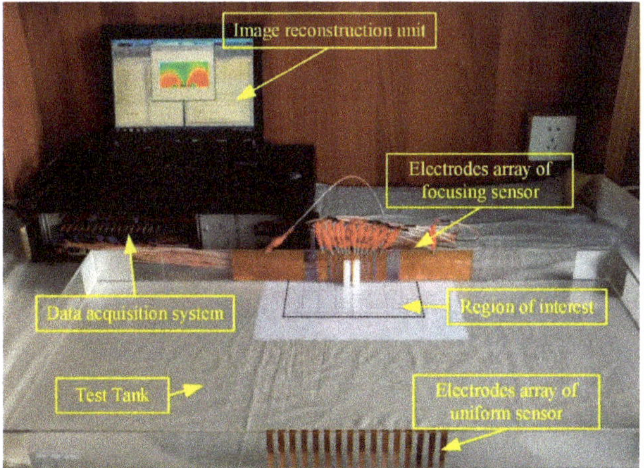


**Figure 15.** The experimental system for OEIT.

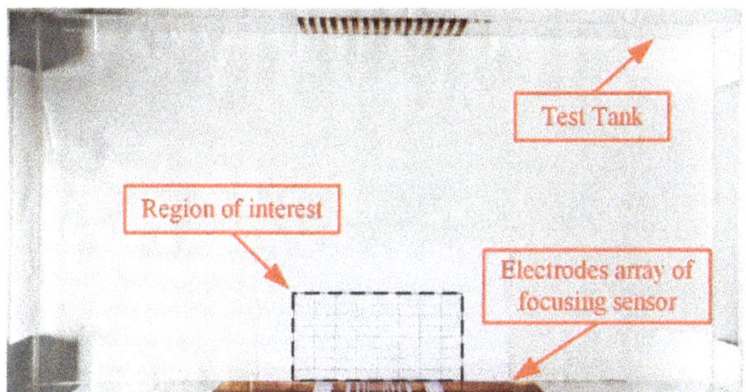


**Figure 16.** The schematic diagram of the region of interest.

**Table 3.** The actual dimensions of the focusing sensor electrodes used in experimental studies.

| Electrode Number | 1 (16) | 2 (15) | 3 (14) | 4 (13) | 5 (12) | 6 (11) | 7 (10) | 8 (9) |
|---|---|---|---|---|---|---|---|---|
| Length (mm) | 110 | 15.5 | 6.0 | 3.5 | 2.6 | 2.0 | 1.7 | 1.5 |

The experimental results of different phantoms based on the proposed focusing sensor and uniform sensor are shown in Figure 17. The inclusion of E1 is an aluminum rod of 10 mm diameter, while E2 is a nylon rod with 3 mm diameter. The two inclusions are set at the region near the central electrode. The test aims to compare the ability of reconstructing the small inclusion as well as the resolution of the two sensors. From the reconstructed results, it can be found that the focusing sensor shows better performance than the uniform sensor and the recognition ability of conventional uniform sensors for small sized inclusions is poor, even if the inclusions are located near the electrode area. For the 10 mm diameter aluminum rod (E1), the focusing sensor can reconstruct the position and shape of the inclusions with high accuracy, while the uniform sensor can only reconstruct the general area of conductivity variation with large image artifacts. When the size of the inclusions is reduced to 3 mm (E2), the uniform sensor cannot reconstruct the conductivity variation, while the focusing sensor can still reconstruct the inclusion clearly. The above results shows that the proposed focusing sensor is more sensitive to small conductivity changes in the near electrode region and the focusing sensor can improve the reconstruction resolution to at least 3% in the near central electrodes region. E3 and E4 show the experimental results for double inclusions, which is to test the ability of different sensors to distinguish multiple inclusions in close proximity. E3 is two same sizes of 10 mm aluminum rods located near the electrodes. The uniform sensor still can only reconstruct the blurred image and the inclusions are not clearly identifiable. However, the two inclusions can be clearly distinguished from the reconstruction image of focusing sensor, although the reconstructed inclusions have slight deformation. E4 shows two same size nylon rods with 5 mm diameter and closer to the electrodes, the reconstruction result of focusing sensor is significantly better than that of the uniform sensor. The uniform sensor cannot recognize small size inclusions, while focusing sensor can not only recognize two small size inclusions that are lose to each other, but also reconstruct them with a high accuracy. Table 4 shows the quantitative analyses of the experimental reconstruction results. The quantitative analyses shows that the results of focusing sensor can reach higher CC and lower RE compared with the results of uniform sensor for all the experimental reconstruction images. The above results show that the proposed focusing sensor has higher sensitivity and reconstruction accuracy to small size inclusions in the near-electrode region. Compared with the uniform sensor, the focusing sensor has better reconstruction quality and higher reconstruction accuracy for small conductivity changes in the focusing region. The experimental results show that the resolution of the focusing sensor can reach 3% in the near center electrode region.

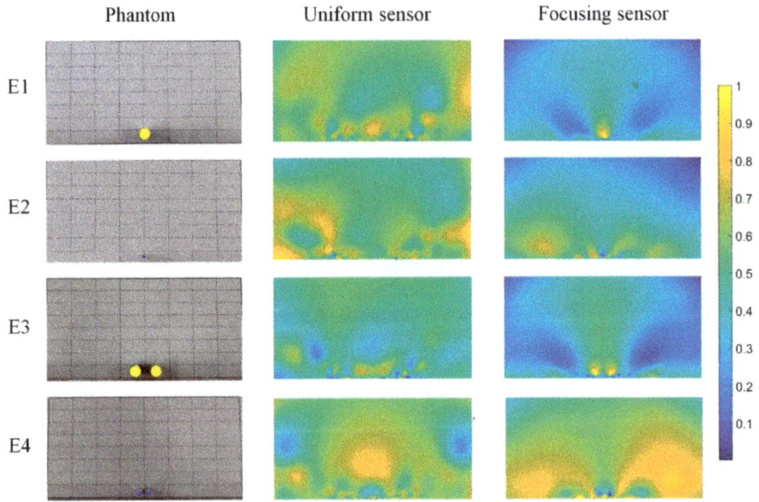

**Figure 17.** Experimental results from proposed focusing sensor and uniform sensor with Tikhonov regularization.

**Table 4.** Quantitative analyses of experimental reconstruction results shown in Figure 17.

|  | E1 | | E2 | | E3 | | E4 | |
|---|---|---|---|---|---|---|---|---|
|  | Uniform | Focusing | Uniform | Focusing | Uniform | Focusing | Uniform | Focusing |
| RE | 0.314 | 0.150 | 0.238 | 0.093 | 0.271 | 0.164 | 0.325 | 0.137 |
| CC | 0.681 | 0.879 | 0.713 | 0.895 | 0.737 | 0.828 | 0.632 | 0.841 |

To further test the performance of the proposed focusing sensor and the two different data fusion methods, an experimental phantom is carried out and compared under different methods. The phantom is two 5 mm nylon rods located near the central electrodes and a 25 mm nylon rod far away from the electrodes. Figure 18 shows the reconstructed results based on each individual sensor and the two data fusion methods. The experimental results are consistent with the simulation results. The focusing sensor has high resolution ability in the focusing region. It can reconstruct small targets in the area near the electrodes, but its reconstructed results for the target far away from the electrodes are worse than the uniform sensor. However, the uniform sensor can hardly reconstruct the two small targets when there has a large inclusion far away from the electrodes. The situation is improved through the data fusion method 1. From the results, Method 1 can basically present small inclusions near the electrode and large inclusions far from the electrode. However, the results of Method 2, which has severe deformation and image artifacts, are relatively poor. This is because in the experimental environment, uncertain noises and errors will increase the difficulty of data fusion under different sensors. Therefore, Method 2 adopts the method of solving the same solution matrix, which will increase the ill-posedness of the solution. Method 1 solves the optimal solution under each sensor and then fuses the image. By this method, the conductivity distribution of the measured field can be reconstructed more accurately. Because the data fusion method proposed in this paper combines the advantages of the two sensors, this method may have the possibility of realization in some application fields. For example, in the surface geological exploration and landmine detection, these two sensors could be used to scan the same static object, successively. Then these two kinds of data could be fused to reconstruct the same object.

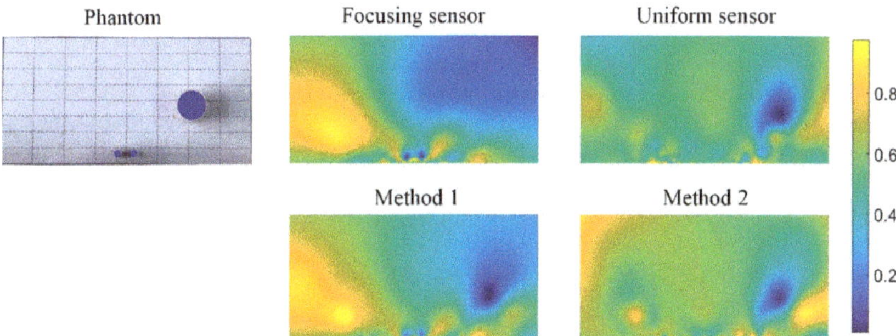


**Figure 18.** Experimental reconstructions from focusing sensor, uniform sensor, and two fusion methods.

## 4. Conclusions

A new sensor design based on conformal transformation for OEIT is proposed to improve the image reconstruction quality and low reconstruction accuracy of OEIT. The proposed focusing sensor is derived from the uniform distribution of electrodes in the conformal circular domain through conformal transformation method. The numerical results show that compared with the conventional uniform sensor, the proposed focusing sensor can achieve high resolution reconstruction of the inclusions near the central electrode with a minimum resolution of 2%. The evolution results roughly determine the high-resolution imaging area of the focusing sensor is [−2.5, 2.5] of $x$-axis and [0, 2.5] of $y$-axis. Moreover, the focusing sensor can distinguish small sized inclusions that are close to each other in the focusing region and reconstruct them with high quality. Qualitative and quantitative results show that the results of the proposed focusing sensor have smaller minimum resolution and higher resolution ability when the position of target is near the central electrodes. The experimental results show that the resolution of the focusing sensor can reach 3% in the near center electrode region. The data fusion methods of the two sensors are also discussed. The experimental results show that the method that solves the optimal solution under each sensor and then fuses the image can lead to better results compared with the method of solving the same solution matrix. However, the proposed method can be used for scanning imaging based on a one-dimensional linear electrode array at present, and the method of constructing two-dimensional focusing electrode arrays is still the main topic of future research. Future work will focus on the construction of two-dimensional electrode array as well as the further improvement of the reconstruction algorithms.

**Author Contributions:** Conceptualization, S.R. and F.D.; methodology, S.R.; software, S.R.; validation, Y.W., S.R. and F.D.; formal analysis, Y.W.; investigation, Y.W.; resources, F.D.; data curation, Y.W.; writing—original draft preparation, Y.W.; writing—review and editing, S.R.; visualization, W.Y.; supervision, S.R.; project administration, F.D.; funding acquisition, F.D.

**Funding:** This research was funded in part by the National Natural Science Foundation of China, grant number 61571321, and in part by the Natural Science Foundation of Tianjin City, grant number 17JCQNJC03500.

**Conflicts of Interest:** The authors declare no conflicts of interest.

## References


1. Holder, D.S. *Electrical Impedance Tomography: Methods, History and Applications*; CRC Press: Boca Raton, FL, USA, 2004.
2. Yao, J.F.; Takei, M. Application of process tomography to multiphase flow measurement in industrial and biomedical fields—A review. *IEEE Sens. J.* **2017**, *17*, 8196–8205. [CrossRef]
3. Xu, Z.F.; Yao, J.F.; Wang, Z.; Liu, Y.L.; Wang, H.; Chen, B.; Wu, H.T. Development of a portable electrical impedance tomography system for biomedical applications. *IEEE Sens. J.* **2018**, *8*, 8117–8124. [CrossRef]
4. Ren, S.J.; Sun, K.; Liu, D.; Dong, F. A statistical shape constrained reconstruction framework for electrical impedance tomography. *IEEE Trans. Med. Imag.* **2019**, in press. [CrossRef]
5. Liu, D.; Kolehmainen, V.; Siltanen, S.; Laukkanen, A.M.; Seppanen, A. Estimation of conductivity changes in a region of interest with electrical impedance tomography. *Inverse Probl. Imag.* **2015**, *9*, 211–229. [CrossRef]
6. Liu, D.; Khambampati, A.K.; Du, J. A parametric level set method for electrical impedance tomography. *IEEE Trans. Med. Imag.* **2018**, *37*, 451–460. [CrossRef] [PubMed]
7. Youssef, A.M.; Elkaliouby, H.; Zabramawi, Y.A. Sinkhole detection using electrical resistivity tomography in Saudi Arabia. *J. Geophys. Eng.* **2012**, *9*, 655–663. [CrossRef]
8. Church, P.; Mcfee, J.E.; Gagnon, S.; Wort, P. Electrical impedance tomographic imaging of buried landmines. *IEEE Trans. Geosci. Remote Sens.* **2006**, *44*, 2407–2420. [CrossRef]
9. Baltopoulos, A.; Polydorides, N.; Pambaguian, L.; Vavouliotis, A.; Kostopoulos, V. Damage identification in carbon fiber reinforced polymer plates using electrical resistance tomography mapping. *J. Compos. Mater.* **2013**, *47*, 3285–3301. [CrossRef]
10. Allen, S.; Allen, L.R. Electrical impedance tomography-based sensing skin for quantitative imaging of damage in concrete. *Smart Mater. Struct.* **2014**, *23*, 085001.
11. Mueller, J.L.; Isaacson, D.; Newell, J.C. Reconstruction of conductivity changes due to ventilation and perfusion from EIT data collected on a rectangular electrode array. *Physiol. Meas.* **2001**, *22*, 97–106. [CrossRef]
12. Borsic, A.; Halter, R.; Wan, Y.; Hartov, A.; Paulsen, K.D. Electrical impedance tomography reconstruction for three-dimensional imaging of the prostate. *Physiol. Meas.* **2010**, *31*, S1-16. [CrossRef]
13. Cherepenin, V.A.; Gulyaev, Y.V.; Korjenevsky, A.V.; Sapetsky, S.A.; Tuykin, T.S. An electrical impedance tomography system for gynecological application GIT with a tiny electrode array. *Physiol. Meas.* **2012**, *33*, 849–862.
14. Aristovich, K.Y.; Santos, G.S.D.; Packham, B.C.; Holder, D.S. A method for reconstructing tomographic images of evoked neural activity with electrical impedance tomography using intracranial planar arrays. *Physiol. Meas.* **2014**, *35*, 1095–1109. [CrossRef] [PubMed]
15. Murphy, E.K.; Mahara, A.; Halter, R.J. A novel regularization technique for microendoscopic electrical impedance tomography. *IEEE Trans. Med. Imag.* **2016**, *35*, 1593–1603. [CrossRef]
16. Perez, H.; Pidcock, M.; Sebu, C. A three-dimensional image reconstruction algorithm for electrical impedance tomography using planar electrode arrays. *Inverse Probl. Sci. Eng.* **2017**, *25*, 471–491. [CrossRef]
17. Liu, J.; Xiong, H.; Lin, L.; Li, G. Evaluation of measurement and stimulation patterns in open electrical impedance tomography with scanning electrode. *Med. Biol. Eng. Comput.* **2015**, *53*, 589–597. [CrossRef]
18. Wang, Y.; Ren, S.J.; Dong, F. A transformation-domain image reconstruction method for open electrical impedance tomography based on conformal mapping. *IEEE Sens. J.* **2018**, *19*, 1873–1883. [CrossRef]
19. Baber, C.C.; Brown, B.H.; Freeston, I.L. Imaging spatial distributions of resistivity using applied potential tomography. *Electron. Lett.* **1983**, *19*, 93–95.

20. Vauhkonen, P.J.; Vauhkonen, M.; Savolainen, T.; Kaipio, J.P. Three-dimensional electrical impedance tomography based on the complete electrode model. *IEEE Trans. Bio-Med. Eng.* **1999**, *46*, 1150–1160. [CrossRef]
21. Ahlfors, L.V. *Complex Analysis*, 3rd ed.; McGraw-Hill Book Company: New York, NY, USA, 1953.
22. Kober, H. *Dictionary of Conformal Representations*, 2nd ed.; Dover Publications: New York, NY, USA, 1957.
23. Adler, A.; Lionheart, W.R. Uses and abuses of EIDORS: An extensible software base for EIT. *Physiol. Meas.* **2006**, *27*, 25–42. [CrossRef] [PubMed]
24. Ren, S.J.; Wang, Y.; Liang, G.H.; Dong, F. A robust inclusion boundary reconstructor for electrical impedance tomography with geometric constraints. *IEEE Trans. Instrum. Meas.* **2019**, *68*, 762–773. [CrossRef]
25. Ren, S.J.; Soleimani, M.; Dong, F. Inclusion boundary reconstruction and sensitivity analysis in electrical impedance tomography. *Inverse Probl. Sci. Eng.* **2018**, *26*, 1037–1061. [CrossRef]
26. Woo, E.J.; Hua, P.; Webster, J.G. Finite-element method in electrical impedance tomography. *Med. Biol. Eng. Comput.* **1994**, *32*, 530–536. [CrossRef] [PubMed]
27. Chittenden, C.T.; Soleimani, M. Planar array capacitive imaging sensor design optimization. *IEEE Sens. J.* **2017**, *17*, 8059–8071. [CrossRef]
28. Vauhkonen, M.; Vadasz, D.; Karjalainen, P.A.; Somersalo, E.; Kaipio, J.P. Tikhonov regularization and prior information in electrical impedance tomography. *IEEE Trans. Med. Imag.* **1998**, *17*, 285–293. [CrossRef]
29. Borsic, A.; Graham, B.M.; Adler, A.; Lionheart, W.R.B. In vivo impedance imaging with total variation regularization. *IEEE Trans. Med. Imag.* **2010**, *29*, 44–54. [CrossRef]
30. Brandstatter, B. Jacobian calculation for electrical impedance tomography based on the reciprocity principle. *IEEE Trans. Magn.* **2003**, *39*, 1309–1312. [CrossRef]
31. Adler, A.; Arnold, J.H.; Bayford, R.; Borsic, A.; Brown, B.; Dixon, P.; Faes, T.J.C.; Frerichs, I.; Gagon, H.; Garber, Y.; et al. GREIT: A unified approach to 2D linear EIT reconstruction of lung images. *Physiol. Meas.* **2009**, *30*, 35–55. [CrossRef]
32. Yang, W.Q.; Peng, L.H. Image reconstruction algorithms for electrical capacitance tomography. *Meas. Sci. Technol.* **2003**, *14*, R1–R13. [CrossRef]
33. Tan, C.; Liu, S.W.; Jia, J.B.; Dong, F. A Wideband Electrical Impedance Tomography System Based on Sensitive Bioimpedance Spectrum Bandwidth. *IEEE Trans. Instrum. Meas.* **2019**, in press. [CrossRef]

*Article*

# Adaptive Selection of Truncation Radius in Calderon's Method for Direct Image Reconstruction in Electrical Capacitance Tomography [†]

**Shijie Sun [1,2,3], Lijun Xu [1,2,*], Zhang Cao [1,2], Jiangtao Sun [1,2] and Wenbin Tian [1,2]**

[1] School of Instrumentation and Optoelectronic Engineering, Beihang University, Beijing 100191, China; sunsj@buaa.edu.cn (S.S.); zh_cao@buaa.edu.cn (Z.C.); jiangtao_sun@buaa.edu.cn (J.S.); wenbin_tian@buaa.edu.cn (W.T.)

[2] Beijing Advanced Innovation Center for Big Data-Based Precision Medicine, Beihang University, Beijing 100191, China

[3] School of Computer Science and Engineering, Beihang University, Beijing 100191, China

[*] Correspondence: lijunxu@buaa.edu.cn; Tel.: +86-010-8231-7325

[†] This paper is an extended version of our conference paper: Sun, S.; Xu, L.; Cao, Z.; Sun, J. Influence of the Integral Parameters in Calderon Method on Image Quality for Electrical Capacitance Tomography. In Proceedings of the 9th World Congress on Industrial Process Tomography, Bath, UK, 2–6 September 2018; pp. 635–640.

Received: 27 March 2019; Accepted: 24 April 2019; Published: 29 April 2019

**Abstract:** Calderon's method has been successfully used for the direct image reconstruction in electrical capacitance tomography. In the method, the truncation radius adopted in numerical integral greatly influences the reconstruction results. In the past, the truncation radius is selected as a constant empirically according to the permittivity distribution pattern and noise level. In this paper, the influence of the truncation radius in Calderon's method on the reconstruction results was first analyzed by numerical simulation. Then, a strategy for adaptive selection of the truncation radius was proposed. The amplitude information of the elements in the scattering transform matrix computed from the Dirichlet-to-Neumann (DN) map was used to determine the range for the truncation radius selection, and the phase information was further used to select a proper truncation radius value within this selection range. Finally, experiments were carried out to verify the strategy. Experimental results showed that small relative image error and good visual effect could be obtained by using the truncation radius selected by the proposed strategy.

**Keywords:** electrical capacitance tomography; industrial process tomography; direct image reconstruction; Calderon's method; truncation radius

---

## 1. Introduction

Electrical capacitance tomography (ECT) is an imaging technique to obtain the permittivity distribution in the region of interest from capacitance measurements on the external boundary [1]. The ECT technique has been developed since the 1980s and widely used for process monitoring in industrial applications, e.g., two-phase flow [2–4], fluidized beds [5] and flame monitoring [6,7]. Compared with other process tomographic techniques based on ultrasound, optics, X-ray and gamma-ray, it has several advantages including low cost, rapid response, good portability and radiation-free [8]. A typical ECT system usually consists of three main parts [9]: (1) An ECT sensor, (2) a data acquisition and processing unit, and (3) a computer for image reconstruction.

With the ECT data acquisition system, the capacitances between every two electrodes of the ECT sensor can be measured [10]. Then the permittivity distribution can be reconstructed by using a suitable reconstruction method. Because of the 'soft-field' property, the image reconstruction in

ECT, i.e., recovering permittivity distribution from surface capacitance measurements, is a non-linear inverse problem and is severely ill-posed [11]. Therefore, the performance of image reconstruction method is crucial to ECT.

In the literature, most image reconstruction methods are implemented based on the sensitivity theorem [12], e.g., the one-step methods including linear back projection (LBP) method [13], Tikhonov method, singular value decomposition (SVD) method [14] etc., and the iterative methods including the Landweber method [15], algebraic reconstruction technique (ART) method [16], etc. These methods have been widely used in many applications. However, the intrinsic 'soft-filed' property of ECT in the sensing region is ignored in these cases [17].

Besides the sensitivity-based methods, another one-step method for direct image reconstruction is introduced by Calderon [18] and has been successfully applied in ECT [17,19,20]. Unlike the sensitivity-based methods, the boundary map matrix is used in the Calderon's method and the two-dimensional Fourier transform is applied to reconstruct the perturbations of the permittivity distribution. Calderon's method can provide the gray value at any pixel of the reconstructed image using a direct and independent approach. It is time-saving as no matrix inversion or iterative process is involved [19].

Several parameters need to be determined when using the Calderon's method, i.e., the number of electrodes, integral nodes and truncation radius used in the numerical integral. (1) If more electrodes are used in ECT, the quality of the reconstructed image can be improved, but the hardware burden will increase. The number of electrodes is generally selected as 12 or 16. (2) More integral nodes will increase the accuracy of the integral process but result in a long computing time. In our previous study, the number of integral nodes was selected as 30 [21]. (3) Increasing the truncation radius will increase the high-frequency components in the reconstructed image, but also increase the image noise. In practical applications, the truncation radius is usually set as a constant empirically at present. However, when we apply the Calderon's method to the monitoring of industrial processes, e.g., two-phase flows [22], the selection of initial truncation radius greatly influences the reconstructed results and the recognition of flow patterns. For example, it is appropriate to choose a small truncation radius for laminar flows and a large truncation radius for core-type flows. Since the true distribution is usually unknown, it is difficult to select a suitable truncation radius. Therefore, it is necessary to find a method to adaptively select the truncation radius with respect to different permittivity distributions.

In this paper, the basic fundamental of the Calderon's method is first presented. Then numerical simulations are carried out to analyze the influences of the truncation radius in the Calderon's method on the reconstructed results with four typical permittivity distributions. A method is proposed to adaptively select the truncation radius using the amplitude and phase information of the scattering transform, which can improve the applicability of the Calderon's method.

## 2. Calderon's Method for Direct Image Reconstruction in ECT

For ECT, the governing equation of the sensing field, denoted by $\Omega$, is:

$$\nabla \cdot \varepsilon(z)\nabla\varphi(z) = 0 \tag{1}$$

where $z$ represents the point with coordinates $(x, y)$. $\varepsilon(z)$ and $\varphi(z)$ are the permittivity and electrical potential at $z$. Figure 1 shows a typical ECT sensor.

According to the divergence theorem, the relationship between the integrals of the spatially varying permittivity and voltage and current measurements on the boundary is:

$$\int_{\Omega} v(z)\nabla \cdot \varepsilon(z)\nabla\varphi(z)dz = \int_{\partial\Omega} v(z)\varepsilon(z)\frac{\partial\varphi(z)}{\partial n}dl - \int_{\Omega} \nabla v(z) \cdot \varepsilon(z)\nabla\varphi(z)dz = 0 \tag{2}$$

where $v(z)$ is an arbitrary continuous function in L2-space and $dl$ is the measured arc length on the boundary.

The Dirichlet-to-Neumann (DN) map takes the given voltage distribution on the boundary to the current density distribution, which is also called the voltage-to-current-density map and can be expressed as:

$$\Lambda_\varepsilon : \varphi(z)|_{\partial\Omega} \rightarrow \varepsilon(z)\frac{\partial\varphi(z)}{\partial n}|_{\partial\Omega} \tag{3}$$

where $\Lambda_\varepsilon$ denotes the DN map when $\Omega$ contains $\varepsilon(z)$. Equation (2) can be rewritten as:

$$\int_\Omega \varepsilon(z)\nabla v(z) \cdot \nabla \varphi(z)dz = \int_{\partial\Omega} v(z)\Lambda_\varepsilon[\varphi(z)]dl \tag{4}$$

For a disturbed permittivity $\varepsilon(z) = 1 + \delta\varepsilon(z)$, if this change only exists in $\Omega$ and $\varphi_{1+\delta\varepsilon}(z) \approx \varphi_1(z)$, we obtain:

$$\int_\Omega \delta\varepsilon(z)\nabla v(z) \cdot \nabla \varphi_1(z)dz \approx \int_{\partial\Omega} v(z)(\Lambda_{1+\delta\varepsilon} - \Lambda_1)[\varphi_1(z)]dl \tag{5}$$

In Calderon's method, we construct that $\varphi_1(z) = e^{ikz}$ and $v(z) = e^{\overline{ikz}}$. The right side of Equation (5) is defined as the scattering transform, $t(k)$, and $k = k_1 + ik_2$. Then Equation (5) becomes:

$$t(k) \approx \int_\Omega \delta\varepsilon(z)\nabla v(z) \cdot \nabla \varphi_1(z)dz = -2(k_1^2 + k_2^2)\iint_\Omega \delta\varepsilon(x,y)e^{-i(-2k_1,2k_2)\cdot(x,y)}dxdy \tag{6}$$

here, the scattering transform can be regarded as the two-dimensional Fourier transform of the permittivity distribution.

Then the permittivity change can be obtained by using inverse Fourier transform,

$$\delta\varepsilon(x,y) \approx \frac{1}{2\pi^2}\iint_{R^2} \frac{t(k_1 + ik_2)}{k_1^2 + k_2^2}e^{i(-2k_1,2k_2)\cdot(x,y)}dk_1dk_2 \tag{7}$$

In the polar coordinates of parameters $r$ and $\theta$, Equation (7) can be rewritten as:

$$\delta\varepsilon(x,y) \approx \frac{1}{2\pi^2}\int_0^R \int_{-\pi}^{\pi} \frac{t(re^{i\theta})}{r}e^{i2r(-\cos\theta,\sin\theta)\cdot(x,y)}d\theta dr \tag{8}$$

where $R$ is the truncation radius of the region used in the numerical integral.

The scattering transform has been approximated and linearized to provide practical and simpler implementations in the D-bar method [23]. Since Calderon's method is an approximation of the D-bar method, the approximation method of scattering transform $t(k)$ is also used here. When we consider the noise in practical measurements, $t(k)$ needs to be truncated, which can be expressed by:

$$t(k) \approx \begin{cases} \int_{\partial\Omega} e^{\overline{ikz}}(\Lambda_{1+\delta\varepsilon} - \Lambda_1)e^{ikz}dl, & |k| \le R \\ 0, & |k| > R \end{cases} \tag{9}$$

The truncation radius, $R$, is the upper limit of the frequencies in the inverse Fourier transform process.

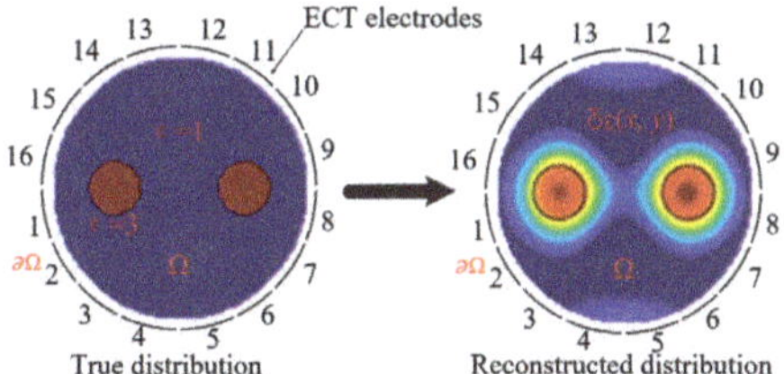


**Figure 1.** A typical electrical capacitance tomography (ECT) sensor.

## 3. Adaptive Selection of Truncation Radius

### 3.1. Influence of the Truncation Radius on Reconstructed Results in Spatial Domain

In Calderon's method, three parameters need to be pre-determined, i.e., the number of electrodes, integral nodes and truncation radius used in the numerical integral. As the integral nodes have been studied and selected as 30 in our previous work, the number of electrodes and truncation radius were analyzed here. In the following, four typical permittivity distributions, i.e., core, multiple objects, annular and stratified, were simulated, as shown in Figure 2a–d. The red color represents the region with high permittivity ($\varepsilon = 3$) and the blue color represents the region with low permittivity ($\varepsilon = 1$). The finite element method (FEM) was used to solve the forward problem, i.e., to obtain the boundary capacitance values from different permittivity distributions.

From Equation (4), increasing the number of electrodes in ECT can improve the computational accuracy of the scattering transform. However, increasing the number of electrodes will increase the hardware burden. To analyze the influences of the number of electrodes on the reconstructed results, the variations in the relative image errors against the number of electrodes for the typical permittivity distributions are shown in Figure 2e. The relative image error between the true permittivity distribution and reconstructed image can be expressed by:

$$err = \frac{\|g_r - g\|}{\|g\|} \tag{10}$$

where $err$ is the relative image error, $g_r$ is the true permittivity distribution and $g$ is the reconstructed permittivity distribution. The truncation radii are selected as 6, 5, 2.4 and 2.4 for the four permittivity distributions, respectively.

Figure 2e shows that the relative image errors decrease significantly if the number of electrodes changes from 8 to 16. When the number of electrodes is larger than 16, the relative image errors only change a little bit. With consideration of the hardware burden and image quality, the number of electrodes is selected as 16 in this paper.

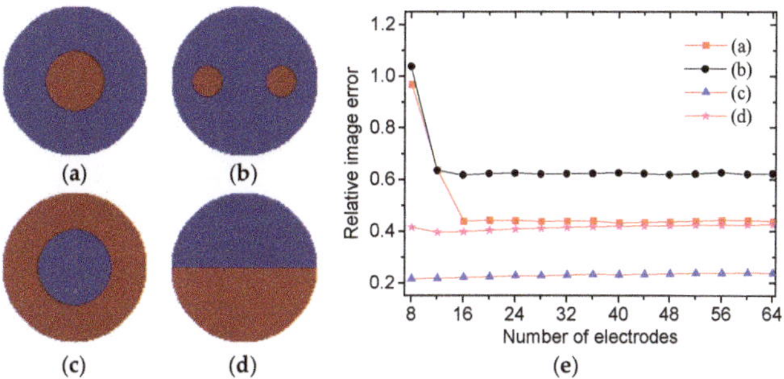


**Figure 2.** Four typical permittivity distributions: (**a**) Core, (**b**) multiple objects, (**c**) annular and (**d**) stratified. (**e**) Relative image errors against number of electrodes for the four distributions.

Then the influences of the truncation radius on the reconstructed results are investigated. The reconstructed images of different permittivity distributions with different truncation radii and without noise are shown in Figures 3a–6a. The black dotted lines represent the edges of the original region with high permittivity. The reconstructed radially symmetric permittivity distributions along the *x*- or *y*-axis are shown in Figures 3b–6b. The reconstructed permittivity is normalized by dividing the maximum permittivity and setting the negative permittivity to 0. The negative permittivities are retained in these figures to show the influence of *R*.

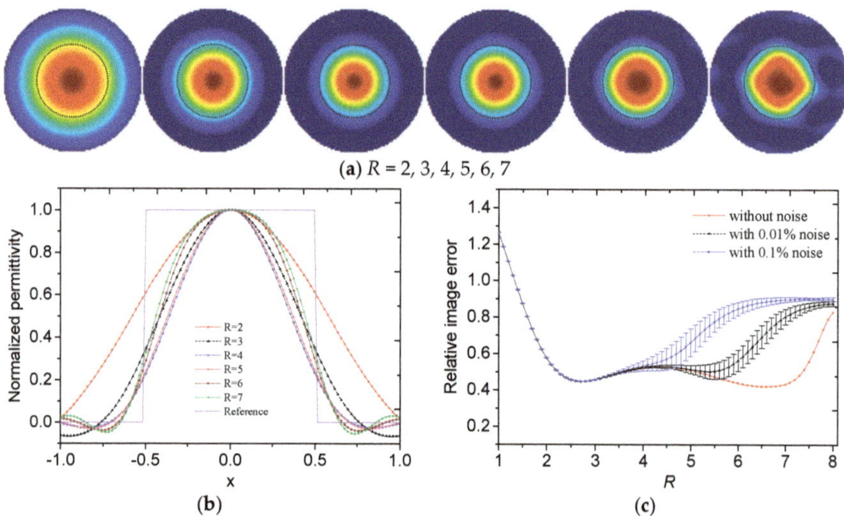

**Figure 3.** (**a**) Reconstructed images, (**b**) reconstructed radially symmetric permittivity distribution along the *x*-axis and (**c**) relative image errors with different truncation radii for core-type distribution.

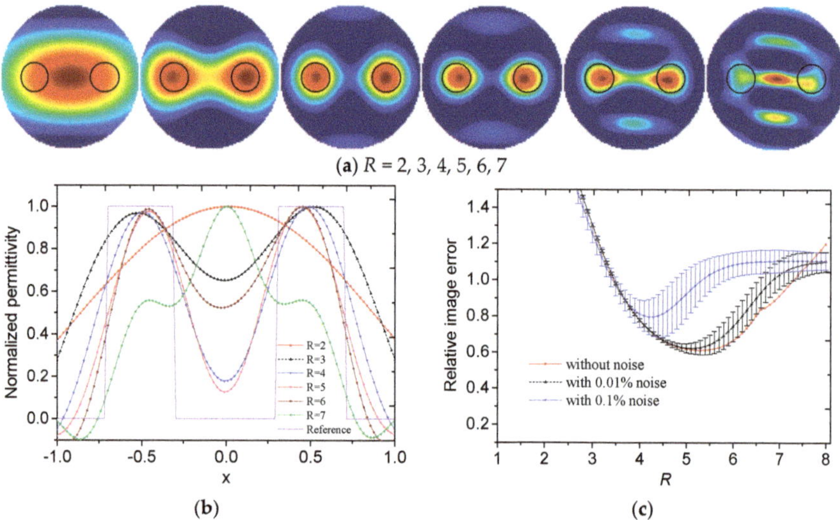

**Figure 4.** (**a**) Reconstructed images, (**b**) reconstructed radially symmetric permittivity distribution along the *x*-axis and (**c**) relative image errors with different truncation radii for multiple-object-type distribution.

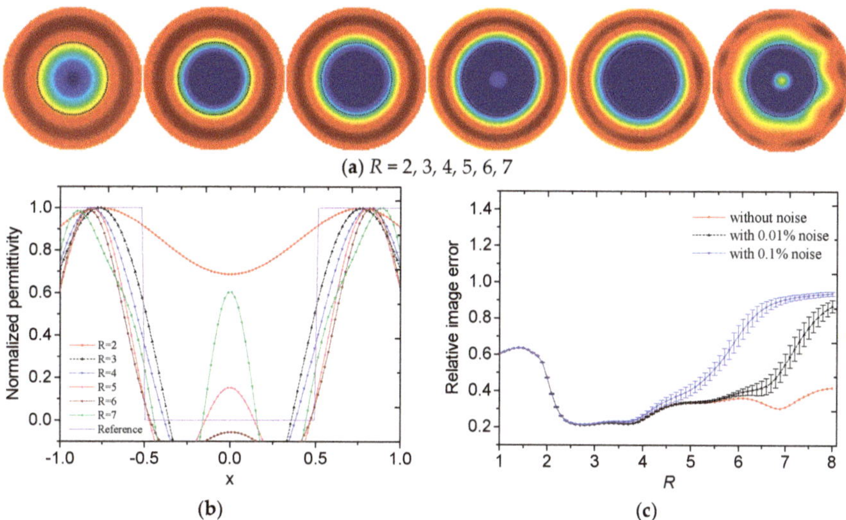

(**a**) $R = 2, 3, 4, 5, 6, 7$

(**b**)

(**c**)

**Figure 5.** (**a**) Reconstructed images, (**b**) reconstructed radially symmetric permittivity distribution along the $x$-axis and (**c**) relative image errors with different truncation radii for annular distribution.

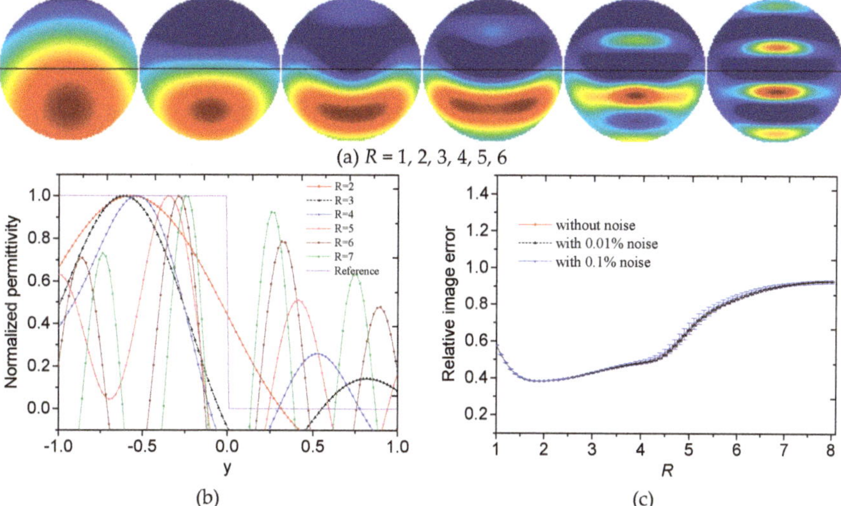

(**a**) $R = 1, 2, 3, 4, 5, 6$

(**b**)

(**c**)

**Figure 6.** (**a**) Reconstructed images, (**b**) reconstructed radially symmetric permittivity distribution along the $y$-axis and (**c**) relative image errors with different truncation radii for stratified distribution.

The relationship between the relative image error and the truncation radius with different permittivity distributions is shown in Figures 3c, 4c, 5c and 6c. The simulated capacitance values are contaminated with noise of relative magnitude 0.01% and 0.1%. The error bars are plotted using the standard deviations, which are calculated by repeating the reconstruction with noise-contaminated capacitance values for 100 times.

In Figures 3a and 4a, a small $R$ means that the reconstructed image contains few high frequency components. In the reconstructed images, the low frequency components contribute to the slowly-changing part, which represents the general overview and contour of the image. A large $R$ means that the reconstructed image contains more high frequency components. More high frequency components can reconstruct sharp changes between adjacent image areas, such as edges and details in the image. However, more high frequency components also give rise to more noise. Therefore, it is important to select an appropriate $R$ to make a trade-off in the Calderon's method.

In Figure 3b, with $R$ increasing, the slope of the normalized permittivity curve increased and the fluctuation of permittivity on the edge increased. In Figure 4b, with $R$ increasing, the centers of the reconstructed circles became closer to the position of $x = 0$. When $R = 4$, the locations of the reconstructed circles were more accurate. By visual inspection, the suitable $R$ was between 4 to 5 for core-type and multiple-object-type distributions.

In Figures 3c and 4c, the truncation radius greatly influenced the relative image error. The truncation radii obtained using the minimum errors without noise for core-type and multiple-object-type distributions were 6.6 and 5.3. When the noise level increased, $R$ with the minimum error decreased.

With consideration of the visual effect of the reconstructed image and the relative image errors, the truncation radius should be selected within a range between 4 and 6 for core-type and multiple-object-type distributions.

Figures 5 and 6 show the reconstructed results of annular and stratified permittivity distributions. Unlike the core-type and multiple-object-type permittivity distributions, the region of high permittivity was on the edge of the sensing region. With consideration of the visual effect of the reconstructed image and the relative image errors, the truncation radius should be selected between 2 and 4, which were lower than that for core-type and multiple-object-type distributions.

The above analysis shows that the truncation radius cannot be set as a constant for ECT in various industrial applications. It is necessary to find a method to adaptively select a suitable $R$.

### 3.2. Influence of the Truncation Radius on Scattering Transform in the Frequency Domain

In this section, $R$ was analyzed in the frequency domain. Equation (9) provides an approximation of $t(k)$ in a grid inside the disc $|k| < R$. $t(k)$ needs to be truncated by selecting a suitable $R$, which acts as a low-pass filter. Since the image reconstruction process, i.e., the calculation of $\delta\varepsilon(x, y)$, can be regarded as the two-dimensional inverse Fourier transform of:

$$F(r, \theta) = \frac{t(re^{i\theta})}{2r} \tag{11}$$

the truncation radius, $R$, plays a role of limiting the frequencies in the inverse Fourier transform process. Increasing $R$ means that the reconstructed images contain more high frequency components and the contours and boundary details of the objects become clearer. However, the noises and artifacts increase with $R$ increasing.

Let us consider the spectral information of the reconstructed image, i.e., amplitude distribution and phase distribution of $F$. The amplitude distributions of $F$ for different permittivity distributions are shown in Figure 7.

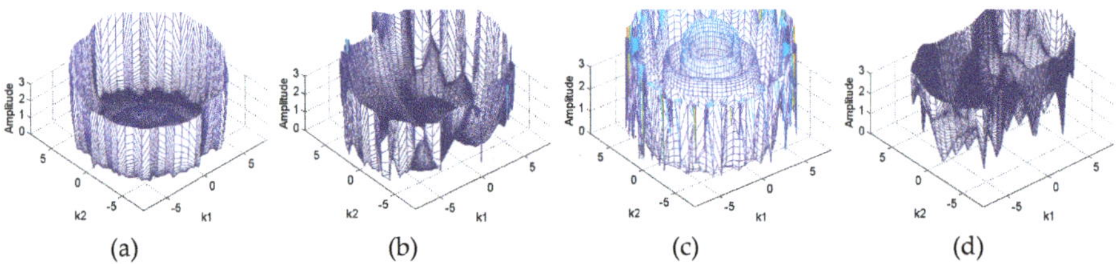

**Figure 7.** Amplitude distributions of $F$ for different permittivity distributions: (**a**) Core, (**b**) multiple objects, (**c**) annular and (**d**) stratified.

To clearly show the variation tendency of the amplitude of *F* in Figure 7, the relationship between the truncation radius and the mean amplitude of *F* is shown in Figure 8. The mean amplitude of *F* is calculated by:

$$\overline{F_a}(r) = \frac{\sum\limits_{\theta=0}^{2\pi} mag(F(r,\theta))}{N} \tag{12}$$

where *mag*() is the amplitude value of a complex number and *N* is the number of $\theta$.

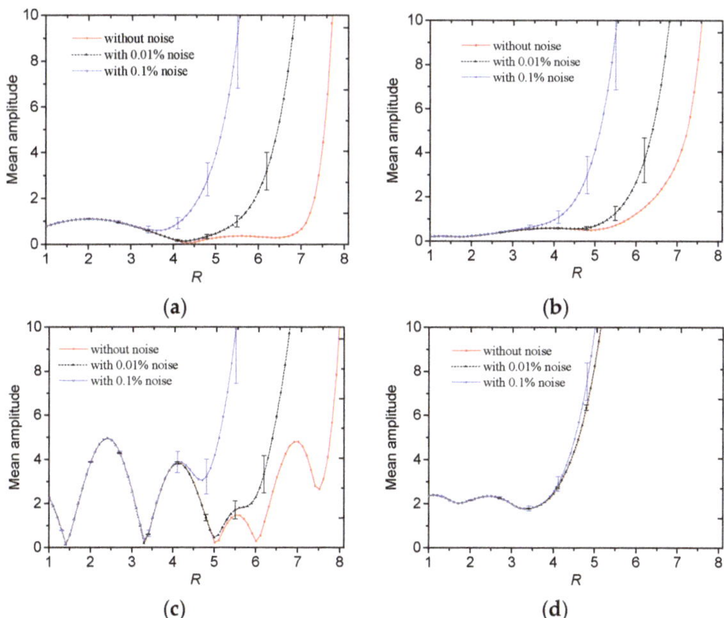

**Figure 8.** Mean amplitude of *F* against the truncation radius for different permittivity distributions: (**a**) Core, (**b**) multiple objects, (**c**) annular and (**d**) stratified.

In Figure 8, the amplitude values without noise increased by a large scale from a fixed value of $\overline{F_a}(r)$, e.g., 5. If the noise level increases, the corresponding truncation radius decreases when $\overline{F_a}(r) = 5$. The values of *F* located in the neighborhood of $R = 0$ were less affected by the noise. Then the maximum *R* could be determined by selecting a fixed mean amplitude value of *F*. For example, if this mean amplitude value was 5, the maximum truncation radii for the four distributions were lower than 8.

Then we considered the absolute phase distributions of *F* for different permittivity distributions, which are shown in Figure 9.

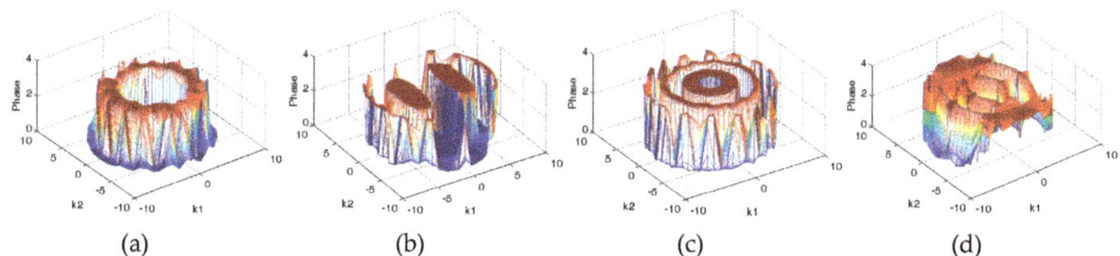

**Figure 9.** Phase distributions of *F* for different permittivity distributions: (**a**) Core, (**b**) multiple objects, (**c**) annular and (**d**) stratified.

Then the mean phase of $F$ is calculated by:

$$\overline{F_p}(r) = \frac{\sum\limits_{\theta=0}^{2\pi} \left| angle(F(r,\theta)) \right|}{N} \tag{13}$$

where *angle*() is the phase value of a complex number. Then the relationship between the truncation radius and mean phase value of $F$ is shown in Figure 10.

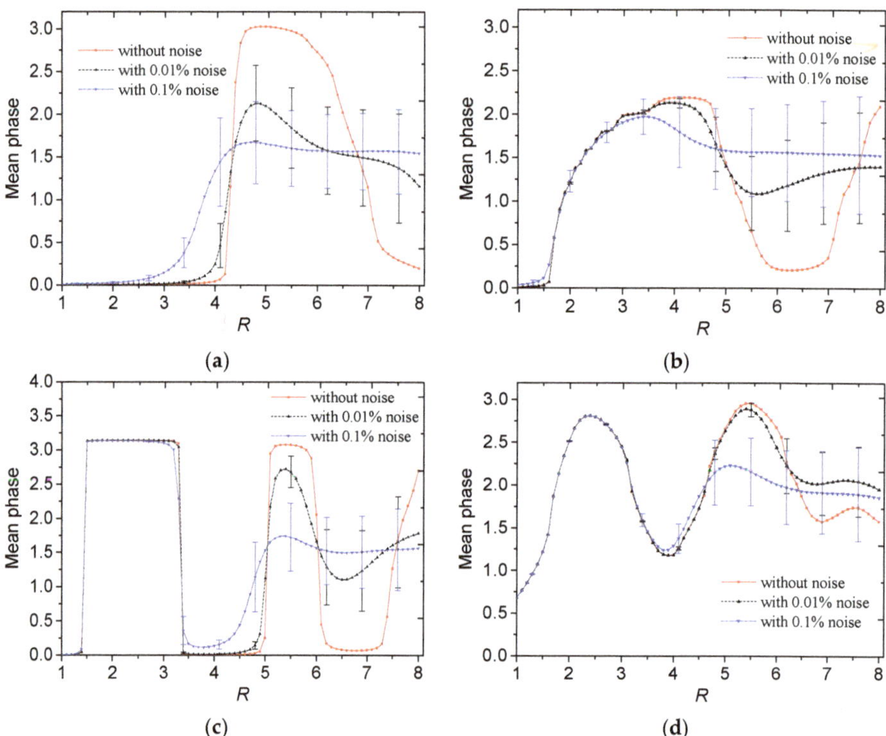

**Figure 10.** Absolute phase value of $F$ against the truncation radius for different permittivity distributions: (**a**) Core, (**b**) multiple objects, (**c**) annular and (**d**) stratified.

In Figure 10a,b, for the core-type and multiple-object-type distributions without noise, there existed one wave crest in the phase curve, and the peak value of phase located in the interval between $R = 4$ to $R = 6$. If the noise level becomes larger, the truncation radius becomes lower to reach the peak value. For the annular and stratified distributions in Figure 10c,d, there existed two wave crests in the phase curve, and the two peak values of phase locate in the intervals between $R = 2$ to 4 and 4 to 6. If the noise level becomes larger, the second wave crest is not obvious but the first wave crest still exists. From the reconstructed results in Section 3, the peak value of the absolute phase curve could be used as an index to select a suitable truncation radius.

### 3.3. Adaptive Selection of the Truncation Radius

$R$ can be theoretically selected as [24]:

$$R = -\frac{1}{10} \log \delta \tag{14}$$

where $\delta$ is the noise level. However, the suitable $R$ for practical applications is generally much larger than the theoretical choice. In electrical impedance tomography for medical applications, $R$ is usually selected between 4 and 6 empirically. Regarding ECT for industrial applications, the permittivity distribution changes dramatically. As described in the previous section, the choice of $R$ in medical EIT is not suitable for annular and stratified distributions any more.

From the above analysis, the amplitude and phase information of $F$ can be combined to determine a suitable truncation radius. Firstly, the scattering transform $t(k)$ is computed from the DN map. Then the truncation radius is selected using the following strategy for adaptive selection of $R$, as shown in Figure 11. When $t(k)$ is obtained, the maximum $R$ can be determined by a fixed mean amplitude of $F$ and the minimum $R$ can be set as a constant, which was selected as 2 in this paper. In this selection range of $R$, the peak value of $\overline{F_p}(r)$ could be detected and used to select the corresponding $R$.

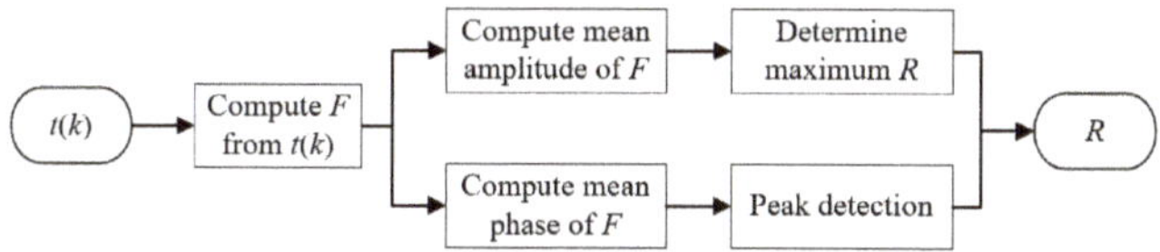


**Figure 11.** Flowchart of the strategy for adaptive selection of truncation radius.

The schematic diagram of Calderon's method with adaptively selected truncation radius is shown in Figure 12. The capacitance values were measured by the ECT data acquisition system and used to construct the DN map. Then $t(k)$ was computed from the DN map and $R$ was selected by the proposed strategy. Finally, the permittivity change could be obtained by using Equation (8) with the selected $R$.

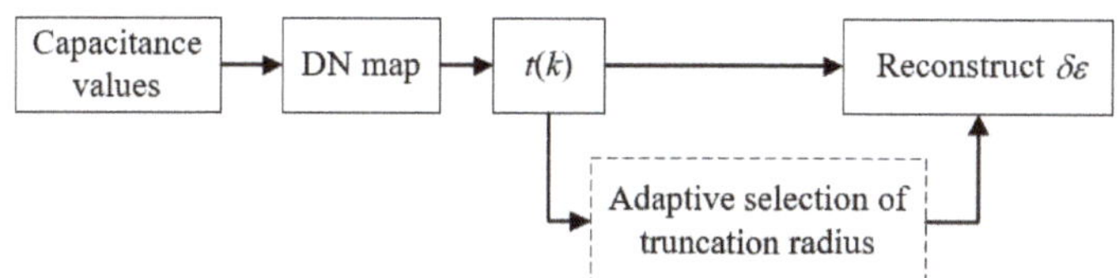


**Figure 12.** Schematic diagram of Calderon's method with adaptively selected truncation radius.

$R$ selected by the proposed method and the corresponding relative errors for the four permittivity distributions are shown in Table 1. The minimum relative errors were obtained using $R$ between 2 to 8 with a step of 0.1 for the four permittivity distributions, and the corresponding $R$ are also shown in Table 1 for comparison. When the simulated capacitance values were contaminated with noises, the truncation radii and relative errors were the mean values from 100 times calculations.

Comparing Figures 3 and 4 and Table 1, the truncation radii selected by the proposed method were in the suitable range between 4 to 6 for the core-type and multiple-object-type distributions, although they brought relatively higher image errors than the actual minimum errors. The truncation radii selected by the proposed method were lower than those with the minimum errors. However, with the selected $R$, fewer artifacts existed in the reconstructed images and the locations of the reconstructed circles were more accurate, which are beneficial for flow pattern recognition. When the noise level increased, the suitable truncation radius decreased.

For the annular and stratified distributions, the truncation radii selected by the proposed method were very close to those obtained according to the minimum errors, which were also in the suitable range between 2 to 4. The noise level had less impact on the selection of truncation radius for the annular and stratified distributions.

**Table 1.** Comparison of the relative errors and truncation radii obtained by the proposed method with those obtained according to the minimum relative error.

|  |  | *R* Selected by the Proposed Method | Relative Error Using the Selected *R* | *R* with Minimum Relative Error | Minimum Relative Error |
|---|---|---|---|---|---|
| Core | without noise | 5 | 50% | 6.6 | 42% |
|  | 0.01% noise | 4.9 | 52% | 3.1 | 45% |
|  | 0.1% noise | 4.7 | 59% | 2.8 | 45% |
| Multiple objects | without noise | 4.1 | 75% | 5.3 | 61% |
|  | 0.01% noise | 4.1 | 80% | 5.2 | 62% |
|  | 0.1% noise | 3.8 | 96% | 4.3 | 78% |
| Annular | without noise | 2.4 | 23% | 2.8 | 21% |
|  | 0.01% noise | 2.4 | 23% | 2.8 | 21% |
|  | 0.1% noise | 2.2 | 30% | 2.7 | 22% |
| Stratified | without noise | 2.4 | 39% | 1.9 | 38% |
|  | 0.01% noise | 2.4 | 39% | 1.9 | 38% |
|  | 0.1% noise | 2.4 | 39% | 1.9 | 38% |

## 4. Experimental Results

In this section, experiments were carried out to evaluate the performance of the proposed method. The number of electrodes in the ECT sensor was selected as 16. The inner diameter of the sensor was 100 mm and the length of the electrode was 120 mm, as shown in Figure 13a. The ECT system developed by Beihang University was used to measure the capacitance values between every two electrodes of the ECT sensor, as shown in Figure 13b. The signal-to-noise ratio (SNR) of the capacitance measurement with the ECT system was larger than 50 dB for the 16-electrodes sensor.

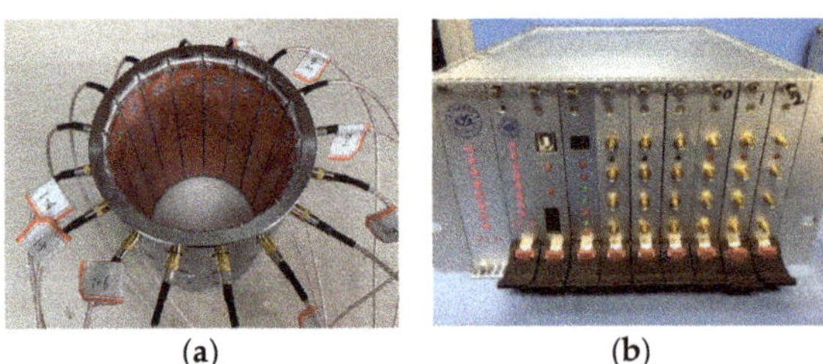

(a)                              (b)

**Figure 13.** (a) ECT sensor and (b) ECT data acquisition system.

Similar to the simulation process, several typical permittivity distributions were constructed. Cylindrical nylon rods were used to construct the core-type and multiple-object-type distributions, i.e., D1 to D3. Several 3D-printed models and a certain amount of fine sand were used to construct the square-shaped, V-shaped, stratified and annular distributions, i.e., D4 to D9. The permittivities of the plastic rods and sand were about 3.

The permittivity distributions were reconstructed using the measured capacitance values and Calderon's method with the proposed strategy for truncation radius selection. The truncation radius selected by the proposed method was denoted as $R\_s$. The photos, cross-sectional views and reconstructed images of the core-type and multiple-object-type distributions are shown in Figure 14.

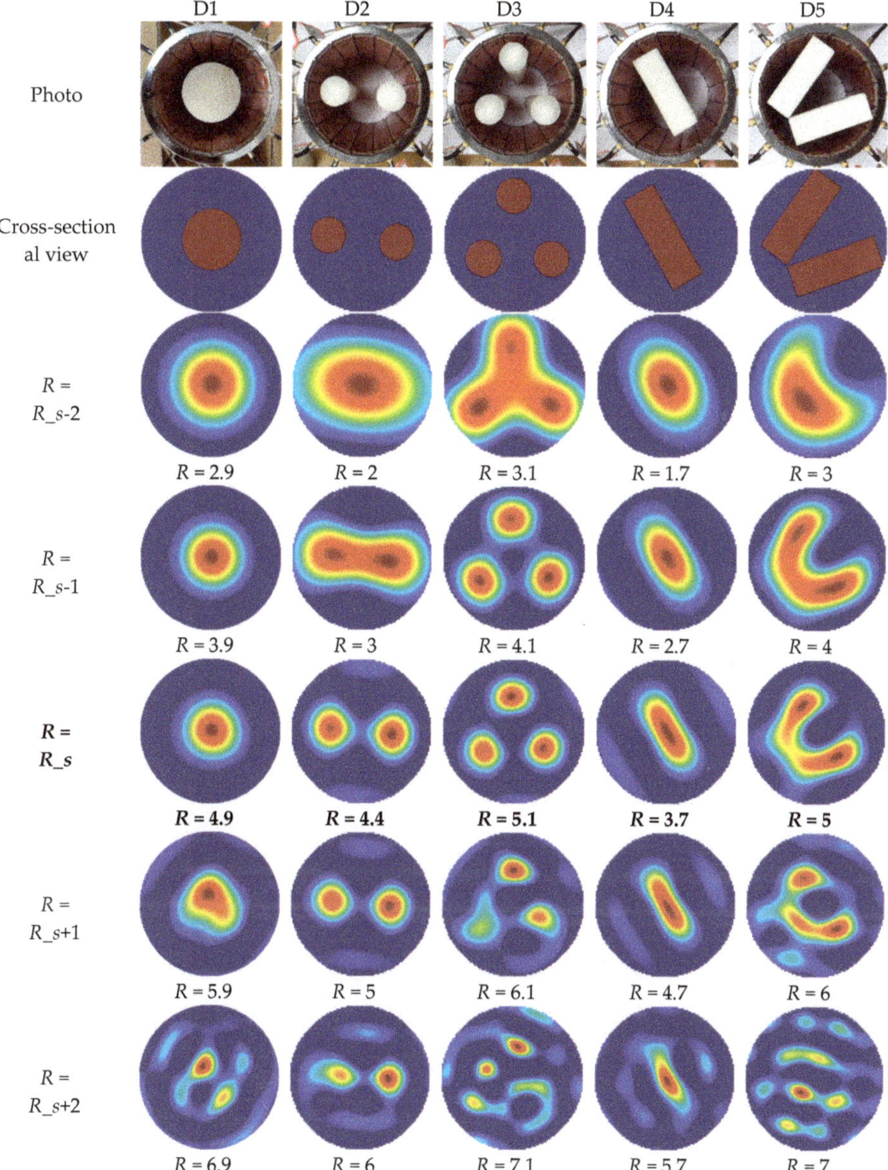

**Figure 14.** Photos, cross-sectional views and reconstructed images of different permittivity distributions from D1 to D5 with different truncation radii.

The selected truncation radii were 4.9, 4.4, 5.1, 3.7 and 5 for the distributions D1 to D5. From the visual point of view, when $R$ was smaller than $R\_s$, the reconstructed high-permittivity regions were excessively large. The low frequency components provided more slow-changing parts. When $R$ was larger than $R\_s$, more artifacts appeared, which meant that the high-frequency noise was introduced with $R$ increasing. The high frequency components brought sharp changes between adjacent image areas.

The photos, cross-sectional views and reconstructed images of the stratified and annular distributions are shown in Figure 15.

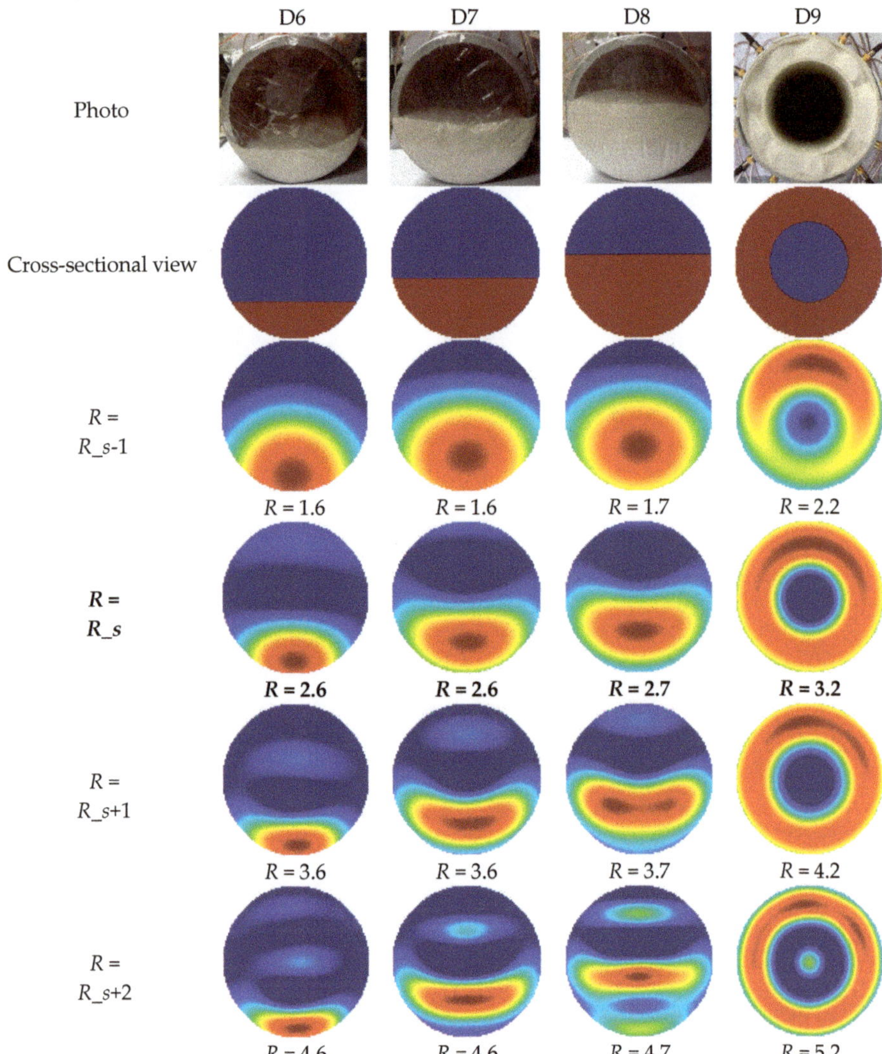

**Figure 15.** Photos, cross-sectional views and reconstructed images of different permittivity distributions from D6 to D9 with different truncation radii.

For the stratified and annular distributions D6 to D9, the selected truncation radii were 2.6, 2.6, 2.7 and 3.2, which were much smaller than those selected for distributions D1 to D5. By visual inspection, the selected truncation radii were also suitable for flow pattern recognition. It should be noted that the reconstructed images were visually good as well when $R$ was a little larger than $R\_s$, e.g., $R\_s + 1$.

To evaluate the reconstructed results with the selected truncation radius quantitatively, the relationship between the relative image errors and the truncation radius are shown in Figure 16.

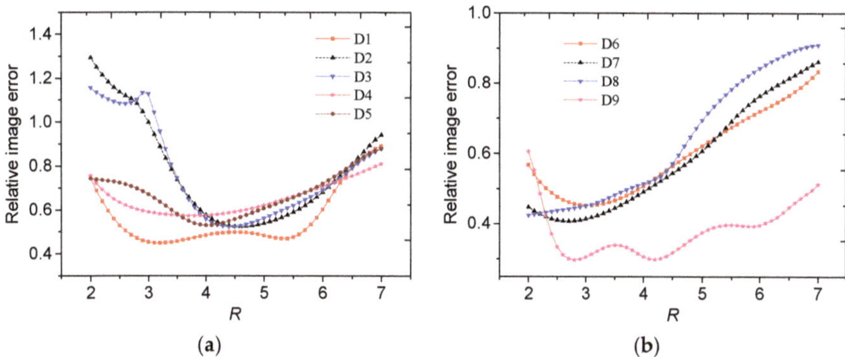

**Figure 16.** Relative image errors against the truncation radius for different permittivity distributions from (**a**) D1 to D5 and (**b**) D6 to D9.

From Figure 16, the relative errors using the selected $R$ could be obtained, as shown in Table 2. Then we could obtain the minimum relative image errors for the permittivity distributions D1 to D9, and the corresponding truncation radii, which are also shown in Table 2 for comparison.

**Table 2.** Comparison of the relative errors and truncation radii obtained by the proposed method with those obtained according to the minimum relative error.

|  | $R$ Selected by the Proposed Method | Relative Errors Using the Selected $R$ | $R$ with Minimum Relative Errors | Minimum Relative Errors |
|---|---|---|---|---|
| D1 | 4.9 | 48% | 3.2 | 45% |
| D2 | 4.4 | 53% | 4.6 | 52% |
| D3 | 5.1 | 57% | 4.4 | 52% |
| D4 | 3.7 | 57% | 3.8 | 57% |
| D5 | 5 | 60% | 4 | 52% |
| D6 | 2.6 | 46% | 3 | 45% |
| D7 | 2.6 | 41% | 2.7 | 41% |
| D8 | 2.7 | 44% | 2 | 42% |
| D9 | 3.2 | 32% | 2.8 | 30% |

For all the distributions except for D3 and D5 in Table 2, the differences between the relative errors using the proposed method and the minimum relative errors were lower than 3%. For the distributions D3 and D5, the images reconstructed by the proposed method could reflect better contours of the objects, compared to the images with the minimum relative errors. There existed some contradictions between the visual effect of the reconstructed image and the relative image errors. The reason may be that there existed geometrical errors in the true distributions used for error calculation. With consideration of a tradeoff between the relative image error and visual effect, the truncation radii selected by the proposed method were in proper selection range.

## 5. Conclusions

A strategy for adaptive selection of the truncation radius was proposed by using the amplitude and phase information of the scattering transform. The following conclusions were drawn.

(1) A small truncation radius means that the reconstructed image contains few high frequency components. The low frequency components contribute to the slowly-changing part in the image. If the truncation radius becomes larger, the images contain more high frequency components and the contours and boundary details of the objects become clearer. However, the noises and artifacts increase;

(2) The permittivity distributions can be divided into two categories, i.e., the core-type and the annular distributions. For the core-type permittivity distributions, i.e., the region of high permittivity is enclosed by a continuous region of low permittivity, the truncation radius should be between 4 and 6 to obtain a better image quality. For annular and stratified distributions, i.e., the region of high permittivity contacts the boundary of the sensing region, the suitable truncation radius should be between 2 and 4, which is much lower than that for core-type distributions. For a higher noise level, a smaller truncation radius should be selected;

(3) The amplitude and phase information of the scattering transform are combined to determine a suitable truncation radius. The experimental results show that small relative image error and good visual effect can be obtained by using the truncation radius selected by the proposed method.

The main contribution of this paper was that a suitable truncation radius in Calderon's method for direct image reconstruction could be adaptively obtained by using the information of scattering transform, without the need of the true distribution pattern. The scattering transform can be directly computed from the measured capacitance values, which makes the proposed method easy to implement and useful in industrial applications.

**Author Contributions:** Conceptualization, S.S.; methodology, S.S. and Z.C.; software, S.S.; formal analysis, L.X.; writing—original draft preparation, S.S.; writing—review and editing, S.S., L.X., J.S. and W.T.; project administration, L.X.; funding acquisition, S.S., L.X. and Z.C.

**Funding:** This research was funded by the National Natural Science Foundation of China, grant number 61801013, 61620106004 and 61522102, the National Postdoctoral Program for Innovative Talents, grant number BX201700021 and the China Postdoctoral Science Foundation, grant number 2018M640039.

**Conflicts of Interest:** The authors declare no conflict of interest.

## References


1. Yang, W.Q. Hardware design of electrical capacitance tomography systems. *Meas. Sci. Technol.* **1996**, *7*, 225–232. [CrossRef]
2. Meribout, M.; Saied, I.M. Real-Time Two-Dimensional Imaging of Solid Contaminants in Gas Pipelines Using an Electrical Capacitance Tomography System. *IEEE Trans. Ind. Electron.* **2017**, *64*, 3989–3996. [CrossRef]
3. Sun, S.; Zhang, W.; Sun, J.; Cao, Z.; Xu, L.; Yan, Y. Real-Time Imaging and Holdup Measurement of Carbon Dioxide Under CCS Conditions Using Electrical Capacitance Tomography. *IEEE Sens. J.* **2018**, *18*, 7551–7559. [CrossRef]
4. Zainal-Mokhtar, K.; Mohamad-Saleh, J. An Oil Fraction Neural Sensor Developed Using Electrical Capacitance Tomography Sensor Data. *Sensors* **2013**, *13*, 11385–11406. [CrossRef] [PubMed]
5. Chandrasekera, T.C.; Li, Y.; Moody, D.; Schnellmann, M.A.; Dennis, J.S.; Holland, D.J. Measurement of bubble sizes in fluidised beds using electrical capacitance tomography. *Chem. Eng. Sci.* **2015**, *126*, 679–687. [CrossRef]
6. Hu, D.; Cao, Z.; Sun, S.; Sun, J.; Xu, L. Dual-Modality Electrical Tomography for Flame Monitoring. *IEEE Sens. J.* **2018**, *18*, 8847–8854. [CrossRef]
7. Sun, S.; Cao, Z.; Huang, A.; Xu, L.; Yang, W. A High-Speed Digital Electrical Capacitance Tomography System Combining Digital Recursive Demodulation and Parallel Capacitance Measurement. *IEEE Sens. J.* **2017**, *17*, 6690–6698. [CrossRef]
8. Olmos, A.; Primicia, J.; Marron, J. Influence of Shielding Arrangement on ECT Sensors. *Sensors* **2006**, *6*, 1118–1127. [CrossRef]
9. Cui, Z.; Wang, H.; Chen, Z.; Xu, Y.; Yang, W. A high-performance digital system for electrical capacitance tomography. *Meas. Sci. Technol.* **2011**, *22*, 055503. [CrossRef]
10. Xu, L.; Sun, S.; Cao, Z.; Yang, W. Performance analysis of a digital capacitance measuring circuit. *Rev. Sci. Instrum.* **2015**, *86*, 054703. [CrossRef]
11. Wang, F.; Marashdeh, Q.; Fan, L.-S.; Warsito, W. Electrical Capacitance Volume Tomography: Design and Applications. *Sensors* **2010**, *10*, 1890–1917. [CrossRef]
12. Kim, Y.S.; Lee, S.H.; Ijaz, U.Z.; Kim, K.Y.; Choi, B.Y. Sensitivity map generation in electrical capacitance tomography using mixed normalization models. *Meas. Sci. Technol.* **2007**, *18*, 2092–2102. [CrossRef]

13. Yang, W.Q.; Peng, L.H. Image reconstruction algorithms for electrical capacitance tomography. *Meas. Sci. Technol.* **2003**, *14*, R1–R13. [CrossRef]

14. Cui, Z.; Wang, Q.; Xue, Q.; Fan, W.; Zhang, L.; Cao, Z.; Sun, B.; Wang, H.; Yang, W. A review on image reconstruction algorithms for electrical capacitance/resistance tomography. *Sens. Rev.* **2016**, *36*, 429–445. [CrossRef]

15. Tian, W.; Sun, J.; Ramli, M.F.; Yang, W. Adaptive Selection of Relaxation Factor in Landweber Iterative Algorithm. *IEEE Sens. J.* **2017**, *17*, 7029–7042. [CrossRef]

16. Lei, J.; Liu, S.; Wang, X.; Liu, Q. An Image Reconstruction Algorithm for Electrical Capacitance Tomography Based on Robust Principle Component Analysis. *Sensors* **2013**, *13*, 2076–2092. [CrossRef]

17. Cao, Z.; Ji, L.; Xu, L. Iterative Reconstruction Algorithm for Electrical Capacitance Tomography Based on Calderon's Method. *IEEE Sens. J.* **2018**, *18*, 8450–8462. [CrossRef]

18. Calderón, A.P. On an inverse boundary value problem. *Comput. Appl. Math.* **2006**, *25*, 133–138. [CrossRef]

19. Cao, Z.; Xu, L.; Wang, H. Image reconstruction technique of electrical capacitance tomography for low-contrast dielectrics using Calderon's method. *Meas. Sci. Technol.* **2009**, *20*, 104027. [CrossRef]

20. Cao, Z.; Xu, L.; Fan, W.; Wang, H. Electrical Capacitance Tomography for Sensors of Square Cross Sections Using Calderon's Method. *IEEE Trans. Instrum. Meas.* **2011**, *60*, 900–907. [CrossRef]

21. Sun, S.; Xu, L.; Cao, Z.; Sun, J. Influence of the Integral Parameters in Calderon Method on Image Quality for Electrical Capacitance Tomography. Available online: https://www.isipt.org/world-congress/9/29273.html (accessed on 27 March 2019).

22. Liang, S.; Ye, J.; Wang, H.; Wu, M.; Yang, W. Influence of the internal wall thickness of electrical capacitance tomography sensors on image quality. *Meas. Sci. Technol.* **2018**, *29*, 035401. [CrossRef]

23. Siltanen, S.; Mueller, J.; Isaacson, D. An implementation of the reconstruction algorithm of A Nachman for the 2D inverse conductivity problem. *Inverse Probl.* **2000**, *16*, 681. [CrossRef]

24. Knudsen, K.; Lassas, M.; Mueller, J.; Siltanen, S. Regularized D-bar method for the inverse conductivity problem. *Inverse Probl. Imaging* **2009**, *3*, 599–624. [CrossRef]

*Article*

# A Lagrange-Newton Method for EIT/UT Dual-Modality Image Reconstruction [†]

**Guanghui Liang [1], Shangjie Ren [1,*], Shu Zhao [2,*] and Feng Dong [1]**

[1]   Tianjin Key Laboratory of Process Measurement and Control, School of Electrical and Information Engineering, Tianjin University, Tianjin 300072, China; ghliang@tju.edu.cn (G.L.); fdong@tju.edu.cn (F.D.)

[2]   Institute of Biomedical Engineering, CAMS & PUMC (Chinese Academy of Medical Sciences and Peking Union Medical College), Tianjin 300192, China

*   Correspondence: rensjie@tju.edu.cn (S.R.); zhaos@bme.cams.cn (S.Z.)

†   This paper is an extended version of our paper published in Liang, G.; Ren, S.; Dong, F. An Inclusion Boundary Reconstruction Method Using Electrical Impedance and Ultrasound Reflection Dual-Modality Tomography. In Proceedings of the 9th World Congress on Industrial Process Tomography, Bath, UK, 2–6 September 2018.

Received: 30 March 2019; Accepted: 23 April 2019; Published: 26 April 2019

**Abstract:** An image reconstruction method is proposed based on Lagrange-Newton method for electrical impedance tomography (EIT) and ultrasound tomography (UT) dual-modality imaging. Since the change in conductivity distribution is usually accompanied with the change in acoustic impedance distribution, the reconstruction targets of EIT and UT are unified to the conductivity difference using the same mesh model. Some background medium distribution information obtained from ultrasound transmission and reflection measurements can be used to construct a hard constraint about the conductivity difference distribution. Then, the EIT/UT dual-modality inverse problem is constructed by an equality constraint equation, and the Lagrange multiplier method combining Newton-Raphson iteration is used to solve the EIT/UT dual-modality inverse problem. The numerical and experimental results show that the proposed dual-modality image reconstruction method has a better performance than the single-modality EIT method and is more robust to the measurement noise.

**Keywords:** electrical impedance tomography; ultrasound tomography; dual-modality imaging; lagrange-newton method

---

## 1. Introduction

Electrical impedance tomography (EIT) is a non-invasive technique to reconstruct the distribution of the media within a closed vessel based on the differences of their conductivities. Due to its advantages of high speed, non-radiation, and low cost, EIT has been widely researched and used in many fields, such as, industrial process imaging [1,2], medical imaging [3,4], geological exploration [5,6], structural health monitoring [7,8], and nondestructive evaluation [9,10]. However, as a soft-field imaging technique, EIT usually suffers from the low spatial resolution and imaging accuracy for its inherent nonlinearity and ill-posedness [11]. It has hindered the further application of EIT in many fields, especially for medical imaging, where a higher imaging accuracy is usually expected [12]. Thus, the study of improving the EIT spatial resolution and imaging accuracy can always attract the attention of the researchers.

Many researchers have devoted to the study of improving the spatial resolution and imaging accuracy of EIT and many methods have been developed. For the nonlinearity of EIT, the directly linearization [13], iterative linearization [14], and directly nonlinear methods [15] have been proposed to formulate the EIT inverse problem. For the ill-posedness of EIT, the regularization methods, such as, Tikhonov regularization [14], Total Variation (TV) regularization [16], and sparse regularization [17], have been proposed to constrain the solution space of EIT inverse problem by introducing the prior information about the targets to be reconstructed. Thus, the construction of the prior information about the target is important for the EIT inverse problem. The prior information is usually predetermined and added into the EIT inverse problem. A commonly used method for constructing the prior information is to use the characteristics of the target itself. For example, in the imaging problems of small targets, the sparse prior constraints are commonly used [17]. For specific target imaging problems, the statistical modeling methods can be used to extract the similarity information of the targets, such as structure or shape, and construct the statistical prior information about the targets. For example, in EIT lung function monitoring, the statistical shape prior extracted from computed tomography (CT) images can be used to alleviate the ill-posedness of EIT inverse problem and improve the EIT imaging accuracy and measurement noise robustness [18].

Although the spatial resolution and imaging accuracy have been improved to some extent by the regularization constraint from the prior information in EIT inverse problem, the acquisition of the prior information about the target usually falls into the problems of lack of the data sources and feature extraction difficulties. In addition, the individual difference of the imaging targets can also invalidate the prior information in some practical applications. Thus, relying on a single-modality imaging technique is difficult to substantially improve the EIT spatial resolution and imaging accuracy because the available prior information is usually limited. Considering the potential information complementarity between different imaging modalities, the dual-modality imaging techniques have attracted more and more attention in the past decade. For example, the positron emission computed tomography (PET)/CT [19], magnetic resonance imaging (MRI)/CT [20] and Ultrasound/MRI [21] have been successfully used in medical diagnosis. The ultrasound tomography (UT)/electrical capacitance tomography (ECT) [22], CT/ECT [23], Gamma Densitometry Tomography (GDT)/EIT [24] have been widely used in multiphase flow process parameters detection. Thus, the dual-modality imaging technique is promising to improve the EIT imaging accuracy by adding the prior information from the other imaging modality.

UT is also a non-invasive technique to reconstruct the media distribution based on the differences of their acoustic impedances and has been widely used in industrial imaging [25] and medical imaging [26]. Different from the EIT, UT is a 'hard-field' imaging technique and the spatial resolution and imaging accuracy of UT is mainly depended on the amount of measured data available. The more measured data means that more ultrasonic transducers are needed in UT [27]. However, the number of the ultrasonic transducer is usually not enough compared to the need for the imaging accuracy in practical applications. Fortunately, UT can obtain some accurate local media distribution information even the number of the ultrasonic transducer is limited [28,29]. Thus, the prior information about the media distribution from UT is promising to improve the imaging accuracy of EIT.

Some studies have been reported about the EIT/UT dual-modality imaging in the past decade. In 2008, Steiner et al. proposed an EIT/ultrasound reflection tomography (URT) dual-modality imaging method for small object detection in biomedical applications, where the measurement information from URT was used to construct the weighted regularization matrix in EIT inverse problem [30]. In 2010, Wan et al. presented a transrectal ultrasound coupled EIT imaging method for transrectal cancer detection, where the contour of the target was captured by ultrasound measurement, and then used to guide the EIT image reconstruction [31]. In 2015, Teniou et al. proposed an electrical resistance tomography (ERT)/URT dual-modality imaging method to detect the small lesions in biological soft tissue, where the measurements from URT was used as hard constraints during the EIT image reconstruction process [32]. In 2017, Pusppanathan et al. proposed an ECT/ultrasound transmission tomography (UTT) dual-modality imaging method for multiphase flow imaging, where the images from the ECT and UTT are fused by the pixel-based fuzzy logic method, and then used to distinguish oil/gas/water three-phase media [33]. In 2018, Liang et al. proposed a shape-based EIT/URT dual-modality imaging method, where some accurate boundaries information about the inclusions detected from URT are used to improve the EIT inclusion boundary reconstruction accuracy [34].

In addition, the emerging impediography can improve the EIT image accuracy by coupling the mechanical effects driven from ultrasound waves [35–37]. Although the EIT/UT dual-modality imaging has made some progresses, the way of information fusion between EIT and UT is still an open issue in EIT/UT dual-modality imaging researches.

In this paper, a dual-modality image reconstruction method combining EIT and UT is proposed. Some prior information about the background medium distribution is determined by the ultrasound reflection and transmission measurements, and then used as a hard constraint added into the EIT inverse problem through an equality constraint equation. Then, the ultrasound constraint EIT image reconstruction problem is reformed by the Lagrange multiplier method and iteratively solved with Newton-Raphson method. To test the performance of the proposed dual-modality imaging method, the numerical and experimental tests are conducted.

## 2. Mathematical Model of EIT and UT

### 2.1. EIT Mathematical Model

EIT mathematical model is derived from Maxwell's electromagnetic theory under the condition of low frequencies. In the quasi-static approximation, the electrical fields can be described in terms of a scalar voltage potential $u$ satisfying the following equation:

$$\nabla \cdot \sigma \nabla u = 0 \quad \text{in} \quad \Omega, \tag{1}$$

where $\sigma$ represents the conductivity in the imaging domain $\Omega$.

Considering the contact impedance between the background medium and electrodes, the complete electrode model (CEM) is used to define the boundary conditions [38]:

$$\begin{cases} \sigma \nabla u \cdot \vec{n} = 0, & \text{on} \quad \partial\Omega \setminus \cup_{l=1}^{L} e_l \\ \int_{e_l} \sigma \nabla u \vec{n} = I_l, & \\ u + z_l \sigma \nabla u \cdot \vec{n} = U_l, & \text{on} \quad e_l \end{cases} \tag{2}$$

where $\partial\Omega$ is the boundary of the imaging domain $\Omega$, $e_l \subset \partial\Omega$ represents the location covered by the $l$-th electrode, $L$ is the total number of the electrodes, $\vec{n}$ is the outward normal vector on $\partial\Omega$, $z_l$ is the contact impedance between the $l$-th electrode and background medium, $I_l$ and $U_l$ are the electric current and electric potential at the $l$-th electrode, respectively. Subject to the law of the conservation of charge, the electric current and voltage in the electrodes satisfy the following equation:

$$\sum_{l=1}^{L} I_l = 0 \quad \text{and} \quad \sum_{l=1}^{L} U_l = 0, \tag{3}$$

Thus, with the definition of EIT mathematical model and the corresponding boundary conditions, the electric potential, conductivity, electric current, and contact impedance can be formulated as the following nonlinear equation:

$$U = u(\sigma, z, I), \tag{4}$$

Considering a small change on the conductivity, the corresponding change on the boundary voltage can be linearly represented using Equation (4) as follows:

$$\Delta U = u(\sigma, z, I) - u(\sigma_0, z, I) = \left. \frac{du(\sigma, z, I)}{d\sigma} \right|_{\sigma=\sigma_0} (\sigma - \sigma_0) + O\left((\sigma - \sigma_0)^2\right)$$

$$= u'(\sigma_0, z, I)\Delta\sigma + O\left((\Delta\sigma)^2\right) \tag{5}$$

If the change of conductivity $\Delta\sigma$ is small enough, the high order terms $O\left((\Delta\sigma)^2\right)$ can be neglected, and Equation (5) will be simplified as follows:

$$\Delta U = u'(\sigma_0, z, I)\Delta\sigma, \tag{6}$$

further, the Equation (6) is usually discretized as follows:

$$y = Jx, \tag{7}$$

where $y \in R^{M\times 1}$ represents the boundary voltage difference, $x \in R^{N\times 1}$ represents the unknown conductivity difference, $J \in R^{M\times N}$ represents the Jacobian matrix, which can be calculated using the Geselowitz's sensitivity theorem [39]:

$$J_{m,n} = -\int_{\Omega_n} \frac{\nabla u\left(I^i\right)}{I^i} \frac{\nabla u\left(I^j\right)}{I^j} dS, \tag{8}$$

where $J_{m,n}$ is the element at the $m$-th row and $n$ column of the Jacobian matrix, $u\left(I^i\right)$ is the electric potential for the $i$-th electrode pair with electric current $I^i$, $u\left(I^j\right)$ is the electric potential for the $j$-th electrode pair with electric current $I^j$, $\Omega_n$ is the region covered by $n$-th pixel.

For a given conductivity difference $x$, the boundary voltage $y$ can be determined by the Equations (1)–(3). This process is usually mentioned as the EIT forward problem, where the Boundary Element Method (BEM) solver developed by Ren et al. is used to calculate the EIT forward problem and obtain the sensitivity matrix by the perturbation method [40,41]. In contrast, given the boundary voltage $y$, the EIT inverse problem is to calculate the conductivity difference $x$ by the Equation (7). Since the sensitivity matrix $J$ is usually singular, the unknown conductivity difference $x$ cannot be directly obtained from the Equation (7) in EIT inverse problem. The regularization methods are commonly used to solve it, which will be introduced in the latter section.

### 2.2. UT Mathematical Model

When the influence of medium heterogeneity on acoustic wave propagation is not considered, the propagation of ultrasonic wave in medium can be approximated as linear propagation model. This model is usually referred as the geometrical/ray acoustic model and widely used in UT image reconstruction. Thus, the mathematical model of UT can be referred to the geometrical optics or ray optics model, and the most commonly considered modes of the ultrasonic wave propagation are transmission and reflection modes, as shown in Figure 1. In addition, the piston transducer is used in the paper, where the energy of the ultrasonic wave is mainly concentrated in a narrow main lobe area [42]. The power of the transmission and reflection wave is mainly affected by the acoustic impedance of the media where the ultrasonic wave pass through, and it can be described as follows:

$$\alpha_R = \left(\frac{P_r}{P_i}\right)^2 = \left(\frac{Z_2 - Z_1}{Z_2 + Z_1}\right)^2, \tag{9}$$

$$\alpha_T = \left(\frac{P_t}{P_i}\right)^2 = \left(\frac{2Z_2}{Z_2 + Z_1}\right)^2, \tag{10}$$

where $P_i$, $P_r$, $P_t$ represent incident, reflection, and transmission sound pressures, respectively. $\alpha_R$ and $\alpha_T$ represent the reflection and transmission coefficients, respectively. $Z_1$ and $Z_2$ are the acoustic impedance of the media and equal to the product of medium density $\rho$ and sound velocity $c$.

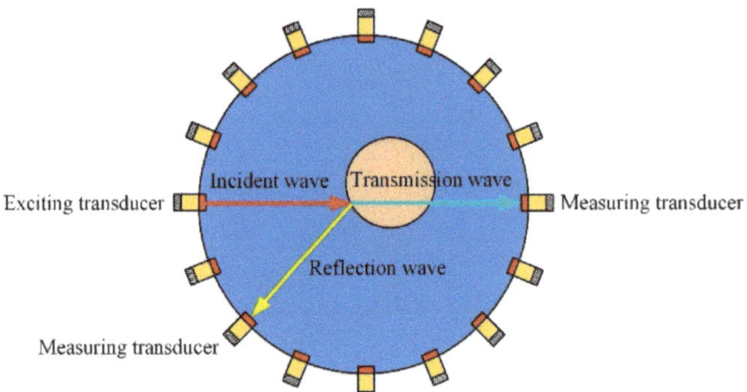


**Figure 1.** Ultrasound transmission and reflection physical model.

### 2.2.1. Transmission Mode Mathematical Model

Under the assumption that the ultrasonic wave is propagating along the straight line, the mathematical model of transmission mode UT is similar to the x-ray CT. The ultrasonic waves sending from the exciting transducer will pass through the media and arrive at the measuring transducers. Consider that the acoustic attenuation effect of the medium, the transmission mode UT usually utilizes the energy attenuation of the ultrasonic waves to invert the medium distribution in the ultrasonic wave propagation paths. Inspired from the Equation (10), we reformulate the sending acoustic energy $I_e$ from the exciting transducer and the receiving acoustic energy $I_m$ from the measuring transducer as the following equation:

$$\iint_{\Omega_k} \mu \mathrm{d}S = -\ln\frac{I_m}{I_e},\tag{11}$$

where $\mu$ is the medium's attenuation coefficient, $\Omega_k$ is the region where the $k$-th ultrasonic wave pass through.

Consider a scenario that medium located in the region $\Omega_k$ changes, the corresponding attenuation coefficient $\mu$ in $\Omega_k$ will be also changing to $\mu + \delta\mu$. Then, the energy attenuation of the ultrasonic wave in the region $\Omega_k$ will change and the Equation (11) can be rewritten as follows:

$$\iint_{\Omega_k} (\delta\mu)\mathrm{d}S = \iint_{\Omega_k} (\mu + \delta\mu)\mathrm{d}S - \iint_{\Omega_k} \mu \mathrm{d}S = -\ln\frac{I_m^a}{I_e} - \left(-\ln\frac{I_m^b}{I_e}\right) = \ln\frac{I_m^b}{I_m^a},\tag{12}$$

where $\delta\mu$ is the variation of the attenuation coefficient in $\Omega_k$, $I_m^b$ and $I_m^a$ represent the acoustic energy before and after the attenuation coefficient changes, respectively.

### 2.2.2. Reflection Mode Mathematical Model

Similar to the propagation of optics ray in the medium, the mathematical model of the reflection mode UT can be derived from the Fermat's principle, where the ultrasonic waves will reflect at the media interface and continue to propagate to the measuring transducer along the shortest propagation path. Thus, the ultrasonic waves sending from the exciting transducer will be reflected at the interface where the acoustic impedances change and continue to propagate in the background medium to the measuring transducer. When the exciting and measuring transducer are determined, the detected points will fall on an elliptical arc. This elliptical arc can be determined by the coordinates of the exciting and measuring transducer and the time of flight (TOF) of the ultrasonic wave, which can be described as follows:

$$\frac{1}{c} \cdot \left[ \sqrt{(x_d - x_e)^2 + (y_d - y_e)^2} + \sqrt{(x_d - x_m)^2 + (y_d - y_m)^2} \right] = t_f, \tag{13}$$

where $c$ is the sound speed in the background medium, $t_f$ is the TOF of the ultrasonic wave in the background medium, $(x_d, y_d)$ is the coordinate of the detected point, which obeys the arc distribution, $(x_e, y_e)$ and $(x_m, y_m)$ are the coordinates of the exciting and measuring transducer, respectively.

### 2.3. Complementarity of EIT and URT Sensitive Field

The boundary measurements of EIT and UT have different sensitivities about the changes of the sensitive parameters in different spaces of the imaging domain, and higher sensitivity means higher detection resolution and accuracy. In EIT, the electric current is injected into the imaging domain through the electrodes attached at the boundary. The electric field lines near the edge of the imaging domain are dense than the central region of the imaging domain. Therefore, the EIT measurements have a higher sensitivity to the changes in conductivity near the edge of the imaging domain. According to the reciprocity principle, the sensitivity of EIT measurements about the changes of conductivity is proportional to the integration of the electric field intensity [43]. As shown in Figure 2a, EIT has higher sensitivity near the edge of the imaging domain. In UT, the ultrasonic waves emitted by the transducers are mainly concentrated in a narrow main lobe range along the radial of the transducer. Only the changes of acoustic impedance occurring at the propagation path of ultrasonic waves will cause the change on reflection or transmission measurements. In the imaging domain, all the ultrasonic waves propagation paths of the transducers are intersected in the central of the imaging domain. Therefore, the UT measurements have a higher sensitivity about the changes in acoustic impedance near the central of the imaging domain [44]. According to the radon transform theory, the sensitivity of UT measurements about the change of acoustic impedance can be formed by the superimposition of all single transducer ultrasonic waves propagation paths. As shown in Figure 2b, UT has higher sensitivity near the central of the imaging domain. Based on the complementarity of EIT and UT sensitivity space distribution, the combination of EIT and UT is expected to improve the imaging resolution and accuracy.

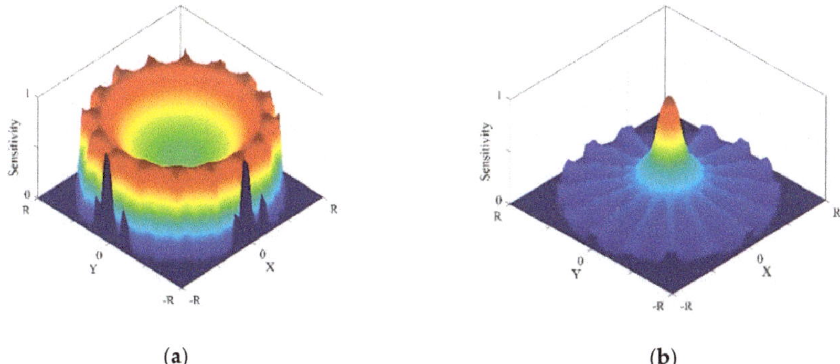


**Figure 2.** 3-D sensitivity distribution: (**a**) Electrical impedance tomography (EIT), (**b**) ultrasound tomography (UT).

## 3. EIT/UT Dual-Modality Image Reconstruction

### 3.1. Unification of Mesh Model and Reconstruction Target

In the paper, the purpose of both EIT and UT is to reconstruct the media distribution based on the change of conductivity or acoustic impedance distribution. In some applications, when the media distribution changes, the conductivity and acoustic impedance will also change. For example, in gas/liquid two-phase flow cross-section reconstruction problem, the presence of bubble will change the distribution of conductivity and acoustic impedance, the targets of EIT and UT are all the reconstruction of the bubble distribution. Thus, the reconstruction target of EIT and UT can be uniformed to the changes of the media distribution, which means that the background medium has the same representation in EIT and UT. In addition, another factor to consider is the mesh model. Based on this demand of reconstructing the same unknown media distribution using EIT and UT, the mesh model for EIT and UT should be also uniformed in the same imaging region. To achieve the purpose, the same mesh model is used in EIT and UT, as shown in Figure 3.

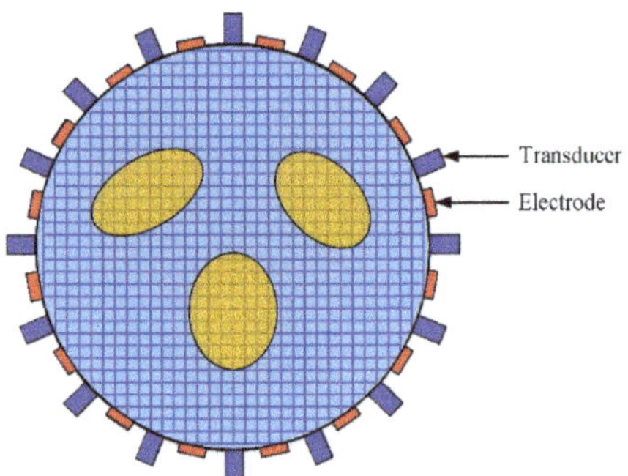


**Figure 3.** Mesh model of EIT/UT dual-modality imaging.

*3.2. Total Variation Regularization for EIT*

It's known that EIT inverse problem is a typical ill-posed inverse problem due to the large condition number of the Jacobian matrix $J$, which means that it's unrealistic to directly solve the unknown $x$ from the Equation (7). To overcome this problem, the regularization methods are usually adopted. First, we convert the EIT inverse problem to find the minimum from the least-square equation $\|Jx - y\|_2^2$. Then, a penalty term, also referred as regularization term, is added to stabilize the solution process by constraining the solution space. This process can be formulated in the following equations:

$$\min_{x \in R^N} f(x), \tag{14}$$

$$f(x) = \frac{1}{2}\|Jx - y\|_2^2 + \beta R(x), \tag{15}$$

where the regularization parameter $\beta$ is a positive constant and balances the residual term $\|Jx - y\|_2^2$ and the penalty term $R(x)$ in the inverse problem solving process. There are many forms of the regularization term based on the difference of the solution space constraints. Here, the Total Variation (TV) regularization term is adopted, and the regularization term can be written as:

$$R(x) = \int_\Omega |\nabla x| dS, \tag{16}$$

However, the Equations (14)–(16) cannot be directly solved due to the non-differentiability of $|\nabla x|$. To overcome this problem, a smooth approximation is usually adopted [45]. Then, the regularization term $R(x)$ can be rewritten as:

$$R(x) = \int_\Omega \sqrt{|\nabla x|^2 + \theta} dS, \tag{17}$$

where $\theta > 0$ is a predetermined small constant and is selected as $1 \times 10^{-6}$ here.

### 3.3. Constraints Information from UT Measurements

As mentioned above, the purpose of EIT and UT is to reconstruct the media distribution based on the change of conductivity or acoustic impedance from the boundary measurements. Since there is no change in the conductivity and acoustic impedance in the background medium region, the image reconstruction of the background medium satisfies the following relationship in EIT and UT:

$$\Omega^b = \Omega^b_{EIT}(\delta\sigma = 0) = \Omega^b_{UT}(\delta Z = 0, \delta\mu = 0), \tag{18}$$

where $\Omega^b$ represents the background medium region, $\Omega^b_{EIT}$ and $\Omega^b_{UT}$ are the background medium region reconstructed from EIT and UT, respectively.

Thus, the background medium region reconstructed from UT can be used as the prior information in EIT image reconstruction. Consider two scenarios whether there is a target on the ultrasonic wave's propagation path, as shown in Figure 4a,b. If there is no target on the ultrasonic wave propagation path, the ultrasonic waves sending from the exciting transducer will continue to propagate to the measuring transducer and the attenuation of the acoustic energy is only caused by the background medium. And then, we can mark the pixels on the transmission ultrasonic waves propagation path as background medium region $\Omega^b_{UT}$. In contrast, if there is a target on the ultrasonic wave propagation path, the ultrasonic waves sending from the exciting transducer will reflect on the target interface, and then continue to propagate to the measuring transducer through the background medium. The time of flight $t_f$ can be extracted from the reflection waves. And then, using the Equation (13) and the main lobe area of the transducer, we can calculate the region covered by the reflection ultrasonic waves propagation paths. After that, we can also mark the pixels on the reflection ultrasonic waves propagation path as background medium region $\Omega^b_{UT}$.

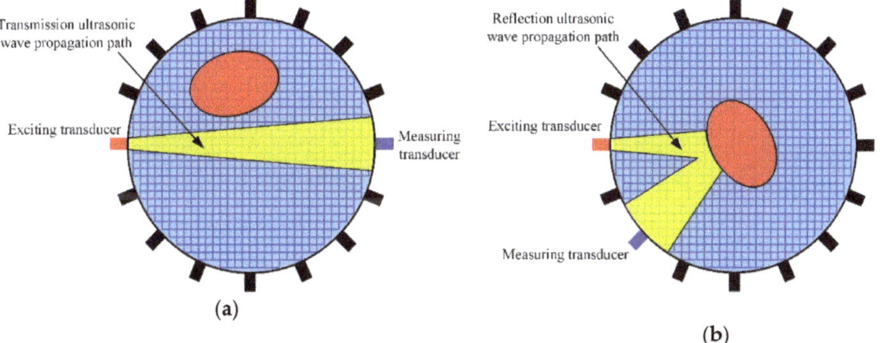


**Figure 4.** Schematic diagram of ultrasonic wave propagation path: (**a**) there is no target on the ultrasonic waves propagation path, (**b**) there is a target on the ultrasonic waves propagation path.

Since the conductivity of the background medium region does not change, we can derive the following equation:

$$h(x) = C\,x = 0, \tag{19}$$

where $h(x)$ represents the constraint function of UT measurements with respect to the conductivity difference $x$, $C$ is the corresponding extend constraint matrix, and it's calculated by the analytical method:

$$C(i,j) = \begin{cases} 1, & i = 1,\dots,M_U, \ j = n(i) \\ 0, & \text{otherwise} \end{cases}, \tag{20}$$

where $C \in R^{M_U \times N}$, $M_U$ is the total number of pixels in region of $\Omega^b_{UT}$, $n(i)$ is the serial number of the pixels located at $\Omega^b_{UT}$ in the set of all pixels.

### 3.4. Lagrange-Newton Solution for Dual-Modality Inverse Problem

In summary, we can convert the media distribution image reconstruction problem into the conductivity change image reconstruction problem for the EIT/UT dual-modality imaging. The EIT/UT dual-modality inverse problem can be described as follows:

$$
\min_{x \in R^{N}} f(x) \\
s.\,t. \quad h(x) = 0
\tag{21}
$$

where $f(x)$ and $h(x)$ are defined by Equations (15) and (19), respectively.

Obviously, the problem (21) is a typical equality constraint optimization problem. The Lagrange function combining with Newton method is commonly used to solve it. First, we formulate the following Lagrange function:

$$
\Psi(x, \lambda) = f(x) - \lambda^{T} h(x) = \frac{1}{2} \| Jx - y \|_{2}^{2} + \beta R(x) - \lambda^{T} h(x),
\tag{22}
$$

where $\lambda$ is the Lagrange multiplier. The stability point of the above Lagrange function corresponds to the optimal solution of the unknown $x$, and satisfies the following conditions:

$$
\begin{cases}
\frac{\partial \Psi}{\partial x} = \nabla f(x) - \nabla h(x)^{T} \lambda = 0 \\
\frac{\partial \Psi}{\partial \lambda} = -h(x) = 0
\end{cases},
\tag{23}
$$

Then, we solve the problem (22) using the Newton-Raphson method, and the iteration solution can be described as $x_{k+1} = x_{k} + (\delta x)_{k}$, where the iteration step length $(\delta x)_{k}$ and the Lagrange multiplier $(\delta \lambda)_{k}$ satisfy the following equation:

$$
\begin{pmatrix} W(x_{k}, \lambda_{k}) & -\nabla h(x_{k})^{T} \\ -\nabla h(x_{k}) & 0 \end{pmatrix} \begin{pmatrix} (\delta x)_{k} \\ (\delta \lambda)_{k} \end{pmatrix} = - \begin{pmatrix} \nabla f(x_{k}) - \nabla h(x_{k})^{T} \lambda_{k} \\ -h(x_{k}) \end{pmatrix},
\tag{24}
$$

where

$$
\begin{cases}
\nabla h(x) &= \nabla(Cx) = C \\
\nabla f(x) &= \nabla\left(\frac{1}{2}\|Jx - y\|_{2}^{2} + \beta R(x)\right) = J^{T}(Jx - y) + \beta L_{\theta}(x)x \\
W(x, \lambda) &= \nabla^{2}(\Psi(x, \lambda)) = \nabla^{2}\left(f(x) - \lambda^{T} h(x)\right) = J^{T} J + \beta L_{\theta}(x) \\
L_{\theta}(x) &= L_{a}^{T}(x) H^{-1}(x) L_{a}(x),
\end{cases}
\tag{25}
$$

where $J^{T}$ is the transpose of $J$, $L_{a} \in R^{Q \times N}$ is a sparse matrix, which can be determined by the topological relationship between the pixels, referring to the literature [45], $H \in R^{Q \times Q}$ is a diagonal matrix whose diagonal entry can be described as follows:

$$
H_{qq}(x) = \sqrt{\|L_{a,q}(x)x\|^{2} + \theta}, \quad q = 1, 2, \ldots, Q,
\tag{26}
$$

where $L_{a,q} \in R^{1 \times N}$ is the $q$-th row element of $L_{a}$, $Q$ is the total number of adjacent element pairs along the horizontal or vertical direction, referring to the literature [45].

To analyze the iteration convergence and determine the iteration termination condition of the problem (21), we define the following penalty function:

$$
G(x, \lambda) = \|\nabla f(x) - \nabla h(x)^{T} \lambda\|_{2}^{2} + \|h(x)\|_{2}^{2},
\tag{27}
$$

In summary, the implement of the proposed EIT/UT dual-modality image reconstruction method can be described as Algorithm 1.

---

**Algorithm 1:** Lagrange-Newton Method for EIT/UT Dual-Modality Image Reconstruction

---

**Step 1: Initialization.**
  Calculate the initial value of $x$ using linear back-projection (LBP) algorithm [13]: $x_0 = J^T y$,
  Given the value of the iteration termination parameter: $\varepsilon \geq 0$, $\varepsilon = 1 \times 10^{-6}$,
  Given the value of the intermediate parameter in iteration: $\kappa \in (0, 1)$, $\kappa = 0.5$, $\eta = 1$.
**Step 2: Termination Condition Judgment.**
  Calculate $G(x_k, \lambda_k)$ by Equation (27), and judge:
  If $G(x_k, \lambda_k) \leq \varepsilon$, stop iteration and return the $x_k$ as the optimal estimation value,
  Otherwise, calculate $(\delta x)_k$ and $(\delta \lambda)_k$ by the Equation (24).
**Step 3: Compute Step Length.**
  Calculate $G(x_k + \eta(\delta x)_k, \lambda_k + \eta(\delta \lambda)_k)$ and $G(x_k, \lambda_k)$ by Equation (27), and judge:
  If $G(x_k + \eta(\delta x)_k, \lambda_k + \eta(\delta \lambda)_k) \leq (1 - \kappa\eta)G(x_k, \lambda_k)$, go to Step 4,
  Otherwise, set $\eta = \eta/4$, continue to the Step 3.
**Step 4: Iteration Update.**
  Update the conductivity difference $x$: $x_{k+1} = x_k + \eta(\delta x)_k$,
  Update the Lagrange multiplier $\lambda$: $\lambda_{k+1} = \lambda_k + \eta(\delta \lambda)_k$,
  Set $k=k+1$ and go to Step 2.

---

## 4. Results

To evaluate the performance of the proposed EIT/UT dual-modality imaging method, a series of numerical and experimental tests are carried out and compared with single-modality EIT method. In the tests, three typical EIT image reconstruction methods, Newton's One-Step Error Reconstructor (NOSER), L1 regularization, and TV regularization, are used in single-modality EIT image reconstruction. Considering that the selection of regularization parameters has a great influence on the imaging results, the regularization parameters in all methods are empirically selected in the range of $10^{-6}$ to $10^{-3}$, and the one that reached the most similar result to the real model was used in the comparison [45]. In EIT, the adjacent electric current excitation and adjacent voltage measurement model is used to obtain the boundary measurements. In UT, the piston transducers are used in numerical and experimental tests, and the acoustic energy can be better concentrated in a narrow beam range. In transmission mode, three transducers directly opposite to the exciting transducer are used to obtain the transmission measurements. In reflection mode, three transducers on each adjacent side of the exciting transducer are used to obtain the reflection measurements. The ultrasound frequency is 1MHz in the numerical and experimental tests. In addition, the reconstruction results from EIT/UT dual-modality method is simplified as UET in the paper.

### 4.1. Numerical Results and Discussion

The numerical tests are conducted by the commercial multi-physical coupling software COMSOL and MATLAB. The simulation model is designed by simulating the discrete bubbles in a conductive liquid. The conductivity of the bubble is $1 \times 10^{-6}$ Sm$^{-1}$ and the background medium conductivity is $1 \times 10^{-2}$ Sm$^{-1}$. The density of the bubble is 1.29 kgm$^{-3}$ and sound velocity is 340 ms$^{-1}$. The density of the background medium is 1000 kgm$^{-3}$ and sound velocity is 1400 ms$^{-1}$. The radius of the imaging domain is 75 mm, and 16 electrodes and 16 ultrasound transducers are staggered evenly and arranged on the periphery of the observation domain, as shown in Figure 3. The piston ultrasound transducers sequentially send and receive the three periods sinusoidal pulse signal in the narrow field range.

### 4.1.1. Quantitative Index

To evaluate the reconstruction results, four quantitative indexes of relative image error ($RE$), correlation coefficient ($CC$), position error ($PE$) and area difference ($AD$), are used in the paper. The $RE$ and $CC$ are usually used to describe the degree of difference of the true conductivity difference and reconstructed conductivity difference. The definition of $RE$ and $CC$ can be expressed as follows:

$$RE = \frac{\|x - x^*\|_2}{\|x^*\|_2},$$ (28)

$$CC = \frac{\sum\limits_{n=1}^{N} (x_n - \bar{x})(x_n^* - \bar{x}^*)}{\sqrt{\sum\limits_{n=1}^{N} (x_n - \bar{x})^2 \sum\limits_{n=1}^{N} (x_n^* - \bar{x}^*)^2}},$$ (29)

where $x$ is the calculated conductivity difference vector, $x^*$ is the real conductivity difference vector, $x_n$ and $x_n^*$ are the $n$-th element of $x$ and $x^*$, respectively, $\bar{x}$ and $\bar{x}^*$ are the mean values of $x$ and $x^*$, respectively. A lower value of $RE$ and a higher value of $CC$ mean a good image reconstruction result.

The $PE$ and $AD$ are respectively used to evaluate the target position and shape reconstructed accuracy by comparing the segmented reconstructed images and the real distribution images. The segmented threshold is selected as 25% of the minimum pixel amplitudes, as follows:

$$[x_q]_n = \begin{cases} 1, & \text{if } [x]_n \leq \frac{1}{4}\min(x) \\ 0, & \text{otherwise} \end{cases},$$ (30)

where $x$ is the calculated image pixel amplitudes (conductivity difference), $x_q$ is the segmented image pixel amplitudes. The $PE$ and $AD$ are defined as follows:

$$PE = \left| c_t - c_q \right|,$$ (31)

$$AD = \frac{\text{sum}\left(x_b \cup x_q\right) - \text{sum}\left(x_b \cap x_q\right)}{N},$$ (32)

where $c_t$ is the center of gravity of the target in real conductivity difference distribution image, $c_q$ is the center of gravity of the target in reconstructed conductivity difference distribution image, $x_b$ is the binary image of the real conductivity difference distribution, $x_q$ is the segmented image of the reconstructed conductivity difference distribution image, $N$ is the total number of the pixels.

### 4.1.2. Image Reconstruction Results Analysis

In numerical tests, five different models are used to compare the proposed EIT/UT dual-modality imaging method (UET) and the single-modality EIT imaging method using NOSER, L1, TV. The image reconstruction results are shown in Figures 5–8 when adding no noise, 40 dB, 25 dB, 20 dB Gaussian noise in EIT measurements, respectively. The real targets boundaries are indicated by the solid black line in the reconstructed images. It can be seen from the Figures 5–8 that the UET has a cleaner background and better imaging performance compared with the other methods, which can be also verified by the quantitative indexes, as shown in Figure 9. For different quantitative indexes, we respectively calculate their mean values in five media distribution models under a certain noise condition. When the noise level increases, the imaging results from single-modality EIT using NOSER, L1 and TV are gradually getting worse, while the proposed UET method can always better reconstruct the media distribution in different noise levels. Thus, the proposed UET method is more robust to the EIT measurement noise. In addition, the expected time of UET is about 0.32 s for per image reconstruction. The expected time of EIT using NOSER, L1, TV are about 0.02 s, 0.2 s, 0.3 s for per image reconstruction, respectively. Thus, the imaging speed of UET is comparable to that of single-modality EIT using TV method, but the imaging accuracy of UET have an obvious improvement.

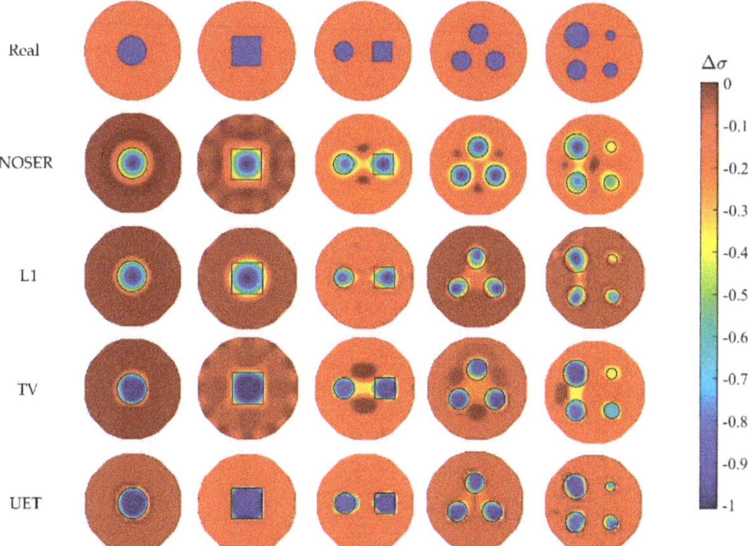

**Figure 5.** Reconstruction results with no noise.

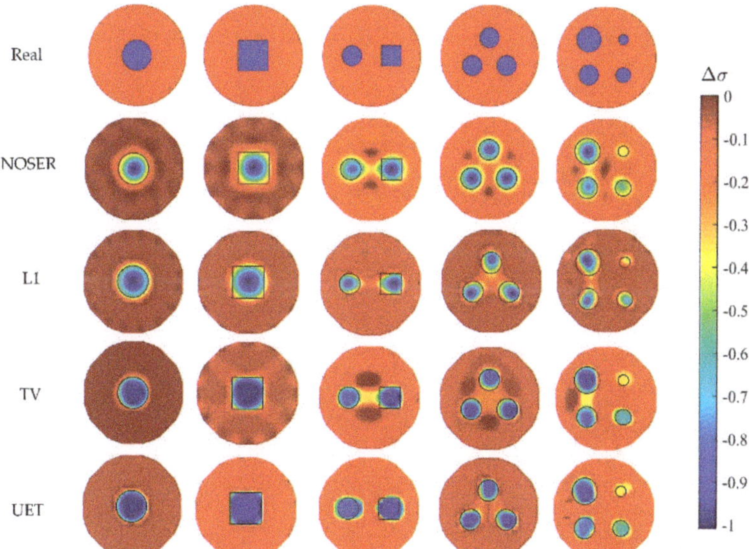

**Figure 6.** Reconstruction results with 40 dB noise.

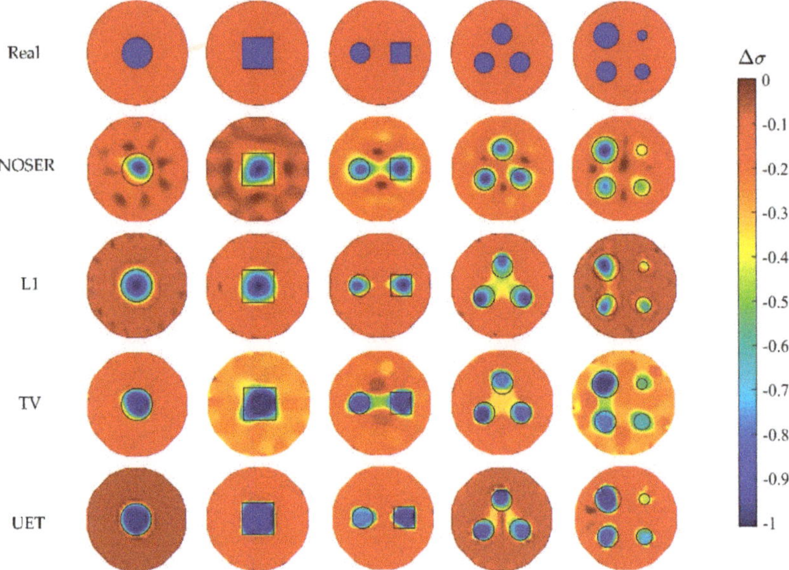

**Figure 7.** Reconstruction results with 25 dB noise.

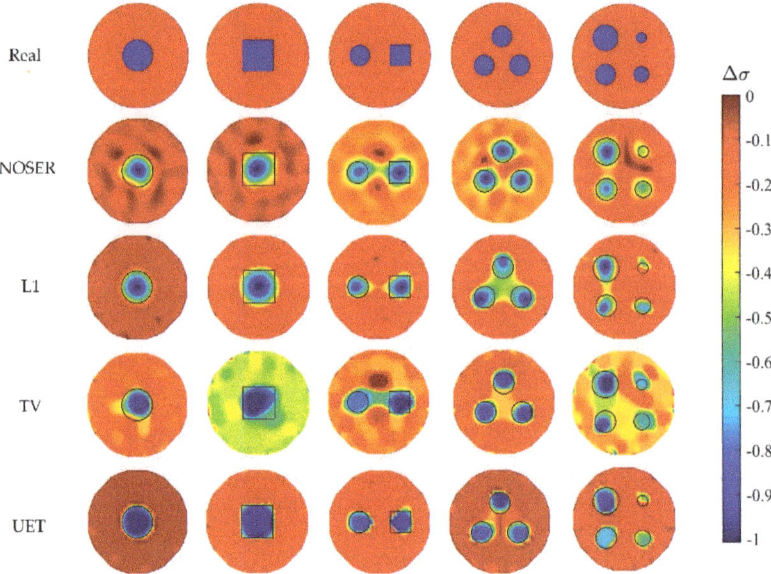

**Figure 8.** Reconstruction results with 20 dB noise.

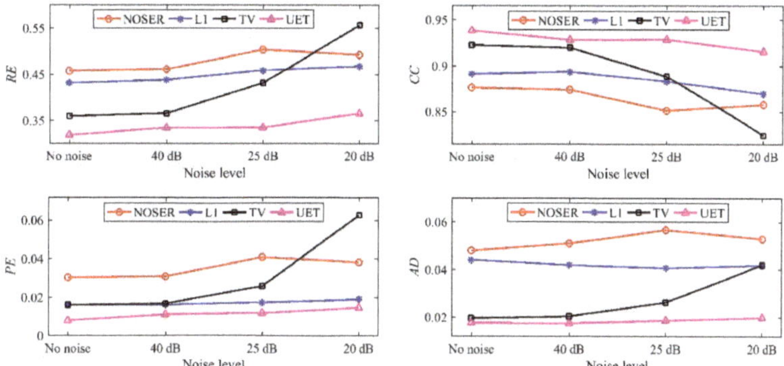

**Figure 9.** Mean value of the quantitative indexes.

### 4.1.3. Comparison of Edge Reconstruction Ability

It can be seen from the numerical results that the UET has a cleaner edge than other methods. To verify this conclusion, we further analyze the numerical results of the model with one circle and one square in Figure 5. First, the pixel values of the reconstructed images are normalized to the interval [0, 1]. Then, the contour of the pixel values along the center horizontal line of the reconstructed images is drawn in Figure 10. It's obvious that the contour of UET reconstructed image is closer to the real contour. Thus, the proposed UET method has a better edge reconstruction ability than the single-modality EIT methods.

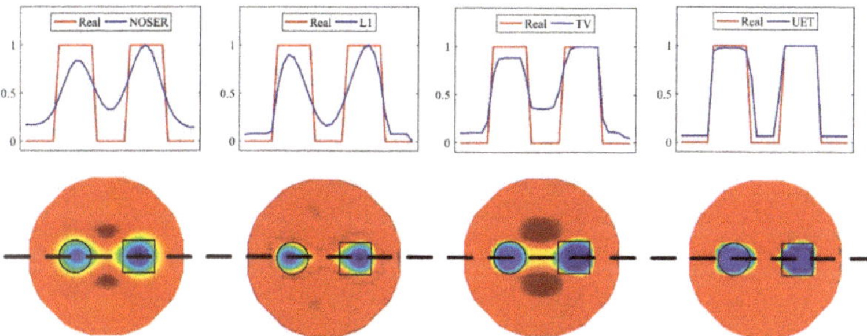

**Figure 10.** Contour of the reconstructed images along the center horizontal line.

### 4.2. Experimental Results and Discussion

To further verify the performance of the proposed method, a set of experimental tests was carried out. The experimental system is shown in Figure 11, which is developed by Tianjin University, China [46]. It mainly contains three parts, the sensor array, the data acquisition and control unit, and the image reconstruction and visualization unit. The radius of the sensor is 75 mm and 16 electrodes and 16 ultrasonic transducers are evenly placed on the periphery of the sensor. The contact resistance between the electrodes and background medium is 37.1 $\Omega/\text{cm}^2$. The background medium consists of $Na_2SO_4$ solution with the conductivity of 1.41 mS/cm. The data acquisition speed of EIT is about 1000-frames/s, and the data acquisition speed of UT is about 17-frames/s. The expected time of the image reconstruction in experimental tests is similar to numerical tests.

The experimental results are shown in Figure 12, where five different distribution models (A, B, C, D, E) of the nylon rod ($10^{-11}$ mS/cm) are used. It can be seen from the results that the proposed UET method has a better imaging performance compared with the single-modality EIT methods. The quantitative analysis of the experimental results is shown in Figure 13, which can also verify the conclusion that the proposed UET method shows better reconstruction performance.

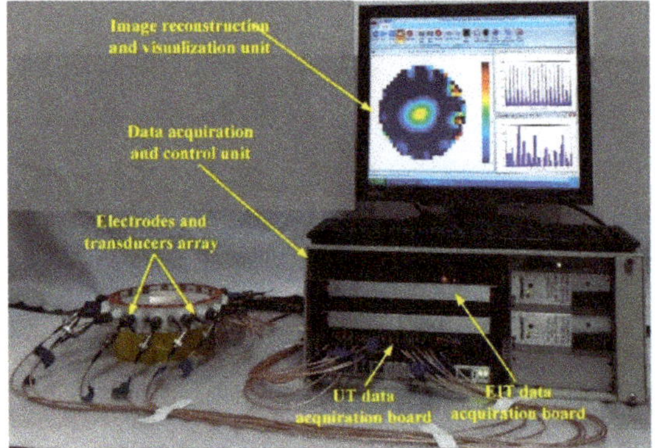


**Figure 11.** EIT/UT dual-modality data acquisition system.

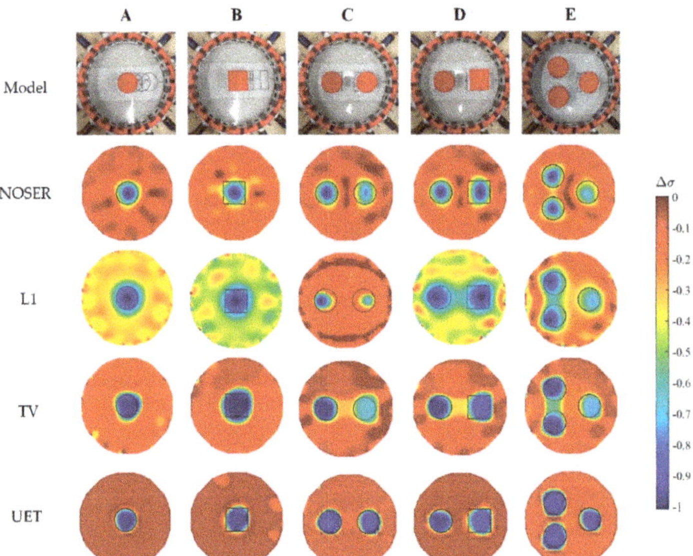

**Figure 12.** Reconstruction results of the experimental tests.

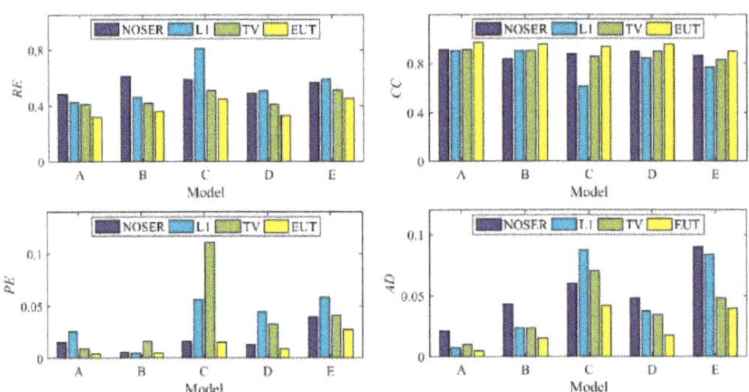

**Figure 13.** Quantitative analysis of the experimental results.

## 5. Conclusions

A novel EIT/UT dual-modality imaging method is proposed, where the measurement information obtained from UT can be used as prior constraint and improve the EIT image reconstruction performance. The dual-modality inverse problem is constructed by the equality constraint equation, and then, solved by the Lagrange-Newton method. The numerical and experimental tests show that the proposed EIT/UT dual-modality imaging method has a better image reconstruction performance than the traditional single-modality EIT methods, where the reconstructed images of UET have clearer targets edges and less artifacts in the background region. In the experimental tests, the average *RE* of UET have reduced 29.7%, 31.1%, 14.9% than single-modality EIT using NOSER, L1, TV, respectively. The average *CC* of UET have improved 7.7%, 16.9%, 7.4% than single-modality EIT using NOSER, L1, TV, respectively. The average *PE* of UET have reduced 32.9%, 68.2%, 71.3% than single-modality EIT using NOSER, L1, TV, respectively. The average *AD* of UET have reduced 55.1%, 50.6%, 36.5% than single-modality EIT using NOSER, L1, TV, respectively. In addition, the UET shows better performance of edge reconstruction, where the contour of UET reconstructed image is closer to the real contour than single-modality EIT methods. Furthermore, the proposed UET method is more robust to the EIT measurement noise, where the quantitative indexes of UET have no significant change with the increase of the measurement noise. The future work will focus on the research of enriching the UT measurements by adding the number of the transducers or increasing the diffusion angle of the transducers, and further improve the image reconstruction performance of EIT/UT dual-modality imaging.

**Author Contributions:** G.L. developed the algorithm and ran the testes; S.R. designed the study; S.Z. interpreted the data and images; F.D. designed the conception; all the authors were involved in writing and editing the manuscript.

**Funding:** This research was funded in part by the National Natural Science Foundation of China, grant number 61571321, and in part by the Natural Science Foundation of Tianjin City, grant number 17JCQNJC03500.

**Conflicts of Interest:** The authors declare no conflicts of interest.

## References

1.  Dong, F.; Jiang, Z.X.; Qiao, X.T.; Xu, L.A. Application of electrical resistance tomography to two-phase pipe flow parameters measurement. *Flow Meas. Instrum.* **2003**, *14*, 183–192. [CrossRef]
2.  Yao, J.F.; Takei, M. Application of process tomography to multiphase flow measurement in industrial and biomedical fields—A review. *IEEE Sens. J.* **2017**, *17*, 8196–8205. [CrossRef]
3.  Liu, D.; Khambampati, A.K.; Du, J.F. A parametric level set method for electrical impedance tomography. *IEEE Trans. Med. Imaging* **2018**, *37*, 451–460. [CrossRef]
4.  Nissinen, A.; Kaipio, J.P.; Vauhkonen, M.; Kolehmainen, V. Contrast enhancement in EIT imaging of the brain. *Physiol. Meas.* **2015**, *37*, 1–24. [CrossRef] [PubMed]
5.  Storz, H.; Storz, W.; Jacobs, F. Electrical resistivity tomography to investigate geological structures of the earth's upper crust. *Geophys. Prospect.* **2000**, *48*, 455–471. [CrossRef]
6.  Loke, M.H.; Chambers, J.E.; Rucker, D.F.; Kuras, O.; Wilkinson, P.B. Recent developments in the direct-current geoelectrical imaging method. *J. Appl. Geophys.* **2013**, *95*, 135–156. [CrossRef]
7.  Tallman, T.N.; Gungor, S.; Wang, K.W.; Bakis, C.E. Damage detection and conductivity evolution in carbon nanofiber epoxy via electrical impedance tomography. *Smart Mater. Struct.* **2014**, *23*, 045034. [CrossRef]
8.  Tallman, T.N.; Semperlotti, F.; Wang, K.W. Enhanced delamination detection in multifunctional composites through nanofiller tailoring. *J. Intell. Mater. Syst. Struct.* **2015**, *26*, 2565–2576. [CrossRef]
9.  Hallaji, M.; Seppänen, A.; Pour-Ghaz, M. Electrical resistance tomography to monitor unsaturated moisture flow in cementitious materials. *Cem. Concr. Res.* **2015**, *69*, 10–18. [CrossRef]
10. Hallaji, M.; Seppänen, A.; Pour-Ghaz, M. Electrical impedance tomography-based sensing skin for quantitative imaging of damage in concrete. *Smart Mater. Struct.* **2014**, *23*, 085001. [CrossRef]
11. Holder, D.S. *Electrical Impedance Tomography: Methods, History and Applications*; Institute of Physics Publishing: London, UK, 2005.

12. Murphy, E.K.; Mahara, A.; Halter, R.J. Absolute reconstructions using rotational electrical impedance tomography for breast cancer imaging. *IEEE Trans. Med. Imaging* **2017**, *36*, 892–903. [CrossRef]

13. Barber, D.C.; Brown, B.H. Applied potential tomography. *J. Phys E Sci. Instrum.* **1984**, *17*, 723. [CrossRef]

14. Vauhkonen, M.; Vadasz, D.; Karjalainen, P.A.; Somersalo, E.; Kaipio, J.P. Tikhonov regularization and prior information in electrical impedance tomography. *IEEE Trans. Med. Imaging* **1998**, *17*, 285–293. [CrossRef]

15. Liu, D.; Kolehmainen, M.; Siltanen, S.; Laukkanen, A.M.; Seppanen, A. Estimation of conductivity changes in a region of interest with electrical impedance tomography. *Inverse Probl. Imag.* **2015**, *9*, 211–229. [CrossRef]

16. Chen, B.; Abascal, J.; Soleimani, M. Electrical resistance tomography for visualization of moving objects using a spatiotemporal total variation regularization algorithm. *Sensors* **2018**, *18*, 1704. [CrossRef]

17. Liu, S.H.; Jia, J.B.; Zhang, Y.D.; Yang, Y.J. Image reconstruction in electrical impedance tomography based on structure-aware sparse bayesian learning. *IEEE Trans. Med. Imaging* **2018**, *37*, 2090–2102. [CrossRef]

18. Ren, S.J.; Sun, K.; Liu, D.; Dong, F. A statistical shape constrained reconstruction framework for electrical impedance tomography. *IEEE Trans. Med. Imaging* **2019**, in press. [CrossRef]

19. Cherry, S.R. Multimodality imaging: beyond PET/CT and SPECT/CT. *Semin. Nucl. Med.* **2009**, *39*, 348–353. [CrossRef]

20. Veninga, T.; Huisman, H.; Maazen, R.W.; Huizenga, H. Clinical validation of the normalized mutual information method for registration of CT and MR images in radiotherapy of brain tumours. *J. Appl. Clin. Med. Phys.* **2004**, *5*, 66–79. [CrossRef]

21. Kaplan, I.; Oldenburg, N.E.; Meskell, P.; Blake, M.; Church, P.; Holupka, E.J. Real time MRI-ultrasound image guided stereotactic prostate biopsy. *Magn. Reson. Imaging* **2002**, *20*, 295–299. [CrossRef]

22. Steiner, G. Sequential fusion of ultrasound and electrical capacitance tomography. *Int. J. Inf. Syst. Sci.* **2006**, *2*, 484–497.

23. Zhang, R.H.; Wang, Q.; Wang, H.X.; Zhang, M.; Li, H.L. Data fusion in dual-mode tomography for imaging oil–gas two-phase flow. *Flow Meas. Instrum.* **2014**, *37*, 1–11. [CrossRef]

24. Johansen, G.A.; Abro, E. A new CdZnTe detector system for low-energy gamma-ray measurement. *Sens. Actuators A Phys.* **1996**, *54*, 493–498. [CrossRef]

25. Liu, S.; Liu, S.; Ren, T. Acoustic tomography reconstruction method for the temperature distribution measurement. *IEEE Trans. Instrum. Meas.* **2017**, *66*, 1936–1945. [CrossRef]

26. Duric, N.; Boyd, N.; Littrup, P.; Sak, M.; Myc, L.; Li, C.; West, E.; Minkin, S.; Martin, L.; Yaffe, M.; et al. Breast density measurements with ultrasound tomography: A comparison with film and digital mammography. *Med. Phys.* **2013**, *40*, 013501. [CrossRef]

27. Langener, S.; Vogt, M.; Ermert, H.; Musch, T. A real-time ultrasound process tomography system using a reflection-mode reconstruction technique. *Flow Meas. Instrum.* **2017**, *53*, 107–115. [CrossRef]

28. Liang, G.H.; Ren, S.J.; Dong, F. An inclusion boundary reconstruction method using electrical impedance and ultrasound reflection dual-modality tomography. In Proceedings of the 9th World Congress on Industrial Process Tomography (WCIPT), Bath, UK, 2–6 September 2018.

29. Liang, G.H.; Ren, S.J.; Dong, F. Ultrasound guided electrical impedance tomography for 2D free-interface reconstruction. *Meas. Sci. Technol.* **2017**, *28*, 074003. [CrossRef]

30. Steiner, G.; Soleimani, M.; Watzenig, D. A bio-electromechanical imaging technique with combined electrical impedance and ultrasound tomography. *Physiol. Meas.* **2008**, *29*, 63–75. [CrossRef] [PubMed]

31. Wan, Y.; Halter, R.; Borsic, A.; Manwaring, P.; Hartov, A.; Paulsen, K. Sensitivity study of an ultrasound coupled transrectal electrical Impedance Tomography system for prostate imaging. *Physiol. Meas.* **2010**, *31*, 17–29. [CrossRef] [PubMed]

32. Teniou, S.; Meribout, M. A multimodal image reconstruction method using ultrasonic waves and electrical resistance tomography. *IEEE Trans. Image Process.* **2015**, *24*, 3512–3521. [CrossRef] [PubMed]

33. Pusppanathan, J.; Rahim, R.A.; Phang, F.A.; Mohamad, E.J.; Seong, C.K. Single-plane dual-modality tomography for multiphase flow imaging by integrating electrical capacitance and ultrasonic sensors. *IEEE Sens. J.* **2017**, *17*, 6368–6377. [CrossRef]

34. Liang, G.H.; Ren, S.J.; Dong, F. An augmented Lagrangian trust region method for inclusion boundary reconstruction using ultrasound/electrical dual-modality tomography. *Meas. Sci. Technol.* **2018**, *29*, 074008. [CrossRef]

35. Hassan, H.; Semperlotti, F.; Wang, K.W.; Tallman, T.N. Enhanced imaging of piezoresistive nanocomposites through the incorporation of nonlocal conductivity changes in electrical impedance tomography. *J. Intell. Mater. Syst. Struct.* **2018**, *29*, 1850–1861. [CrossRef]
36. Zhao, L.; Yang, J.; Wang, K.W.; Semperlotti, F. An application of impediography to the high sensitivity and high resolution identification of structural damage. *Smart Mater. Struct.* **2015**, *24*, 065044. [CrossRef]
37. Ammari, H.; Bonnetier, E.; Capdeboscq, Y.; Tanter, M.; Fink, M. Electrical impedance tomography by elastic deformation. *SIAM J. Appl. Math.* **2008**, *68*, 1557–1573. [CrossRef]
38. Somersalo, E.; Cheney, M.; Isaacson, D. Existence and uniqueness for electrode models for electric current computed tomography. *SIAM J. Appl. Math.* **1992**, *52*, 1023–1040. [CrossRef]
39. Geselowitz, D. An application of electrocardiographic lead theory to impedance plethysmography. *IEEE Trans. Biomed. Eng.* **1971**, *18*, 38–41. [CrossRef] [PubMed]
40. Ren, S.J.; Soleimani, M.; Dong, F. Inclusion boundary reconstruction and sensitivity analysis in electrical impedance tomography. *Inverse Probl. Sci. Eng.* **2018**, *26*, 1037–1061. [CrossRef]
41. Ren, S.J.; Wang, Y.; Liang, G.H.; Dong, F. A robust inclusion boundary reconstructor for electrical impedance tomography with geometric constraints. *IEEE Trans. Instrum. Meas.* **2019**, *68*, 762–773. [CrossRef]
42. Li, N.; Cao, M.C.; Xu, K.; Jia, J.B.; Du, H.B. Ultrasonic transmission tomography sensor design for bubble identification in gas-liquid bubble column reactors. *Sensors* **2018**, *18*, 4256. [CrossRef] [PubMed]
43. Brandstatter, B. Jacobian calculation for electrical impedance tomography based on the reciprocity principle. *IEEE Trans. Magn.* **2003**, *39*, 1309–1312. [CrossRef]
44. Rahiman, M.H.F.; Rahim, R.A.; Rahim, H.A.; Zakaria, Z.; Pusppanathan, J. A study on forward and inverse problems for an ultrasonic tomography. *J. Teknol.* **2014**, *70*, 113–117. [CrossRef]
45. Song, X.Z.; Xu, Y.B.; Dong, F. A spatially adaptive total variation regularization method for electrical resistance tomography. *Meas. Sci. Technol.* **2015**, *26*, 125401. [CrossRef]
46. Dong, F.; Xu, C.; Zhang, Z.Q.; Ren, S.J. Design of parallel electrical resistance tomography system for measuring multiphase flow. *Chin. J. Chem. Eng.* **2012**, *20*, 368–379. [CrossRef]

 *sensors*

*Article*

# Comparison of Selected Machine Learning Algorithms for Industrial Electrical Tomography

**Tomasz Rymarczyk [1,2], Grzegorz Kłosowski [3,*], Edward Kozłowski [3] and Paweł Tchórzewski [2]**

[1]   University of Economics and Innovation in Lublin, 20-209 Lublin, Poland; tomasz@rymarczyk.com
[2]   Research & Development Centre Netrix S.A., 20-704 Lublin, Poland; pawel.tchorzewski@netrix.com.pl
[3]   Faculty of Management, Lublin University of Technology, 20-618 Lublin, Poland; e.kozlovski@pollub.pl
*   Correspondence: g.klosowski@pollub.pl

Received: 28 January 2019; Accepted: 25 March 2019; Published: 28 March 2019

**Abstract:** The main goal of this work was to compare the selected machine learning methods with the classic deterministic method in the industrial field of electrical impedance tomography. The research focused on the development and comparison of algorithms and models for the analysis and reconstruction of data using electrical tomography. The novelty was the use of original machine learning algorithms. Their characteristic feature is the use of many separately trained subsystems, each of which generates a single pixel of the output image. Artificial Neural Network (ANN), LARS and Elastic net methods were used to solve the inverse problem. These algorithms have been modified by a corresponding increase in equations (multiply) for electrical impedance tomography using the finite element method grid. The Gauss-Newton method was used as a reference to machine learning methods. The algorithms were trained using learning data obtained through computer simulation based on real models. The results of the experiments showed that in the considered cases the best quality of reconstructions was achieved by ANN. At the same time, ANN was the slowest in terms of both the training process and the speed of image generation. Other machine learning methods were comparable with the deterministic Gauss-Newton method and with each other.

**Keywords:** machine learning; inverse problem; electrical impedance tomography; image reconstruction; industrial tomography

---

## 1. Introduction

This article presents the results of research on the use of tomographic sensors for the analysis of industrial processes with the use of dedicated measuring devices and image reconstruction algorithms.

Electrical impedance tomography (EIT) is a non-invasive, high-potential application imaging method. It is suitable for continuous real-time visualization of the dynamic distribution of electrical conductivity inside the tested object [1]. To perform EIT reconstructions, we use weak alternating currents (1–5 mA) with low frequency (1–100 kHz) and measure the appropriate peripheral voltages by means of a set of electrodes attached to the object's surface [2]. A cross-sectional image of internal spatial conductivity is obtained from voltage measurements gained from different electrode pairs. Despite its relatively low spatial resolution, the EIT is now a widely accepted tomographic imaging technique that is widely used in many areas, such as monitoring industrial processes [3–5], geophysical research [6–8] and biomedical diagnosis [2,9,10]. Mathematical reconstruction of conductor maps in the EIT is about solving a non-linear and ill-posed inverse problem from noisy data [11]. Regulatory techniques can be used to mitigate the instability of solutions. One of the most commonly used methods is a one-step approach to Gauss-Newton reconstruction (GN) [12], which allows the use of sophisticated, regulated models to describe the problem of the inverse EIT through a heuristic determined predecessor [13]. Landweber iteration is a modification of the steepest gradient descent approach and is also widely used in EIT [14]. The algebraic reconstruction technique (ART) is a valid method of reconstructing the computed tomography images that can be used in the EIT [15]. Other important methods include: regularization using total variation (TV) [16], which allows image reconstruction while preserving the edge, split augmented Lagrangian shrinkage algorithm [17] and the generalized vector sampled pattern matching method (GVSPM) [18].

Because deep learning is good for mapping complicated nonlinear functions, attempts are increasingly being made to apply deep learning methods based on convolutional neural networks (CNNs) for EIT/ERT (electrical resistivity tomography) image reconstruction [11]. Among the CNN, the deep D-bar methods are also used [19]. D-bar methods are based on a rigorous mathematical analysis. They provide robust direct reconstructions by using a low-pass filtering of the associated nonlinear Fourier data [9].

In the EIT tomography, algorithms belonging to machine learning methods can be successfully used. Typical examples of this kind of method are: Lasso (least absolute shrinkage and selection operator), Elastic net, least-angle regression (LARS) [6], artificial neural networks [20] and convolutional neural networks [11], multivariate adaptive regression splines (MARS), k-nearest neighbors (KNN), random forest (RF), gradient boosting machine (GBM) [21], Principal Component and Partial Least Square Regression [22].

The current development of EIT algorithms is largely focused on the use of machine learning methods [23]. Hence the need to verify whether such algorithms are in fact better than the classical, known deterministic methods to which the Gauss-Newton method belongs [12,24].

In comparison to other known imaging methods used in industry [25], electrical impedance tomography (EIT) has a number of advantages. These include, among others: higher time resolution, lower costs, opportunities for wider use, etc. However, reconstruction of the EIT may be unstable and has a fundamental disadvantage resulting from the need to solve the inverse problem [26]. The sensitivity of the EIT solutions to measurement, numerical and model errors entails the need to adjust the model parameters to specific cases. Many such methods have been developed over the years. These serious constraints on the EIT therefore favor the development of more sophisticated algorithms [27–29]. It is worth mentioning that most 2D reconstruction methods are also applicable in 3D situations with minor modifications [30].

The authors of the article developed three original variants of known algorithms based on machine learning techniques, and then compared them to the deterministic method as well as to each other. In order to make a precise assessment enabling a reliable comparison, universal evaluation metrics were used: Mean Squared Error (MSE), Relative Image Error (RIE) and Image Correlation Coefficient (ICC).

Advanced automation and control of production processes play a key role in enterprises [31,32]. Technological equipment and production lines can be considered the heart of industrial production, while information technologies and control systems are its brain. Tomographic imaging of objects creates a unique opportunity to discover the complexity of the structure without the need to invade the object. There is a growing need for information on how internal flows behave in the process equipment. This should be performed non-invasively by tomographic instrumentation [33].

Sensor technologies are mainly based on electrical tomography (ET) [34–38], which includes electrical capacitance tomography (ECT) [39–45] and electrical resistance tomography (ERT) [7,46,47]. It allows reconstruction of the image by the distribution of conductivity or permittivity of the object from electrical measurements at the edge of the object.

The results of the reconstruction of individual algorithms with different measurement models were compared. The tests were carried out for real data obtained from real laboratory measurements. The electronic devices for measuring the material values and to collect data from the measurement sensors were designed and made by the authors.

The main novelty of the presented method is a machine learning approach based on learning many separate subsystems (ANN, LARS, Elastic net), while each subsystem is dedicated to a single pixel of the output image (Figure 1). The deterministic method, Gauss-Newton with Laplace regularization should be treated as a reference, enabling objective comparison of standard techniques with machine learning methods.

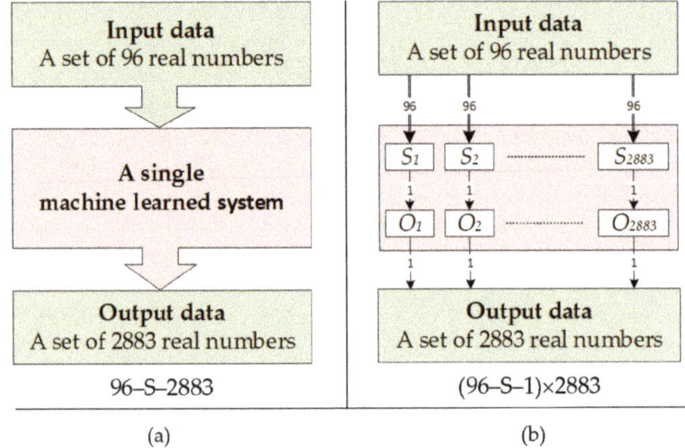


**Figure 1.** Comparison of the traditional concept with the improved concept: (**a**) a single prediction system with 96 predictors and 2883 responses; (**b**) multiple prediction system composed of 2883 separately trained subsystems, each of which has 96 predictors and 1 response.

Figure 1a shows the traditional machine-trained algorithm. It consists of a singular predictive (regression) system with many outputs. The input vector includes 96 measurements of voltage drops measured on individual electrode pairs. The predictive system has 2883 outputs, which makes its training difficult. The large number of system outputs is the main reason for the unsatisfactory quality of reconstructed tomographic images.

Figure 1b shows the scheme of the novel multiple system. Its characteristic feature is that on the basis of the same 96-element vector of predictors, 2883 separate prediction subsystems ($S_1$, $S_2$, ... , $S_{2883}$) were trained. Each of the subsystems generates only one independent output (response), which is the value of a single pixel of the reconstructed image ($O_1$, $O_2$, ... , $O_{2883}$). Thanks to this approach, each pixel of the output image is the result of the operation of a single-output prediction subsystem. Subsystems with one output are easier to train than subsystems with multiple outputs. Thanks to this, the results obtained using the presented concept (96—S—1) × 2883 are better than those obtained using the traditional concept 96—S—2883.

The article consists of four sections. The measurement models, machine learning methods and descriptions of algorithms were presented in Section 2. The results of the research work in the form of reconstruction of images for measurement data are shown in Section 3. In Section 4, the results obtained are discussed. It also summarizes the presented research.

## 2. Materials and Methods

This section presents the tomographic methods, process tomography, measuring devices, laboratory systems, mathematical algorithms and measurement models used in image reconstruction based on synthetic data and real measurements. Laboratory equipment, tomography devices designed at Research & Development Centre Netrix SA, the Eidors toolbox [48], Microsoft tools, Matlab, Python and R language were used during the research.

### 2.1. Electrical Tomography

Electrical tomography is an imaging technique that uses different electrical properties of different types of materials, including biological tissues. In this method, the power or voltage source is connected to the object, followed by the emergence of current flows or the distribution of voltage at the edge of the object. The collected information is processed by an algorithm that reconstructs the image. This tomography is characterized by a relatively low image resolution. Difficulties in obtaining high resolution result mainly from a limited number of measurements, nonlinear current flow through a given medium and too-low sensitivity of measured voltages depending on changes in conductivity inside the area. Electrical tomography has historically been divided into electrical capacitive tomography for systems dominated by dielectrics, and electrical resistance tomography. The basic theory can be obtained from Maxwell's equations.

A complex "admittivity" can be defined as follows:

$$\gamma = \sigma + i\omega\varepsilon \tag{1}$$

where $\varepsilon$ is the permittivity, $\sigma$ is the electrical conductivity, and $\omega$ is the angular frequency.

In the case of the electric field strength (**E**), the current density (**J**) in the test area will be related to Ohm's law:

$$\mathbf{J} = \gamma\mathbf{E} \tag{2}$$

The gradient of the potential distribution ($u$) has the form:

$$\mathbf{E} = -\nabla u \tag{3}$$

Due to the fact that there are no sources from the Ampère law in the studied region, we have:

$$\nabla{\cdot}\mathbf{J} = 0 \tag{4}$$

Potential distribution in a heterogeneous, isotropic area:

$$\nabla{\cdot}(\gamma\nabla u) = 0 \tag{5}$$

where $u$ is the potential.

Where the capacitance or resistance dominates, the equation factor should be simplified to the form:

$$\nabla{\cdot}(\sigma\nabla u) = 0 \; for \; \frac{\omega\varepsilon}{\sigma} \ll 1 (ERT) \tag{6}$$

$$\nabla{\cdot}(\varepsilon\nabla u) = 0 \; for \; \frac{\omega\varepsilon}{\sigma} \gg 1 (ECT) \tag{7}$$

By solving the inverse problem, we obtain the distribution of material coefficients in the studied area.

Electrical resistance tomography in a process tomography can be interchangeably called electrical impedance tomography (EIT). In the following part of this work, we will mainly use the name, EIT [49–51].

The inverse method and neighboring method in EIT for collecting data from potential measurements at the edge of an object for 16 electrodes is shown in Figure 2.

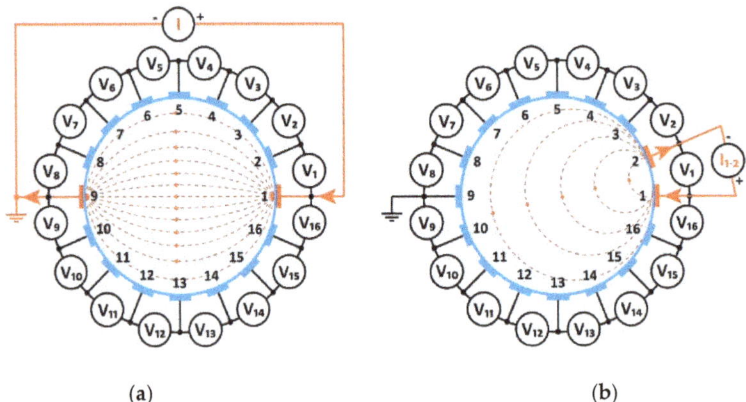

(a)           (b)

**Figure 2.** Measurement model in electrical impedance tomography: (a) opposite, (b) neighboring method.

## 2.2. Measurement Models

In order to test the effectiveness of algorithms for the analysis of processes in industrial tomography, three real measuring models were implemented. Electrical tomography was implemented for the analysis. Figure 3a presents the EIT measuring device (hybrid tomograph), which was made by the Netrix S.A. Research and Development Center. A bucket with electrodes was used as the tank or industrial reactor model (Figure 3b,c).

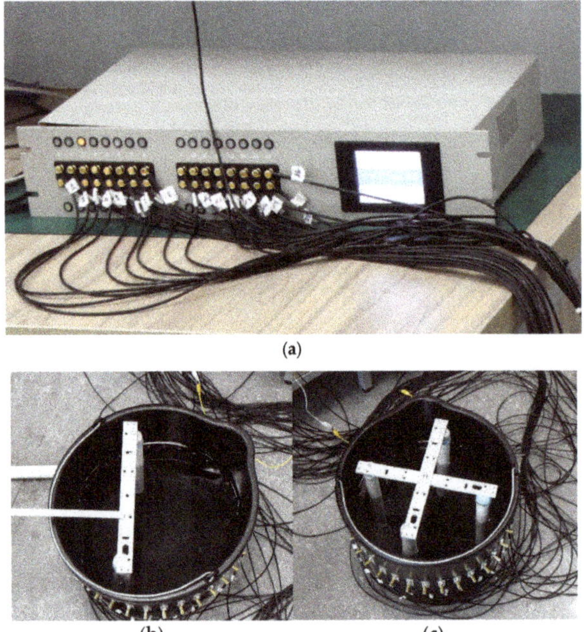

(a)

(b)           (c)

**Figure 3.** The test stand: (a) the measurement device—a hybrid tomograph made by the Netrix S.A. Research and Development Center, (b) tank with 2 phantoms, (c) tank with 4 phantoms.

The arrangement of phantoms inside the investigated object is presented in Figure 4. This is a plan view that corresponds with the pictures of the tank shown in Figure 3.

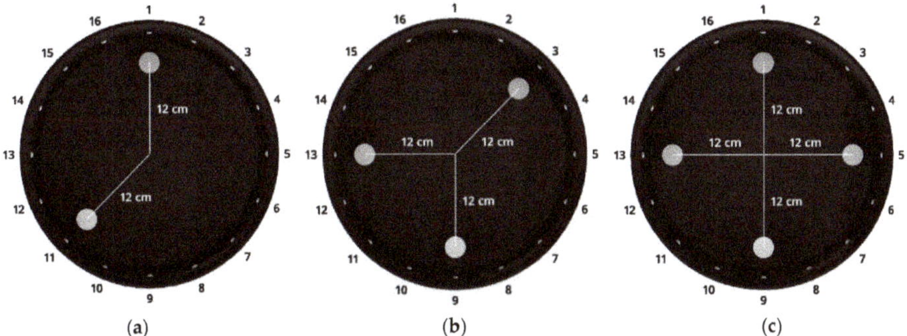

**Figure 4.** Three variants of the arrangement of phantoms in the tested tank with 16 electrodes: (**a**) 2 phantoms, (**b**) 3 phantoms, (**c**) 4 phantoms.

Figure 5 shows a side view of the dimensioned model of the EIT tested tank. On the left side, a tube immersed in the tank with its diameter can be seen.

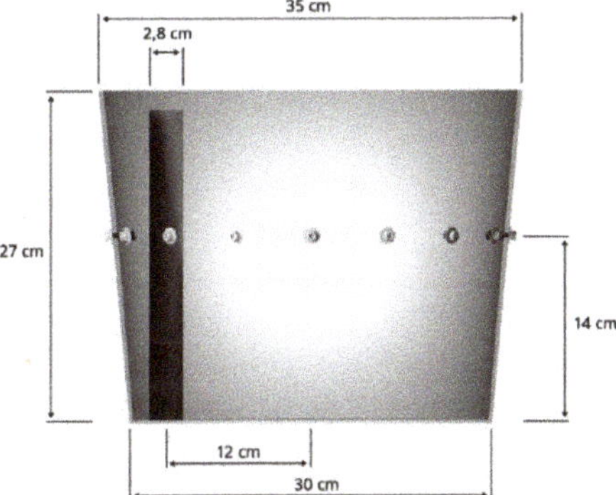

**Figure 5.** Dimensioned model of the EIT tested tank.

Based on the above physical models, a special simulation algorithm was developed to generate learning cases used during the training process of the machine learning systems. Each training case was generated in accordance with the following procedure. First of all, we assume a homogeneous distribution of electrical conductivity. Then, we randomly select the number of internal inclusions. We assume that as a result of the draw we receive a maximum of five objects, each with a circular shape. The radius and conductivity are such that they correspond to the actual tests carried out by the EIT. During the next stage of calculations, the center of each internal object is drawn. For the obtained conductivity distribution, measurement voltages are determined using the finite element method.

Figure 6 shows one of the 50,000 generated cases used to train the predictive system. The cross section of the tank contains 5 randomly arranged artifacts, which corresponds to the 96 vector voltage measurements. Because the polarization of the electrodes changes during individual measurements, the voltage varies during the interval $(-0.06; +0.06)$.

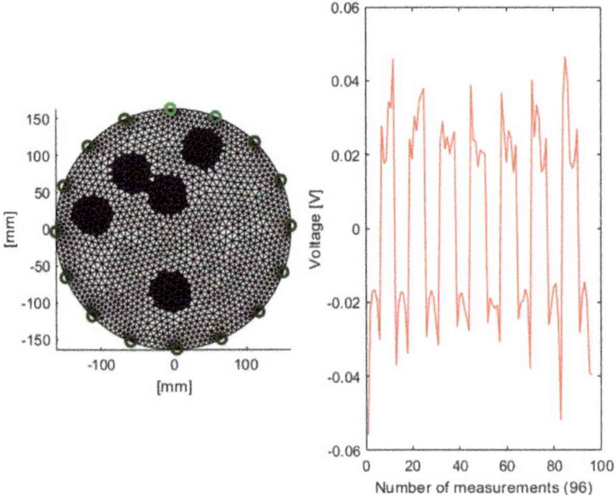

**Figure 6.** A training case generated with the simulation method with a graph showing the voltages.

Based on the dimensions of the physical model, output images (reconstructions) for 3 cases of arrangement of the tubes were also generated (see Figure 4). The background pixel values are 1 and in the reconstructive images are marked in white. In turn, the spots (pixels) of the occurrence of the tubes have a value close to 0 and are colored dark blue.

Algorithm 1 shows the pseudo code used to generate training cases. The script generating the simulation data of the measurements on the electrodes included artificial noise (line *8* in Algorithm 1). For this purpose, a random number generator with a normal distribution was used, with an expected value of 0 and a corresponding standard deviation. In addition, the voltages determined on the basis of numerical simulation are always subject to a certain error, especially when, as in the described case, we make calculations on a grid consisting of a relatively small number of finite elements.

---

**Algorithm 1.** The pseudo code to generate learning cases

| | | |
|---|---|---|
| *1.* | N = 50000; | % The number of cases |
| *2.* | **for** 1: N | |
| *3.* | random selection of the number of objects; | % set of NumberOfObjects variable |
| *4.* | **for** 1: NumberOfObjects | |
| *5.* | random selection of the object's location; | % center and radius |
| *6.* | **end** | |
| *7.* | adding an output image to the set of training cases; | % saving response data |
| *8.* | determination of voltages and adding Gaussian noise; | % Gaussian noise = randn(1, 96) $\times$ 5 $\times$ 10$^{-5}$ |
| *9.* | saving the values of voltages to the training set; | % saving input data |
| *10.* | **end** | |

---

### 2.3. Algorithms and Methods

There are many methods and algorithms used in optimization problems [52–57]. In this article, the authors chose deterministic algorithms based on the Gauss-Newton method as a reference to the machine learning methods. The Gauss-Newton method is often used in electrical tomography because it is quite effective. The next algorithms were based on machine learning methods [8,58], in which an innovative approach to tomographic problems was presented.

### 2.3.1. Image Reconstruction

Process tomography also belongs to the problems of the inverse electromagnetic field. The inverse problem is the process of optimization, identification, or synthesis in which the parameters describing a given field are determined based on the possession of information specific to this field. Such issues are difficult to analyze. They do not have unambiguous solutions and are ill-conditioned due to too little or too much information. They are sometimes contradictory or linearly dependent. Knowledge of the process can make image reconstruction more resistant to incomplete or damaged data. The numerical analysis of the problem was carried out using the finite element method.

The colors of individual pixels on the image correspond to the conductance of the examined cross-section parts. An approach in which each of the separately trained subsystem generates only one output, that is, the value of a single pixel of the output image allows for better mapping of the values of electrical measurements.

To confirm the above thesis, a number of experiments were carried out using three neural networks differing in structure and number of outputs. Three types of ANN with the following structures were trained: 96—10—1 (96 predictors, 10 hidden neurons, 1 response), 96—10—10 (96 predictors, 10 hidden neurons, 10 responses) and 96—20—10 (96 predictors, 20 hidden neurons, 10 responses). The smaller the Mean Square Error (*MSE*) and the bigger the regression (*R*), the better is the quality of ANN. Responses (output pixels) were chosen randomly. The set of 50,000 cases was randomly divided into 3 subsets: training, validating and testing in the proportion of 70:15:15. The results of the experiments are presented in Table 1.

**Table 1.** Comparison of the neural networks with 1 and 10 responses.

| Quality Indicators for Testing Set | ANN Type | | |
|:---:|:---:|:---:|:---:|
| | **96—10—1** | **96—10—10** | **96—20—10** |
| $\overline{MSE}$ | 0.0069 | 0.0087 | 0.0086 |
| $\overline{R}$ | 0.7548 | 0.6994 | 0.6897 |

Only the testing set was used to assess the network quality. To increase the objectivity of experiments, the indicators in Table 1 ($\overline{MSE}$, $\overline{R}$) were the arithmetic mean of 10 experiments performed for each of the three types of ANN.

As you can see, the best results were obtained for the ANN with a single output (96—10—1). A more complex 10-output network (96—20—10) was better than the simpler 96—10—10 ANN having 10 neurons in the hidden layer. However, both neural networks with 10 responses turned out to be worse than ANN with a single response. The abovementioned tests proved that the variant of ANN with one output turned out to be the best. For this reason, in the research multiple LARS, Elastic net and ANN systems were used, in which each of the subsystems generated only one response value.

Figure 7 presents an outline of a machine learning system that was applied to all 3 methods: LARS, Elastic net and ANN. A distinguishing feature of the presented concept is the separate training of each of the 2883 machine learning subsystems. Their number is equal to the resolution of the image output grid (2883 pixels).

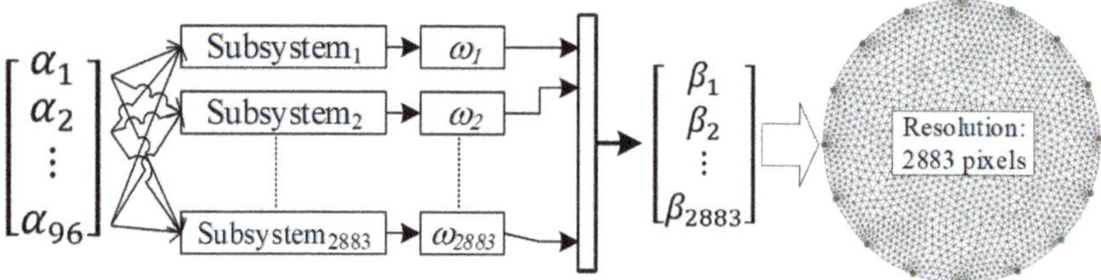

**Figure 7.** A mathematical neural model for converting electrical signals into images.

### 2.3.2. Gauss-Newton Method

The Gauss-Newton method is an effective approach to solve inverse problem in the electrical impedance tomography. It is worth emphasizing that such a problem is nonlinear and ill-posed. In difference imaging, the Gauss-Newton method can be used to minimize differences between reference and inhomogeneous data [12,59].

In general cases, image reconstruction involves determining the global minimum of the objective function, which can be defined as follows:

$$F(\sigma) = \frac{1}{2}\left\{\|\mathbf{U}_m - \mathbf{U}_s(\sigma)\|^2 + \lambda^2\|\mathbf{L}(\sigma - \sigma^*)\|^2\right\} \tag{8}$$

where:

$\mathbf{U}_m$—voltages obtained as a result of the measurements
$\mathbf{U}_s(\sigma)$—voltages received by numerical calculations (FEM) for given conductivity $\sigma$
$\sigma^*$—conductivity represents known properties
$\lambda$—regularization parameter (positive real number)
$\mathbf{L}$—regularization matrix.

Using appropriate approximations, it can be shown that the conductivity in the iteration denoted by $k + 1$ is given by the following formula:

$$\sigma_{k+1} = \sigma_k + \alpha_k\left(\mathbf{J}_k^T\mathbf{W}\mathbf{J}_k + \lambda^2\mathbf{L}^T\mathbf{L}\right)^{-1}\left[\mathbf{J}_k^T\mathbf{W}(\mathbf{U}_m - \mathbf{U}_s(\sigma_k)) - \lambda^2\mathbf{L}^T\mathbf{L}(\sigma_k - \sigma^*)\right] \tag{9}$$

where: $\mathbf{W}$—weighting matrix (it is usually a unit matrix), $\mathbf{J}_k$—Jacobian matrix calculated in $k$-th step, $\alpha_k$—step length. The Gauss-Newton method with Laplace regularization was implemented in our research.

### 2.3.3. LARS

Machine learning is related to the ability of the software to generalize based on previous experience. The important thing is that these generalizations are designed to answer questions about both previously collected data and new information. Using statistical methods with different regression models was presented in [60]. This approach enables quick diagnosis combining low cost and high efficiency. The selection of variables and the detection of data anomalies are not separate problems. To use the variables and outliers at the same time, the low angle regression (LARS) algorithm is used. While it is prudent to be cautious about the generalization of a small set of simulation results, it seems that LARS combined with dummy variables or row samples can provide computationally efficient, robust selection procedures. The proposed multiply LARS algorithm calculates all possible Lasso estimates for a given problem using an order of magnitude of less computing time. Another variation of LARS implements the linear regression of forward stagewise, this combination explains similar numerical results previously observed for Lasso and Stagewise and helps to understand the properties of both methods. A simple approximation of LARS degrees of freedom is available, from which the estimated prediction error value is taken.

If the regression data has only additional outliers, then we can start with a simple regression model:

$$Y = X\beta + \varepsilon \tag{10}$$

where $Y \in R^n, X \in R^{n \times (k+1)}$ denote the observation matrices of response and input variables, respectively, and $\beta \in R^{k+1}$ denotes the vector of unknown parameters. The object, $\varepsilon \in R^n$ presents a sequence of disturbances. The Least Angle Regression algorithm selects the subset of appropriate variables from entire set of available input variables. The linear model is built by employing the forward stepwise regression, where at each step the best variable is inserted to model.

The algorithm of Least Angle Regression is applied as follows:

1. standardize input variables;
2. select the most correlated input variable with the output variable. Add input variable to the linear model;
3. determine the residual from the obtained model;
4. add a variable which is the most correlated with the residual to the model;
5. move coefficient $\beta$ towards its least-squares coefficient;

Repeat steps 2–5 for the suitable number of iterations.

### 2.3.4. Elastic Net

Elastic net is a regularized regression method that linearly combines the L1 and L2 penalties of the Lasso and ridge methods [61–63]. Lasso is a regularization technique. The implemented multiply method can be used to reduce the number of predictors in a regression model or it selects among redundant predictors.

The equation is used to determine the linear regression:

$$\min_{(\beta_0, \beta') \in R^{k+1}} \frac{1}{2n} \sum_{i=1}^{n} (y_i - \beta_0 - x_i \beta')^2 + \lambda P_\alpha (\beta') \tag{11}$$

where $x_i = (x_{i1}, \ldots, x_{ik})$, $\beta' = (\beta_1, \ldots, \beta_k)$ for $1 \leq i \leq n$ and $P_\alpha$ is an Elastic net penalty
$P_\alpha$ is defined as:

$$P_\alpha (\beta') = (1 - \alpha) \frac{1}{2} \|\beta'\|_{L_2} + \alpha \|\beta'\|_{L_1} = \sum_{j=1}^{k} \left( \frac{1 - \alpha}{2} \beta_j^2 + \alpha |\beta_j| \right) \tag{12}$$

We see that the punishment is a linear combination of norms $L_1$ and $L_2$ of unknown parameters $\beta'$. The introduction of the parameter-dependent penalty function to the objective function reduces the estimators of unknown parameters.

### 2.3.5. Multiply Neural Network

This chapter presents the neuronal model enabling efficient reconstruction of tomographic images. Effective use of multiply artificial neural networks in tomography is possible, but the effectiveness of this tool depends on many conditions. First of all, ANNs (artificial neural networks) are able to effectively visualize objects, many of which are already known. Each subsystem means one neural network. All neural networks were trained based on a set of 50,000 simulation-generated cases.

A serious problem limiting the ability to generalize ANNs is overfitting. A good technique to reduce overfitting is to fundamentally limit the capacity of the model. These approaches are called regularization techniques. Among them, the following techniques can be distinguished: parameter norm penalties, early stopping, dropout, and transfer learning. In the case described, the technique of early stopping was used [64].

This technique tries to stop an estimator's training phase prematurely, at the point where it has learned to extract all meaningful associations from the data, before beginning to model its noise. This is done by monitoring MSE (Mean Squared Error) of the validation set and terminating the training phase when this metric stops falling. This way, the estimator has enough time to learn the useful information but not enough to learn from the noise.

All cases were randomly divided into 3 sets: training, validating and testing in 70:15:15 proportions. The training set (35,000 cases) was used to properly train each of the subsystems. The validation set (7500 cases) was used to determine the moment of stopping the iterative training process. The condition for stopping the learning process was a non-decreasing MSE for the validation set for the next 6 iterations. The test set (7500 cases) can be used for independent assessment of network quality after the learning process (MSE, R). The structure of each of the neural networks can be described as follows: 96—10—1. This means that each ANN was a multi-layered perceptron with 96 predictors, one hidden layer with 10 neurons and the output layer with one neuron. Logistic functions were used as the activation functions. All ANNs were trained using the Levenberg-Marquardt algorithm.

Algorithm 2 in the form of Matlab code represents the iterative process of training the multiple neural network shown in Figure 5. In a single structural variable called *nets_for_pixels*, all 2883 neural networks were recorded.

---

**Algorithm 2.** The Matlab code for training multiple ANN system

---

```
% X'- input matrix 96 × 50000 of training cases
% Y'- output matrix 2883 × 50000 of training cases
% Choose a Training Function
trainFcn = 'trainlm';   % In this case Levenberg-Marquardt backpropagation was chosen
hiddenLayerSize = 10;   % Choose a number of hidden layers
net = fitnet(hiddenLayerSize,trainFcn);   % Create a fitting network under variable 'net'
% Choose input and output pre/post-processing functions
% 'removeconstantrows' - remove matrix rows with constant values
% 'mapminmax' - map matrix row minimum and maximum values to [−1 1]
net.input.processFcns = {'removeconstantrows','mapminmax'};
net.output.processFcns = {'removeconstantrows','mapminmax'};
% Setup division of data for training, validation, testing
net.divideFcn = 'dividerand';   % Divide data randomly
net.divideMode = 'sample';   % Divide up every sample
net.divideParam.trainRatio = 70/100;   % 70% of cases is allocated for training
net.divideParam.valRatio = 15/100;   % 15% of cases is allocated for validation
net.divideParam.testRatio = 15/100;   % 15% of cases is allocated for testing
net.performFcn = 'mse';   % Mean Squared Error will be used for performance evaluation
x = X';
y = Y';
N=2883;   % The resolution of output picture grid
parfor i=1:N   % Start 'for' loop with parallel computing
    % Assign an i-th row of reference cases to the variable t. Each of the 2883 lines corresponds
    % to one pixel of the output image
    t = y(i,:);
    % Train the network. The variable 'nets_for_pixels' is a structure that consists of 2883
    % separately trained neural networks.
    [nets_for_pixels{i},~] = train(net,x,t);
end   % End 'parfor' loop
```

---

It should be emphasized that the algorithms used for the multiply Elastic net and multiply LARS methods, although created in R programming language, have an analogical logical structure. Therefore, they are not included in this paper.

## 3. Results

This chapter presents the results of image reconstruction based on designed numerical models and laboratory measurements. Data analysis is an important part of the diagnosis of the process based on tomography. Knowledge of the process can make the image reconstruction better. Inside the tested object, as its cross-section, a mesh of finite elements is generated. As a result of the calculations it obtains a reconstructed image. The inverse problem was solved using both deterministic and machine learning methods.

Figures 8–15 present the results of reconstruction of images based on laboratory measurements of the examined objects. These are not reconstructions based on artificial measurements obtained from a simulation generator. The reconstructions presented below are the result of real measurements generated using a physical model (Figure 3). They contain natural noises and other imperfections, caused by disturbances of the EIT system and the measurement process. As a result, the tomographic images presented below constitute an appropriate comparative basis, enabling objective evaluation of individual reconstruction EIT algorithms.

The systems with 16 and 32 electrodes were used here. Previous research proves that deterministic methods effectively reconstruct the image based on real measurements. The results obtained using multiply neural networks depend primarily on the quality of the training set. In the presented experiments, the data set for ANN was 10 times larger than for LARS or Elastic net and included more cases both in terms of the number of objects (tubes) and their distribution. It is possible that this fact caused the higher quality of the ANN reconstruction. The multiply LARS method is quite sensitive, while multiply Elastic net is quite universal, because by selecting the appropriate regularization parameters you can get enough good reconstructions on the actual data.

All reconstructions presented in this section refer to the three variants of tube arrangements presented in Section 2.2 (Figure 4). Reconstructions were obtained on the basis of test cases generated using the appropriate script. The white color means background. The objects are blue. The colors of the image are correlated with the conductance of the area represented by particular points on the mesh of a given cross-section. All reconstructed images were not improved by data filtering or denoising.

### 3.1. Gauss-Newton Method

The Gauss-Newton with Laplace regularization method was used to reconstruct the image in the electrical tomography for the 16 and 32 electrode systems (Figures 8 and 9). The reconstructions were obtained using the Gauss-Newton method using Laplace regularization. The numerical algorithm operated on a differential basis. So, in this case, we solve the inverse problem after the first iteration. The regularization parameter was 0.08. The reconstructed images illustrate variants with 2, 3 and 4 artifacts.

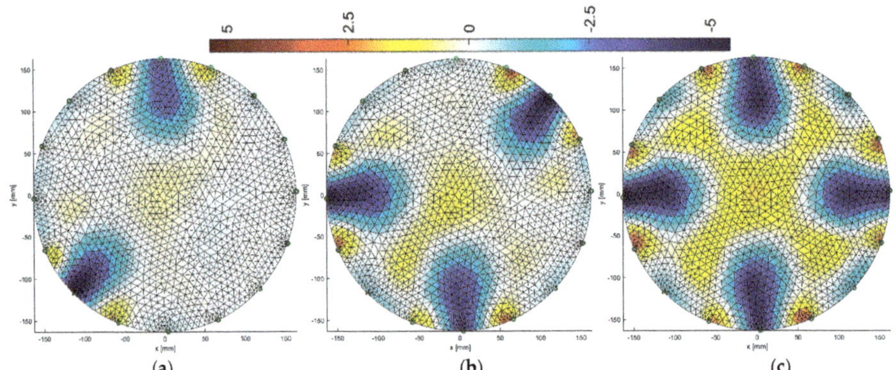

**Figure 8.** Image reconstruction for 16 measurement electrodes by the Gauss-Newton method with Laplace regularization: (**a**) 2 objects, (**b**) 3 objects, (**c**) 4 objects.

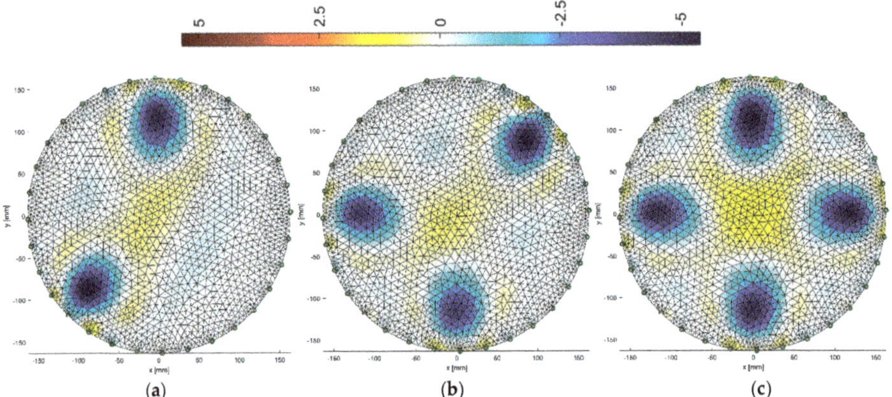

**Figure 9.** Image reconstruction for 32 measurement electrodes by the Gauss-Newton method with Laplace regularization: (**a**) 2 objects, (**b**) 3 objects, (**c**) 4 objects.

By comparing the reconstructive images from Figures 8 and 9 to Figure 4 from Section 2.2, it can be seen that the visual mapping of the position of the objects is correct, but their diameters are larger than in the reference images. The background noise is also visible because it should be uniformly white. There are also significant differences in the quality of images obtained from 16 and 32 electrode systems. The use of 32 electrodes gives much better results in this case.

### 3.2. Multiply Neural Networks

Image reconstruction in the case of multiply neural networks depends largely on the training set. An interesting observation is that the use of 32 electrodes (Figure 11) with respect to 16 (Figure 10) does not affect the visual quality of the imaging.

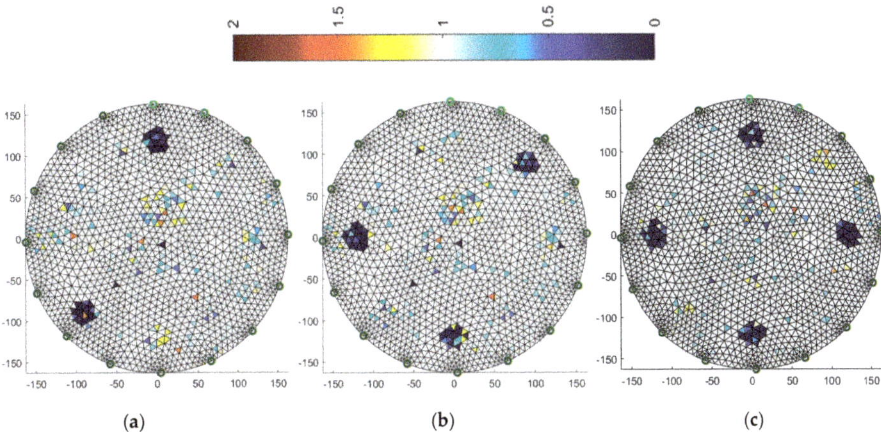

**Figure 10.** Image reconstruction for 16 measurement electrodes by Multiply Neural Networks: (**a**) 2 objects, (**b**) 3 objects, (**c**) 4 objects.

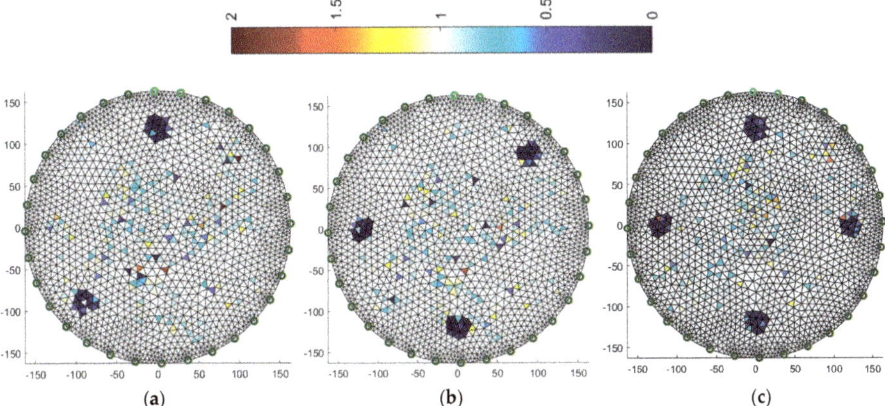

**Figure 11.** Image reconstruction for 32 measurement electrodes by Multiply Neural Networks: (**a**) 2 objects, (**b**) 3 objects, (**c**) 4 objects.

Comparing the reconstructive images of Figures 10 and 11 to Figure 4, it can be seen that the visual representation of the position and also the size of the objects is clearly better than for the Gauss-Newton method. Noise is visible, but it is relatively small and rather point-like.

### 3.3. Multiply LARS

Another algorithm was based on the multiply LARS method. A training set of 5000 elements was used here. In this case, the obtained results for a system with 16 electrodes (Figure 12) are slightly worse than for a system with 32 electrodes (Figure 13). The key element in this method is the separation of a group of independent measurements. The visual mapping of the position of the objects is correct, but their diameters are larger than in the reference images.

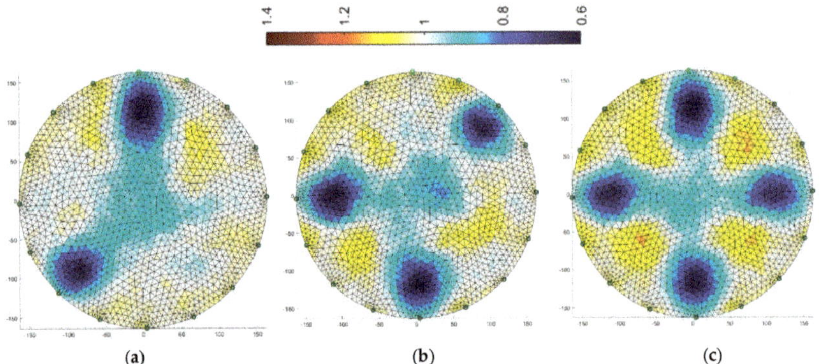

**Figure 12.** Image reconstruction for 16 measurement electrodes by multiply LARS: (**a**) 2 objects, (**b**) 3 objects, (**c**) 4 objects.

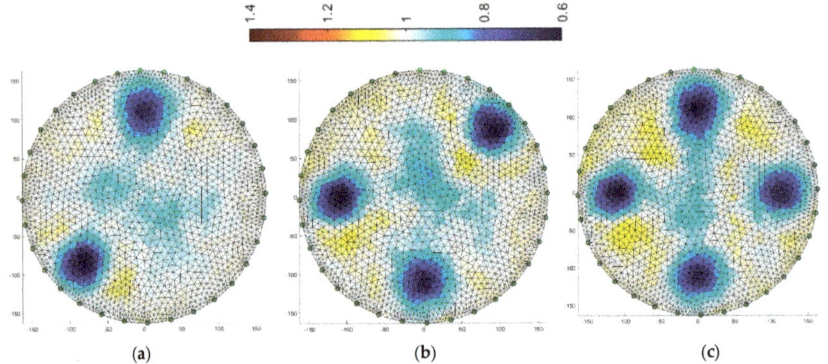

**Figure 13.** Image reconstruction for 32 measurement electrodes by multiply LARS: (**a**) 2 objects, (**b**) 3 objects, (**c**) 4 objects.

*3.4. Multiply Elastic Net*

The final algorithm is multiply Elastic net. It is more universal due to its character and gives quite precise results.

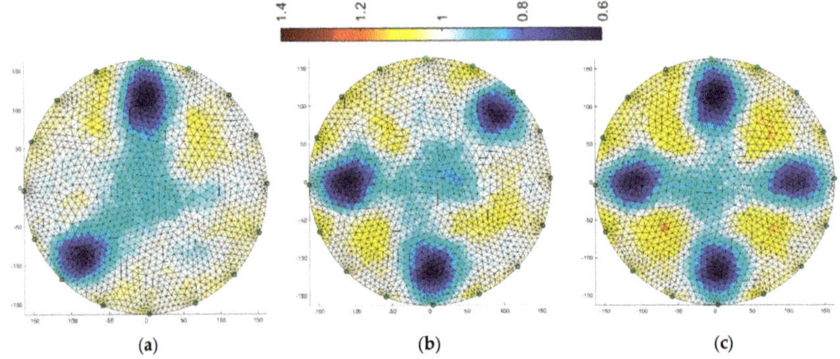

**Figure 14.** Image reconstruction for 16 measurement electrodes by multiply Elastic net: (**a**) 2 objects, (**b**) 3 objects, (**c**) 4 objects.

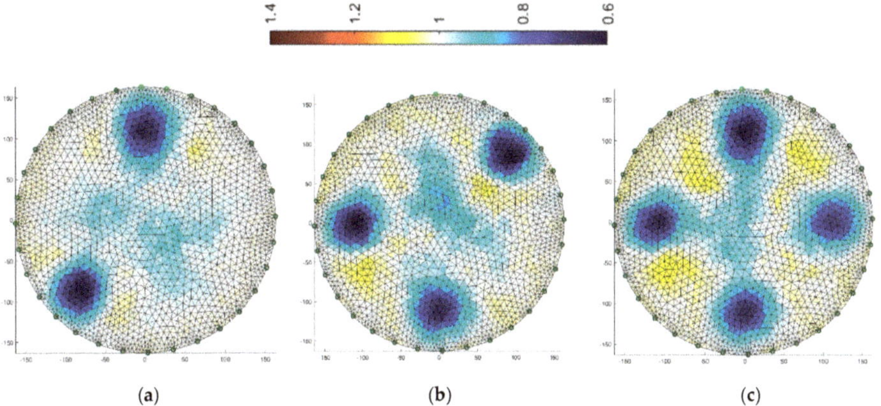

**Figure 15.** Image reconstruction for 32 measurement electrodes by multiply Elastic net: (**a**) 2 objects, (**b**) 3 objects, (**c**) 4 objects.

The same training set was used as for the previous method. Reconstructions for systems with 16 and 32 electrodes are shown in Figures 14 and 15, respectively. The diameters of inclusions are larger than the reference ones, however, the two-fold increase in the number of electrodes gives significantly better results. The accuracy of the mapping increases and the amount of background noise decreases.

*3.5. Comparison of Image Reconstructions*

Visual comparison of individual methods (Gauss-Newton, multiply Neural Networks, multiply LARS and multiply Elastic net) is not very precise. In order to increase the fairness of the comparison, special indicators calculated using mathematical methods were applied. To make this possible using the simulation method, reference images (vectors) were designed for all 6 cases examined. The dimensions of the physical model presented in Section 2.2 were used for this purpose. It is quite easy because in all tested variants the pixels of the background are white (value 1) and the identified objects (tubes) have a dark blue color (value 0) on the cross-section.

In order to compare the methods, the following evaluation metrics were used: Mean Squared Error (MSE), Relative Image Error (RIE), Image Correlation Coefficient (ICC) and the Expected Time of Image Reconstruction expressed in seconds. MSE is evaluated according to (13)

$$\text{MSE} = \frac{1}{n}\sum_{i=1}^{n}\left(y'_i - y^*_i\right)^2 \tag{13}$$

where $n$ is the output image resolution, $y'_i$ is the value of $i$ reference pixel, and $y^*_i$ is the value of $i$ reconstructed pixel.

RIE is evaluated according to (14).

$$\text{RIE} = \frac{\|y' - y^*\|}{\|y'\|} \tag{14}$$

ICC is evaluated according to (15).

$$\text{ICC} = \frac{\sum_{i=1}^{n}\left(y^*_i - \overline{y}^*\right)\left(y'_i - \overline{y}'\right)}{\sqrt{\sum_{i=1}^{n}\left(y^*_i - \overline{y}^*\right)^2 \sum_{i=1}^{n}\left(y'_i - \overline{y}'\right)^2}} \tag{15}$$

where: $\overline{y}'$ is the mean value for reference pixels; $\overline{y}^*$ is the mean value for reconstructed pixels.

The smaller the MSE and RIE, the better the reconstruction quality. ICC is vice versa, the closer to 1, the better the reconstruction.

Table 2 presents the analysis of the quality of image reconstruction for individual methods. The column headers contain information about the number of electrodes in the measurement system (E16 or E32) and the number of hidden objects (O2, O3, O4). For example, E16_O2 means a measuring system with 16 electrodes applied to a case with 2 objects hidden inside the tested tank.

**Table 2.** Comparison of image reconstruction indicators.

| Methods | Evaluation Metrics | Tested Cases | | | | | |
|---|---|---|---|---|---|---|---|
| | | E16_O2 | E16_O3 | E16_O4 | E32_O2 | E32_O3 | E32_O4 |
| ANN | MSE | 0.0074 | 0.0086 | 0.0076 | 0.0060 | 0.0061 | 0.0058 |
| | RIE | 0.0869 | 0.0936 | 0.0886 | 0.0782 | 0.0785 | 0.0771 |
| | ICC | 0.7356 | 0.7371 | 0.8218 | 0.7484 | 0.8163 | 0.7946 |
| | Expected time of image reconstruction [s] | 0.1501 | 0.1578 | 0.1574 | 0.2776 | 0.2785 | 0.2787 |
| Elastic net | MSE | 0.0111 | 0.0148 | 0.0197 | 0.0081 | 0.0131 | 0.0174 |
| | RIE | 0.2466 | 0.3499 | 0.3451 | 0.2120 | 0.2661 | 0.3300 |
| | ICC | 0.5024 | 0.4651 | 0.4535 | 0.5090 | 0.4785 | 0.4702 |
| | Expected time of image reconstruction [s] | 0.00062 | 0.00066 | 0.00071 | 0.0013 | 0.0014 | 0.0014 |
| LARS | MSE | 0.0115 | 0.0153 | 0.0203 | 0.0074 | 0.0121 | 0.0160 |
| | RIE | 0.1053 | 0.1216 | 0.1402 | 0.0871 | 0.1113 | 0.1280 |
| | ICC | 0.4658 | 0.4586 | 0.4438 | 0.5261 | 0.5072 | 0.5082 |
| | Expected time of image reconstruction [s] | 0.00041 | 0.00095 | 0.00092 | 0.0019 | 0.0018 | 0.0018 |
| Gauss-Newton with Laplace regulari-zation | MSE | 0.0199 | 0.0267 | 0.0351 | 0.0110 | 0.0164 | 0.0225 |
| | RIE | 0.1661 | 0.2524 | 0.3415 | 0.1563 | 0.1755 | 0.2402 |
| | ICC | 0.5290 | 0.4643 | 0.4181 | 0.5853 | 0.5984 | 0.5412 |
| | Expected time of image reconstruction [s] | 0.01248 | 0.01010 | 0.00940 | 0.01159 | 0.01229 | 0.01197 |

Analyzing the indicators in Table 2, it can be noticed that for all 6 tested variants and 3 indicators the best quality of reconstruction was obtained with the multiply ANN. The rest of the methods differ in relation to both variants and indicators. For example, in the reconstruction of E16_O2, the best MSE was with Elastic net (MSE = 0.0111), the best RIE was for LARS (RIE = 0.1053) and the best ICC was for Gauss-Newton with Laplace regularization (ICC = 0.5290). So, it is impossible to unambiguously determine the best method from the multiply Elastic net, multiply LARS and Gauss-Newton, but indisputably, the best results were obtained using ANN in the tested cases. At the same time, it can be noticed that multiply ANN is the slowest method among all the tested algorithms, while the fastest methods proved to be multiply Elastic net and multiply LARS.

The learning times of tomographic algorithms based on the analyzed methods belonging to the machine learning group depend on a lot of factors. For example, the training time of the multiply ANN for 16 electrodes, by employing one central processing unit (CPU) core took about 27 hours but with 24 cores it was 4.4 hours. The multiply Elastic net and multiply LARS methods are much faster than multiply ANN. With one core it was about 90 seconds and with 24 cores the training time was about 25 seconds. In the case of 32 electrodes, the training times are about 13% longer for ANN and 5 times longer for Elastic net and LARS.

## 4. Conclusions

The monitoring systems are aimed at automation, analysis and optimization of technological processes using industrial tomography, which allows for analysis of processes taking place in a facility without interfering with its interior. Such solutions enable better understanding and monitoring of industrial processes and facilitate process control in real time. The collected information is processed by an algorithm that reconstructs the image. This type of tomography is characterized by a relatively low image resolution. Difficulties in obtaining high resolution result mainly from a limited number of measurements, non-linear current flow through a given medium and too-low sensitivity of measured voltages depending on changes in conductivity in the area. The main challenge in this area is to design precise measuring devices and algorithms for image reconstruction.

Data analysis is an important part of the diagnosis of the process based on tomography. The inverse problem is the process of optimization, identification or synthesis, in which the parameters describing the field are determined based on the possession of information relevant to this field. Such problems are difficult to analyze. They do not have unambiguous solutions and are misunderstood due to too little or too much information. Knowledge of the process can make image reconstruction more resistant to incomplete information. In the article, the authors used the deterministic method based on the Gauss-Newton method with Laplace regularization as a reference for the selected machine learning methods.

In the process based on electrical tomography, there is no ideal method for reconstructing and analyzing data. Methods and models need to be properly selected depending on the specificity of the problem that needs to be solved. Deterministic methods are usually more awkward with many hidden objects requiring reconstruction. Multiply Neural Networks give better results, but this is mostly dependent on the quantity and quality of the training set. With a large training set, especially for smaller objects, they are really effective. Machine learning methods based on multiply LARS, especially multiply Elastic net, seem to be less accurate, especially for real measurement data, but they are much faster than multiply ANN. The disadvantage of ANN is the long training time and the relatively long reconstruction time. The obtained results were illustrated graphically, which gives the possibility of visual analysis of the processes taking place inside the object, as well as with the use of numerical indicators. The proposed algorithms and gained knowledge should bring benefits to various economic and industrial sectors.

Further works will be focused on improving the methods of image reconstruction using deep learning and the development of measuring devices for both electrical tomography and ultrasound tomography.

**Author Contributions:** T.R. has developed system concepts, research methods and implementation of solutions in industrial tomography of the presented techniques in this article. G.K. has implemented the neural network method. E.K. carried out research especially in the field of statistical methods. P.T. worked on deterministic algorithm and grids in Matlab.

**Acknowledgments:** The authors would like to thank the authorities and employees of the Institute of Mathematics, Maria Curie-Skłodowska University, Lublin, Poland for sharing supercomputing resources.

**Conflicts of Interest:** The authors declare no conflict of interest.

## References

1. González, G.; Huttunen, J.M.J.; Kolehmainen, V.; Seppänen, A.; Vauhkonen, M. Experimental evaluation of 3D electrical impedance tomography with total variation prior. *Inverse Probl. Sci. Eng.* **2016**, *24*, 1411–1431. [CrossRef]
2. Liu, S.; Jia, J.; Zhang, Y.D.; Yang, Y. Image reconstruction in electrical impedance tomography based on structure-aware sparse Bayesian learning. *IEEE Trans. Med. Imaging* **2018**, *37*, 2090–2102. [CrossRef]
3. Kang, S.I.; Khambampati, A.K.; Kim, B.S.; Kim, K.Y. EIT image reconstruction for two-phase flow monitoring using a sub-domain based regularization method. *Flow Meas. Instrum.* **2017**, *53*, 28–38. [CrossRef]
4. Ren, S.; Wang, Y.; Liang, G.; Dong, F. A Robust Inclusion Boundary Reconstructor for Electrical Impedance Tomography with Geometric Constraints. *IEEE Trans. Instrum. Meas.* **2018**, *99*, 1–12. [CrossRef]
5. Yang, Y.; Jia, J. An image reconstruction algorithm for electrical impedance tomography using adaptive group sparsity constraint. *IEEE Trans. Instrum. Meas.* **2017**, *66*, 2295–2305. [CrossRef]
6. Liu, D.; Zhao, Y.; Khambampati, A.K.; Seppänen, A.; Du, J. A Parametric Level set Method for Imaging Multiphase Conductivity Using Electrical Impedance Tomography. *IEEE Trans. Comput. Imaging* **2018**, *4*, 552–561. [CrossRef]
7. Rymarczyk, T. Using electrical impedance tomography to monitoring flood banks. *Int. J. Appl. Electromagn. Mech.* **2014**, *45*, 489–494. [CrossRef]

8. Rymarczyk, T.; Kłosowski, G. Application of neural reconstruction of tomographic images in the problem of reliability of flood protection facilities. *Eksploatacja I Niezawodnosc* **2018**, *20*, 425–434. [CrossRef]

9. Hamilton, S.J.; Hauptmann, A. Deep D-Bar: Real-Time Electrical Impedance Tomography Imaging with Deep Neural Networks. *IEEE Trans. Med. Imaging* **2018**, *37*, 2367–2377. [CrossRef]

10. Tavares, R.S.; Sato, A.K.; Martins, T.C.; Lima, R.G.; Tsuzuki, M.S.G. GPU acceleration of absolute EIT image reconstruction using simulated annealing. *Biomed. Signal Process. Control* **2017**. [CrossRef]

11. Tan, C.; Lv, S.; Dong, F.; Takei, M. Image Reconstruction Based on Convolutional Neural Network for Electrical Resistance Tomography. *IEEE Sens. J.* **2019**, *19*, 196–204. [CrossRef]

12. Farha, M. Combined Algorithm of Total Variation and Gauss-Newton for Image Reconstruction in Two-Dimensional Electrical Impedance Tomography (EIT). In Proceedings of the 2017 International Seminar on Sensor, Instrumentation, Measurement and Metrology (ISSIMM), Surabaya, Indonesia, 25–26 August 2017.

13. Yang, Y.; Jia, J.; Polydorides, N.; McCann, H. Effect of structured packing on EIT image reconstruction. In Proceedings of the 2014 IEEE International Conference on Imaging Systems and Techniques (IST) Proceedings, Santorini, Greece, 14–17 October 2014; pp. 53–58.

14. Wang, H.; Wang, C.; Yin, W. A pre-iteration method for the inverse problem in electrical impedance tomography. *IEEE Trans. Instrum. Meas.* **2004**, *53*, 1093–1096. [CrossRef]

15. Li, T.; Kao, T.J.; Isaacson, D.; Newell, J.C.; Saulnier, G.J. Adaptive Kaczmarz method for image reconstruction in electrical impedance tomography. *Physiol. Meas.* **2013**, *34*, 595–608. [CrossRef]

16. González, G.; Kolehmainen, V.; Seppänen, A. Isotropic and anisotropic total variation regularization in electrical impedance tomography. *Comput. Math. Appl.* **2017**, *74*, 564–576. [CrossRef]

17. Zhou, Y.; Li, X. A real-time EIT imaging system based on the split augmented Lagrangian shrinkage algorithm. *Measurement* **2017**, *110*, 27–42. [CrossRef]

18. Liu, X.; Yao, J.; Zhao, T.; Obara, H.; Cui, Y.; Takei, M. Image reconstruction under contact impedance effect in micro electrical impedance tomography sensors. *IEEE Trans. Biomed. Circuits Syst.* **2018**, *12*, 623–631. [CrossRef]

19. Alsaker, M.; Hamilton, S.J.; Hauptmann, A. A direct D-bar method for partial boundary data electrical impedance tomography with a priori information. *Inverse Probl. Imaging* **2017**, *11*, 427–454. [CrossRef]

20. Fernández-Fuentes, X.; Mera, D.; Gómez, A.; Vidal-Franco, I. Towards a Fast and Accurate EIT Inverse Problem Solver: A Machine Learning Approach. *Electronics* **2018**, *7*, 422. [CrossRef]

21. Brillante, L.; Bois, B.; Mathieu, O.; Lévêque, J. Electrical imaging of soil water availability to grapevine: A benchmark experiment of several machine-learning techniques. *Precis. Agric.* **2016**, *17*, 637–658. [CrossRef]

22. Rymarczyk, T.; Kozłowski, E. Using Statistical Algorithms for Image Reconstruction in EIT. In Proceedings of the MATEC Web Conferences, Majorca, Spain, 14–17 July 2018; Volume 210, p. 02017.

23. Rymarczyk, T.; Kłosowski, G.; Kozłowski, E. Non-Destructive System Based on Electrical Tomography and Machine Learning to Analyze Moisture of Buildings. *Sensors* **2018**, *18*, 2285. [CrossRef]

24. Hoyle, B.S. IPT in Industry—Application Need to Technology Design. In Proceedings of the ISIPT 8th World Congress in Industrial Process Tomography, Igaussu Falls, Brazil, 26–29 September 2016; pp. 1–7.

25. Rymarczyk, T.; Adamkiewicz, P.; Polakowski, K.; Sikora, J. Effective ultrasound and radio tomography imaging algorithm for two-dimensional problems. *Przegląd Elektrotechniczny* **2018**, *94*, 62–69.

26. Romanowski, A. Contextual Processing of Electrical Capacitance Tomography Measurement Data for Temporal Modeling of Pneumatic Conveying Process. In Proceedings of the 2018 Federated Conference on Computer Science and Information Systems (FedCSIS), Poznan, Poland, 9–12 September 2018; pp. 283–286.

27. Rymarczyk, T.; Kłosowski, G.; Gola, A. The Use of Artificial Neural Networks in Tomographic Reconstruction of Soil Embankments. In Proceedings of the International Symposium on Distributed Computing and Artificial Intelligence, Toledo, Spain, 20–22 June 2018; Springer: Cham, Switzerland, 2018; pp. 104–112.

28. Rymarczyk, T. New Methods to Determine Moisture Areas by Electrical Impedance Tomography. *Int. J. Appl. Electromagn. Mech.* **2016**, *52*, 79–87. [CrossRef]

29. Szczęsny, A.; Korzeniewska, E. Selection of the method for the earthing resistance measurement. *Przegląd Elektrotechniczny* **2018**, *94*, 178–181.

30. Liu, S.; Wu, H.; Huang, Y.; Yang, Y.; Jia, J. Accelerated Structure-Aware Sparse Bayesian Learning for 3D Electrical Impedance Tomography. *IEEE Trans. Ind. Inform.* **2019**. [CrossRef]

31. Kozłowski, E.; Mazurkiewicz, D.; Kowalska, B.; Kowalski, D. Binary linear programming as a decision-making aid for water intake operators. In Proceedings of the International Conference on Intelligent Systems in Production Engineering and Maintenance, Wroclaw, Poland, 28–29 September 2017; Springer: Cham, Switzerland, 2017; pp. 199–208.

32. Kłosowski, G.; Kozłowski, E.; Gola, A. Integer linear programming in optimization of waste after cutting in the furniture manufacturing. *Adv. Intell. Syst. Comput.* **2018**, *637*, 260–270.

33. Wang, M. *Industrial Tomography: Systems and Applications*; Woodhead Publishing: Sawston, UK, 2015.

34. Holder, D. *Introduction to Biomedical Electrical Impedance Tomography Electrical Impedance Tomography Methods, History and Applications*; Institute of Physics: Bristol, UK, 2005.

35. Karhunen, K.; Seppänen, A.; Kaipio, J.P. Adaptive meshing approach to identification of cracks with electrical impedance tomography. *Inverse Probl. Imaging* **2014**, *8*, 127–148. [CrossRef]

36. Rymarczyk, T.; Adamkiewicz, P.; Duda, K.; Szumowski, J.; Sikora, J. New electrical tomographic method to determine dampness in historical buildings. *Arch. Electr. Eng.* **2016**, *65*, 273–283. [CrossRef]

37. Al Hosani, E.; Soleimani, M. Multiphase permittivity imaging using absolute value electrical capacitance tomography data and a level set algorithm. *Philos. Trans. R. Soc. A* **2016**, *374*, 20150332. [CrossRef]

38. Kryszyn, J.; Wanta, D.; Smolik, W. Gain Adjustment for Signal-to-Noise Ratio Improvement in Electrical Capacitance Tomography System EVT4. *IEEE Sens. J.* **2017**, *17*, 8107–8116. [CrossRef]

39. Banasiak, R.; Wajman, R.; Jaworski, T.; Fiderek, P.; Fidos, H.; Nowakowski, J. Study on two-phase flow regime visualization and identification using 3D electrical capacitance tomography and fuzzy-logic classification. *Int. J. Multiphase Flow* **2014**, *58*, 1–14. [CrossRef]

40. Garbaa, H.; Jackowska-Strumiłło, L.; Grudzień, K.; Romanowski, A. Application of electrical capacitance tomography and artificial neural networks to rapid estimation of cylindrical shape parameters of industrial flow structure. *Arch. Electr. Eng.* **2016**, *65*, 657–669. [CrossRef]

41. Kryszyn, J.; Smolik, W. Toolbox for 3d modelling and image reconstruction in electrical capacitance tomography. *Informatyka Automatyka Pomiary w Gospodarce i Ochronie Środowiska (IAPGOŚ)* **2017**, *7*, 137–145.

42. Soleimani, M.; Mitchell, C.N.; Banasiak, R.; Wajman, R.; Adler, A. Four-dimensional electrical capacitance tomography imaging using experimental data. *Prog. Electromagn. Res.* **2009**, *90*, 171–186. [CrossRef]

43. Ye, Z.; Banasiak, R.; Soleimani, M. Planar array 3D electrical capacitance tomography. *Insight-Non-Destr. Test. Cond. Monit.* **2013**, *55*, 675–680. [CrossRef]

44. Wajman, R.; Fiderek, P.; Fidos, H.; Sankowski, D.; Banasiak, R. Metrological evaluation of a 3D electrical capacitance tomography measurement system for two-phase flow fraction determination. *Meas. Sci. Technol.* **2013**, *24*, 065302. [CrossRef]

45. Romanowski, A. Big Data-Driven Contextual Processing Methods for Electrical Capacitance Tomography. *IEEE Trans. Ind. Inform.* **2019**, *15*, 1609–1618. [CrossRef]

46. Kłosowski, G.; Rymarczyk, T.; Gola, A. Increasing the Reliability of Flood Embankments with Neural Imaging Method. *Appl. Sci.* **2018**, *8*, 1457. [CrossRef]

47. Demidenko, E.; Hartov, A.; Paulsen, K. Statistical estimation of Resistance/Conductance by electrical impedance tomography measurements. *IEEE Trans. Med. Imaging* **2004**, *23*, 829–838. [CrossRef]

48. Adler, A.; Lionheart, W.R. Uses and abuses of EIDORS: An extensible software base for EIT. *Physiol. Meas.* **2006**, *27*, S25. [CrossRef] [PubMed]

49. Dušek, J.; Hladký, D.; Mikulka, J. Electrical Impedance Tomography Methods and Algorithms Processed with a GPU. In Proceedings of the 2017 Progress In Electromagnetics Research Symposium-Spring (PIERS), St. Petersburg, Russia, 22–25 May 2017; pp. 1710–1714.

50. Rymarczyk, T.; Sikora, J. Applying industrial tomography to control and optimization flow systems. *Open Phys.* **2018**, *16*, 332–345. [CrossRef]

51. Voutilainen, A.; Lehikoinen, A.; Vauhkonen, M.; Kaipio, J. Three-dimensional nonstationary electrical impedance tomography with a single electrode layer. *Meas. Sci. Technol.* **2010**, *21*, 035107. [CrossRef]

52. Babout, L.; Grudzień, K.; Wiącek, J.; Niedostatkiewicz, M.; Karpiński, B.; Szkodo, M. Selection of material for X-ray tomography analysis and DEM simulations: Comparison between granular materials of biological and non-biological origins. *Granul. Matter* **2018**, *20*, 20–38. [CrossRef]

53. Mikulka, J. GPU—Accelerated Reconstruction of T2 Maps in Magnetic Resonance Imaging. *Meas. Sci. Rev.* **2015**, *4*, 210–218. [CrossRef]

54. Bartušek, K.; Fiala, P.; Mikulka, J. Numerical Modeling of Magnetic Field Deformation as Related to Susceptibility Measured with an MR System. *Radioengineering* **2008**, *17*, 113–118.

55. Lopato, P.; Chady, T.; Sikora, R.; Ziolkowski, M. Full wave numerical modelling of terahertz systems for nondestructive evaluation of dielectric structures. *COMPEL* **2013**, *32*, 736–749. [CrossRef]

56. Vališ, D.; Mazurkiewicz, D. Application of selected Levy processes for degradation modelling of long range mine belt using real-time data. *Arch. Civil Mech. Eng.* **2018**, *18*, 1430–1440. [CrossRef]

57. Ziolkowski, M.; Gratkowski, S.; Zywica, A.R. Analytical and numerical models of the magnetoacoustic tomography with magnetic induction. *COMPEL* **2018**, *37*, 538–548. [CrossRef]

58. Hastie, T.; Tibshirani, R.; Friedman, J. *The Elements of Statistical Learning Data Mining, Inference, and Prediction*; Springer: New York, NY, USA, 2009.

59. Madsen, K.; Nielsen, H.; Tingleff, O. *Methods for Non-Linear Least Squares Problems*, 2nd ed.; Informatics and Mathematical Modelling, Technical University of Denmark: Lyngby, Denmark, 2004; p. 60.

60. Fonseca, T.; Goliatt, L.; Campos, L.; Bastos, F.; Barra, L.; Santos, R. Machine Learning Approaches to Estimate Simulated Cardiac Ejection Fraction from Electrical Impedance Tomography. In Proceedings of the Ibero-American Conference on Artificial Intelligence (IBERAMIA 2016), LNAI 10022, San José, Costa Rica, 23–25 November 2016; pp. 235–246.

61. Zou, H.; Hastie, T. Regularization and variable selection via the elastic net. *J. R. Stat. Soc. Ser. B* **2005**, *2*, 301–320. [CrossRef]

62. Tibshirani, R. Regression shrinkage and selection via the lasso. *J. R. Stat. Soc. Ser. B* **1996**, *58*, 267–288. [CrossRef]

63. Wang, J.; Han, B.; Wang, W. Elastic-net regularization for nonlinear electrical impedance tomography with a splitting approach. *Appl. Anal.* **2018**, 1–17. [CrossRef]

64. Raskutti, G.; Wainwright, M.J.; Yu, B. Early stopping and non-parametric regression: An optimal data-dependent stopping rule. *J. Mach. Learn. Res.* **2014**, *15*, 335–366.

*sensors* 

*Article*

# Multiple Wire-Mesh Sensors Applied to the Characterization of Two-Phase Flow inside a Cyclonic Flow Distribution System [†]

César Y. Ofuchi [1,2,*], Henrique K. Eidt [2,3], Carolina C. Rodrigues [2,3], Eduardo N. dos Santos [1,2], Paulo H. D. dos Santos [2,3], Marco J. da Silva [1,2], Flávio Neves, Jr. [1], Paulo Vinicius S. R. Domingos [4] and Rigoberto E. M. Morales [2,3]

[1]   Graduate Program in Electrical and Computer Engineering (CPGEI), Federal University of Technology-PR, Curitiba 80230-901, Brazil; e.n.santos@ieee.org (E.N.d.S.); mdasilva@utfpr.edu.br (M.J.d.S.); neves@utfpr.edu.br (F.N.J.)
[2]   Multiphase Flow Research Center (NUEM), Federal University of Technology-PR, Curitiba 80230-901, Brazil; hkeidt@gmail.com (H.K.E.); carolcimarelli@gmail.com (C.C.R.); psantos@utfpr.edu.br (P.H.D.d.S.); rmorales@utfpr.edu.br (R.E.M.M.)
[3]   Graduate Program in Mechanical and Materials Engineering (PPGEM), Federal University of Technology-PR, Curitiba 80230-901, Brazil
[4]   Petróleo Brasileiro S.A. (Petrobras), Rio de Janeiro 21941-915, Brazil; paulodomingos@petrobras.com.br
*   Correspondence: ofuchi@utfpr.edu.br
†   This paper is an extended version of our paper: César Yutaka Ofuchi, Eduardo N. dos Santos, Eduardo Henrique Eidt, Carolina Cimarelli, Paulo Henrique Santos, Marco José da Silva, Flávio Neves, Jr., Rigoberto E. M. Morales. Investigation of the Gas-Liquid Flow inside a Cyclonic Flow Distribution System Using Wire-Mesh Sensors. In Proceedings of the 9th World Congress on Industrial Process Tomography, Bath, UK, 2–6 September 2018.

Received: 25 November 2018; Accepted: 21 December 2018; Published: 7 January 2019

**Abstract:** Wire-mesh sensors are used to determine the phase fraction of gas–liquid two-phase flow in many industrial applications. In this paper, we report the use of the sensor to study the flow behavior inside an offshore oil and gas industry device for subsea phase separation. The study focused on the behavior of gas–liquid slug flow inside a flow distribution device with four outlets, which is part of the subsea phase separator system. The void fraction profile and the flow symmetry across the outlets were investigated using tomographic wire-mesh sensors and a camera. Results showed an ascendant liquid film in the cyclonic chamber with the gas phase at the center of the pipe generating a symmetrical flow. Dispersed bubbles coalesced into a gas vortex due to the centrifugal force inside the cyclonic chamber. The behavior favored the separation of smaller bubbles from the liquid bulk, which was an important parameter for gas-liquid separator sizing. The void fraction analysis of the outlets showed an even flow distribution with less than 10% difference, which was a satisfactorily result that may contribute to a reduction on the subsea gas–liquid separators size. From the outcomes of this study, detailed information regarding this type of flow distribution system was extracted. Thereby, wire-mesh sensors were successfully applied to investigate a new type of equipment for the offshore oil and gas industry.

**Keywords:** wire-mesh; flow distribution; subsea gas–liquid separation; two-phase flow; cyclonic chamber

---

## 1. Introduction

Sensing technology for two-phase flow monitoring has evolved from simple visualization techniques and global parameters measurement (such as pressure drop and temperature) to use of tomographic and imaging techniques to discover details of flow behavior in pipes and equipment.

The current use of computational power to simulate complex flow and to predict its behavior has also pushed the development of measurement techniques to measure flow parameters with greater detail, i.e., high spatial and temporal resolution [1]. Among tomographic techniques based on a variety of measuring principles—such as gamma-ray, X-ray, impedance, ultrasound and others—a technique known as the wire-mesh sensor has emerged as a very competitive alternative, due to its high spatial and temporal resolution (up to few millimeters and few kilohertz range). It has been applied in a number of pilot plant studies around the world [2–6]. Hence, in this paper, the wire-mesh sensor was applied for the first time to characterize the flow inside reduced-scale subsea equipment from the oil industry.

Offshore deep-water discoveries have driven the interest of the industry in new subsea separation technologies. According to the International Energy Agency, the petroleum withdrawn on platforms represents 30% of all world production. The crude oil of offshore reservoirs is generally mixed with many components, such as liquid water, gas and solids. In this way, devices are used to separate those components, e.g., gravitational separator devices in the offshore platform [7].

In the past years, there is a trend to move the processing unit from the topside to the seabed level, to optimize oil production in deep water environments [8]. Separation of liquid and gas phases at the wellhead level can mitigate the hydrate risk after the separation and increase liquid boosting of submersible centrifugal pumps, or any artificial lift process used, among other advantages. Subsea separation also allows the debottlenecking of topside water treatment (water/hydrocarbon separation and subsea water reinjection) [9]. Gravitational and centrifugal separators are generally used for this purpose due to their high separation efficiency. However, they require large dimensions, which makes the construction, installation and maintenance of this equipment difficult at offshore deep-water applications.

Numerous alternatives to conventional separators based on gravity and centrifugal flow have been proposed. Separators that use the cyclone concepts as VASPS (Vertical Annular Separation and Pumping System) [10], include the CS (Cyclone Separator) [11,12] and the GLCC© (Gas–Liquid-Cylindrical Cyclone) [13,14]. The concept is based on tangential inlets in vertical pipes to provide the swirling motion to the incoming mixture. The resulting centrifugal force enhances phase separation since the liquid phase flows near the wall and the gas phase flows in the middle of the pipe, which is induced by the difference of density between the phases. Compact GLCC separators are smaller than conventional separators and could reduce the costs of the development of an oilfield. However, reducing the size of separators also reduces the separator robustness to handle fluctuations in the flow rate and composition. Hence, some authors proposed the combination of cyclone devices as pre-separators or as partial separators [15]. Another alternative to solve the problems of conventional cyclonic separators is to reduce its dimensions by distributing the flow to smaller vessels with reduced wall thickness [16]. The distributor ideally will produce symmetrical flow rates across all outlets to optimize the separator dimensions project.

In this context, a novel design of a flow distributor system, proposed by the authors, is experimentally investigated. Tomographic instrumentation and a camera were applied to evaluate the two-phase flow behavior inside the distributor device presented in Figure 1. A cyclonic chamber, in which the entries are tangentially located at the bottom and outlets at the top, was used as a pre-distribution/separation device. Due to the positioning of the inputs being tangential in the distributor, an ascendant liquid film flow, driven by a centrifugal field, results when the liquid-gas mixture enters the cyclonic chamber. This flow has the characteristic, in which a thin liquid film flows near the wall under the action of centrifugal and gravitational fields until the outlets. The device could be coupled with gas-liquid separators such as the VASPS. The study focused on the slug flow pattern, since it is the worst-case scenario, as the gas and the liquid flow could be unevenly distributed to different separators. Wire-mesh sensors were used to identify the flow pattern and to measure the void fraction at the input, inside the cyclonic chamber and at the four outputs. A camera was also used to capture the flow behavior through the transparent pipeline.

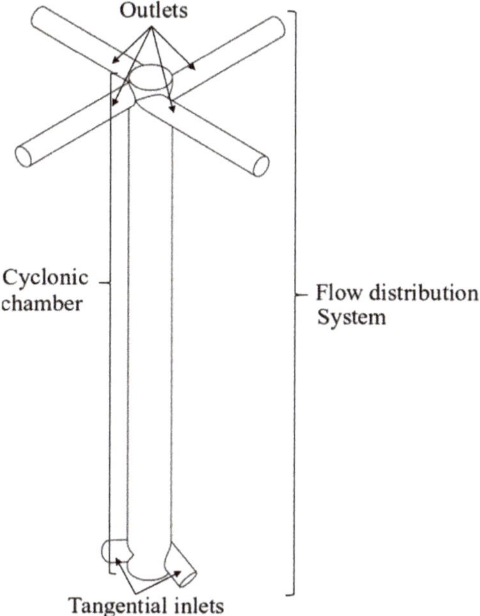



**Figure 1.** Schematic representation of the flow distribution system.

## 2. Experimental Setup

Figure 2 shows a schematic representation of the experimental flow loop used in this work. The water-air mixture flows through an acrylic pipe of 26 mm (1 inch) internal diameter and 7 m in length. Tap water circulates throughout a closed loop using a pump, and air flow is produced by a compressor and stored in a vessel. Air and water flow rates are independently measured by a rotameter and a Coriolis flow meter. The pipe ends in a vertical direction, where the flow is divided into two channels and tangentially enters the cyclonic chamber. The flow is distributed through four outputs with a 13 mm diameter (1/2 inch). The rotameter readings are compensated by the pressure difference at the pipe entrance and the pressure at the measurement point. Temperatures, pressures and flow rates are monitored using industrial sensors connected by Foundation Fieldbus. The flow rate is also controlled by a frequency inverter, by National Instrument System acquisition, using LabVIEW. We wanted to evaluate the slug flow pattern at the input, so we used a flow map (Figure 3), which is a simple method of determining flow regimes based upon a known range of phase velocities. We followed the flow map elaborated by [17] for the setup of gas and liquid superficial velocities (The superficial velocity is normally defined as $Q/A$, where $Q$ is the volumetric flow rate (e.g., m$^3$/s) of the fluid, and $A$ is the cross-sectional area (e.g., m$^2$)), as it had the same vertical, 1 inch, water-air two-phase flow as in our experiment. The highlighted rectangle comprises a test setup with superficial gas velocities ($J_G$) from 0.5 to 1.0 m/s, and superficial liquid velocities ($J_L$) from 0.5 to 2.0 m/s with a 0.5 m/s step.

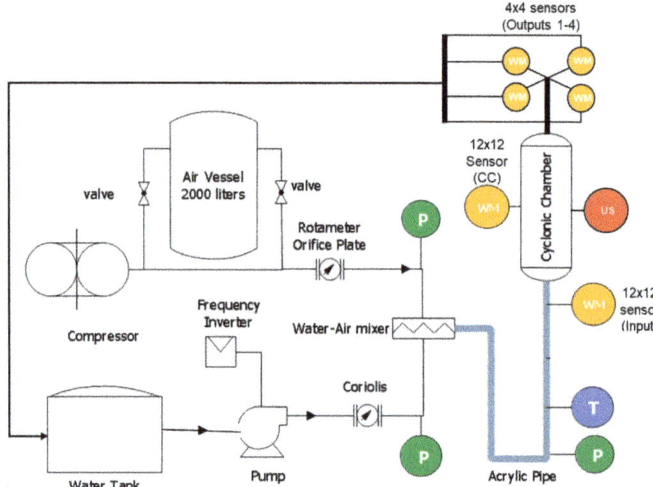


**Figure 2.** Schematic representation of the measurement plant with the wire-mesh sensors (WMS), ultrasound (US), pressure and temperature sensors (P) and (T).

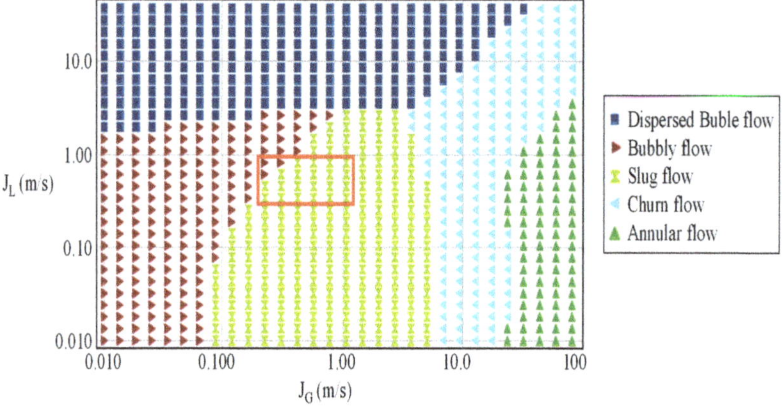


**Figure 3.** Flow map test setup for slug flow based on [17].

To investigate the behavior of the liquid-gas flow inside the distribution system, we analyzed the variation of the void fraction through the system. A wire-mesh sensor (WMS) was used to measure the void fraction [18]. This technique is a reliable flow visualization tool with high spatial and temporal resolution. The sensor consists of transmitter and receiver wires measuring the electrical properties of the flow within its slightly spaced crossing points. The transmitter electrodes are consecutively activated while keeping all other transmitter electrodes at ground potential. The pair of electrodes measured the electrical permittivity of the surrounding flow phase at the crossing point and then converted it to phase fraction, by calibration routines taken before the measurements. More details on the capacitive wire-mesh sensor can be found at [19]. The electronics, commercialized by HZDR Innovation GmbH, was configured with a frame-rate of 1000 Hz. The gas-liquid mixture flows through a 12 × 12 WMS at the system input, then into a 12 × 12 WMS inside the cyclonic chamber, and finally through the four 4 × 4 WMS at the system output, as described in Figure 2. The number of wires is proportional to the pipeline diameter: 1 inch uses 12 × 12 wires and 1/2 inch uses 4 × 4 wires. A simple camera was used to register the flow image at the same wire-mesh acquisition points as a comparison.

The gas fraction $\alpha$ is achieved using the acquired voltages $V$ for all crossing points. The values are calibrated using the sensor covered with a high permittivity material $V_H$ (liquid water) and a low permittivity material $V_L$ (air). Equation (1) shows the relation between the variables. Data are stored in a three-dimensional matrix $\alpha(i,j,k)$, where $i$ and $j$ denote the wire indices and $k$ is the temporal sampling point index (Figure 4).

$$\alpha(i,j,k) = \frac{V(i,j,k) - V_L(i,j)}{V_H(i,j) - V_L(i,j)} \tag{1}$$

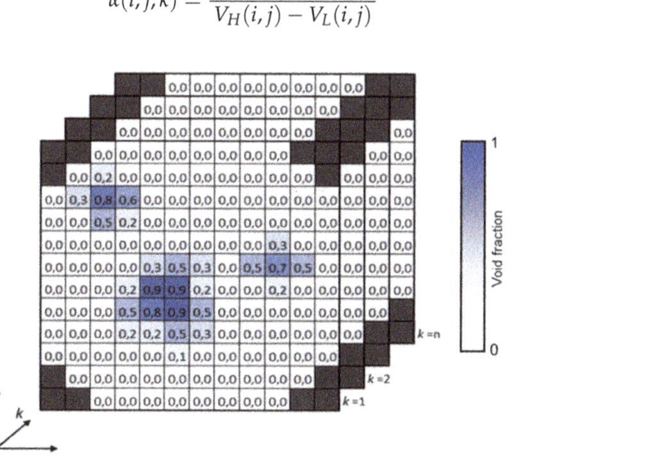

**Figure 4.** Spatial coordinates of the wire-mesh processing data. Void fraction is represented in the color scale.

Based on the $\alpha$ matrix, one can process the void fraction data to analyze the flow behavior. Figure 5 depicts a different data representation of wire-mesh sensor readings, which are explained by Equations (2)–(4). More details can be found on [19].

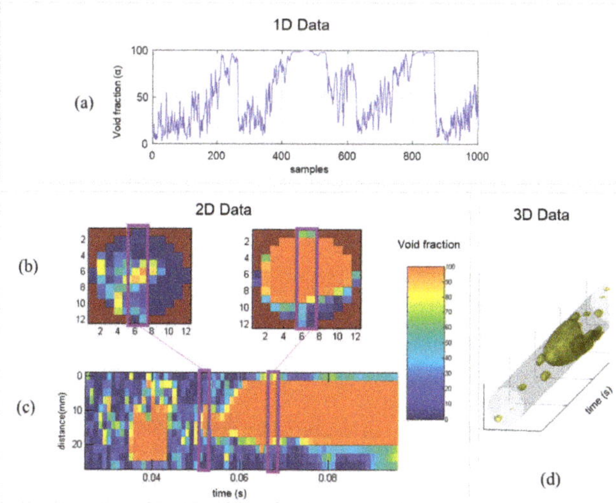

**Figure 5.** Signal processing of water/gas slug flow acquired with the wire-mesh sensor: (**a**) Averaged void fraction time series, (**b**) cross-sectional and (**c**) axial cut slice images, and (**d**) three-dimensional isosurface plot to view the gas-phase boundaries.

We evaluate the overall void fraction behavior of the flow by using the one-dimensional data of the averaged void fraction ($\bar{\alpha}$) time-series, presented in Figure 5a and Equation (2):

$$\bar{\alpha}(k) = \frac{1}{N_i N_j} \sum_{i}^{N_i} \sum_{j}^{N_j} \alpha(i,j,k),$$

(2)

where $N_i$ and $N_j$ are the total number of samples in each axis.

To analyze the flow structure in detail we extract a two-dimensional void fraction data of the cross-section $\bar{\alpha}_{cross}$, represented by Figure 5b and Equation (3):

$$\bar{\alpha}_{cross}(i,j) = \frac{1}{N_k} \sum_{k}^{N_k} \alpha(i,j,k),$$

(3)

where $N_k$ is the total number of time samples. Additional 2D data are obtained by selecting a specific chord, $i_{chord}$ (which in general, is the central chord), to get an axial cut of the void fraction matrix, as described in Figure 5c and Equation (4):

$$\alpha(j,k) = \alpha(i_{chord}, j, k).$$

(4)

In contrast to making slices, a different way of exploring volumetric data is to view the three-dimensional boundaries. In Figure 5d, we create a polygon using the isosurface data from the $\alpha$ matrix and an isovalue (threshold). We use it to view the void fraction structure at the input to confirm the flow pattern. In addition, we can view the flow structure inside the cyclonic chamber, which will be clear in the Results section.

## 3. Results and Discussion

In the first test, wire-mesh sensors were installed at the input, inside the cyclonic chamber, and at two outlets. We proposed a qualitative analysis of the void fraction flow pattern in all test sections. We used a camera in the same position as the WMS, installed in the cyclonic chamber afterward to compare the flow structure. In the second test, WMS were installed at the input and the four outlets. The flow distribution rate at each outlet was quantitatively analyzed using a set of slug flow patterns at the input. Table 1 shows the wire-mesh setup in each experiment.

**Table 1.** Arrangement of wire-mesh sensors in each test.

| Test | Input | Cyclonic Chamber | Outlet |
| --- | --- | --- | --- |
| Flow pattern analysis | 12 × 12 | 12 × 12 | Two 4 × 4 |
| Flow distribution analysis | 12 × 12 | – | Four 4 × 4 |

*3.1. Flow Pattern Analysis*

In this test setup, we evaluated the void fraction of the intermittent flow through the distribution system. The first result was a comparison between the slug flow at the input and inside the cyclonic chamber. Camera and 3D wire-mesh data images are presented in Figure 6.

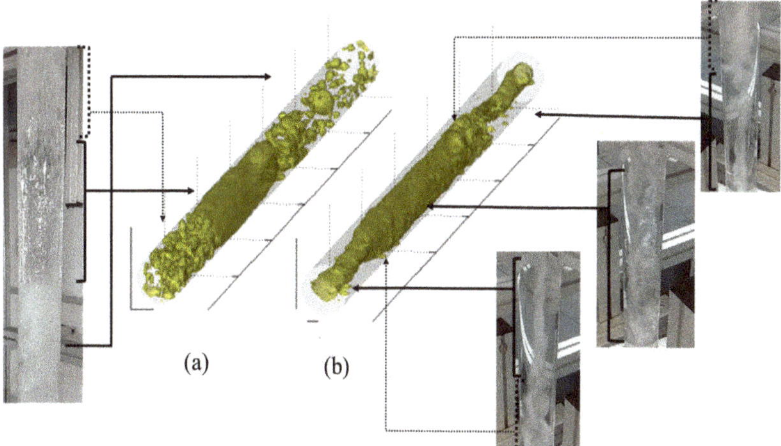

**Figure 6.** Slug flow images with wire-mesh and camera: (**a**) Input and (**b**) inside the cyclonic chamber.

Figure 6a shows the slug flow pattern at the input for superficial gas velocities ($J_G$) of 1.5 m/s and liquid velocities ($J_L$) of 1.0 m/s, which had a dispersed bubbly gas phase followed by a Taylor bubble. The direct tomographic data from the wire-mesh sensor enabled a better view inside the mixture. Figure 6b depicts the flow pattern inside the cyclonic chamber. Centrifugal forces pushed the liquid phase onto the wall of the pipeline with a formation of a gas core in the center. The intermittent gas-liquid flow pattern changed to a rotating flow, where the dispersed bubbles coalesced in a gas vortex.

The same behavior occurred in another test, with only dispersed bubbles ($J_G$ = 0.5 m/s and $J_L$ = 1.5 m/s), as presented in Figure 7. Wire-mesh data at the input showed the bubbles at the input and the resulting gas vortex inside the cyclonic chamber. This behavior was highly desirable in the gas-liquid separator system context. Most separators are sized to provide enough retention time to allow gas bubbles to form and separate out [20]. Hence, the cyclonic chamber may reduce the separator size, in addition to the flow distribution advantage.

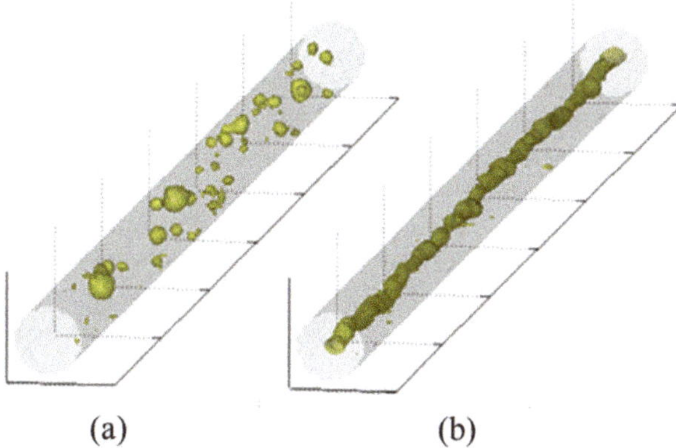

**Figure 7.** Tomographic wire-mesh data in a dispersed bubbly flow with ($J_G$) = 0.5 m/s and ($J_L$) = 1.5 m/s. (**a**) Dispersed bubble at the input and (**b**) gas vortex inside the cyclonic chamber.

Flow distribution can be measured based on the instrumented outlets. Figure 8 depicts a detailed axial and longitudinal view of the void fraction measured at the input, inside the cyclonic chamber and at two outlets with the WMS. Original data were synchronized, as the measurement occurs in different test points.

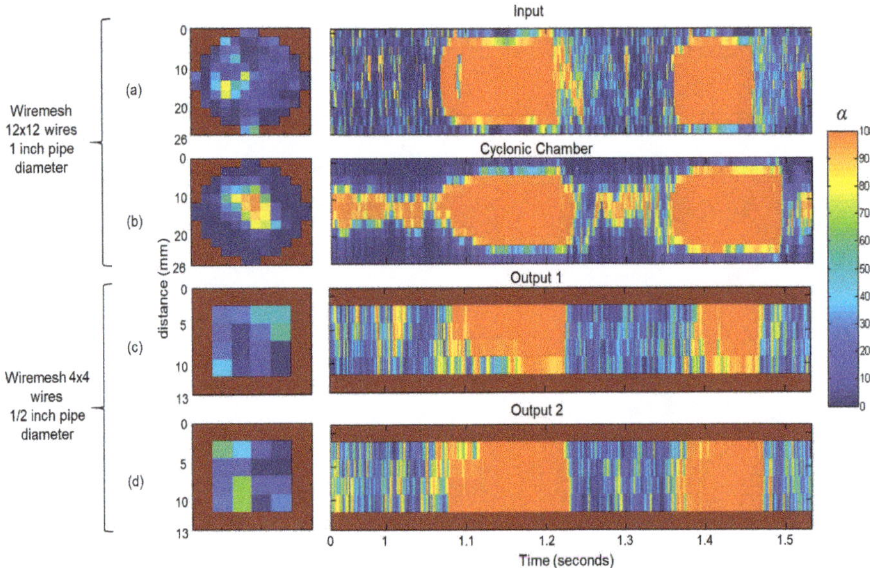



**Figure 8.** Axial and longitudinal view of void fraction using the WMS in two-phase flow with $J_L = 1.0$ m/s and $J_G = 1.0$ m/s. (**a**) Input, (**b**) cyclonic chamber, (**c**) outlet 1 and (**d**) outlet 2.

Figure 8a shows the slug flow pattern at the input. Large gas bubbles were separated by liquid slugs and smaller bubbles were mixed in the liquid slug. The changed flow pattern inside the cyclonic chamber is presented in Figure 8b. Gas and liquid phases maintained the same intermittency as the input. The main difference was the dispersion of the bubbles, inside the liquid slug, which concentrated in the middle of the pipeline due to the centrifugal force. The void fraction at the outlets also followed the slug flow pattern and were equally distributed (Figure 8c,d). Besides the symmetry in flow distribution, the synchronized comparison also showed that the intermittent behavior propagated from the input to the outlets. We also observed the mean void fraction behavior in the input, inside the cyclonic chamber and at the two outlets as the liquid and gas superficial velocities increased.

Figure 9 shows the mean void fraction measured by the WMS for three sets of superficial velocities: $J_L = 0.5/J_G = 0.5$, $J_L = 1.0/J_G = 1.0$ and $J_L = 1.0/J_G = 2.0$. Figure 9a depicts the void fraction at the input and Figure 9b inside the cyclonic chamber using $12 \times 12$ WMS. The mean gas phase was dispersed at the input and more concentrated in the center of the pipeline, inside the cyclonic chamber with a liquid film around. The effect was better observed in lower gas velocities. In higher gas velocities, the gas phase almost occupied the entire pipeline. With $J_L = 1.0$ and $J_G = 2.0$, the mean void fraction profile at the input was almost the same as inside the cyclonic chamber. Figure 9c,d presents the void fraction in the outlets. As they were horizontally oriented, the liquid phase stayed at the bottom of the pipeline. As already observed, both outlets had the same void fraction profile.

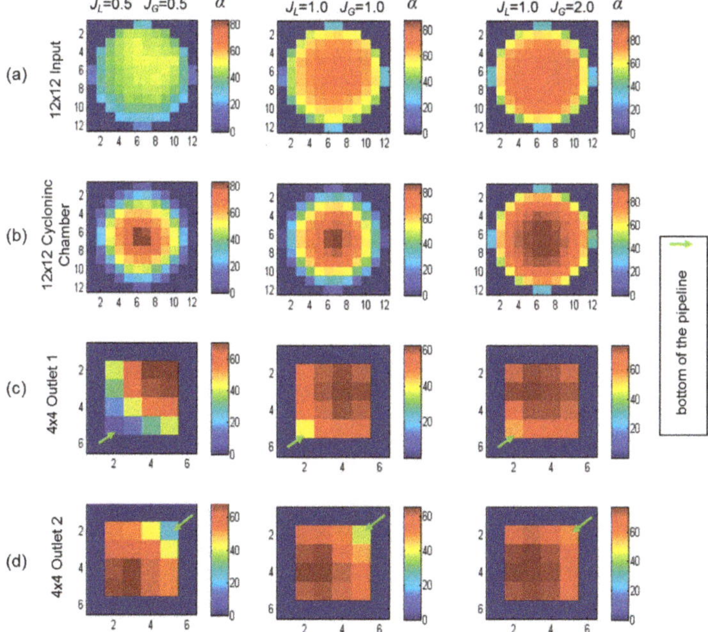

**Figure 9.** Axial view of the mean void fraction by WMS for three sets of gas and liquid superficial velocities: (**a**) Input using 12 × 12 WMS, (**b**) inside the cyclonic chamber using 12 × 12 WMS, (**c**) outlet 1 using 4 × 4 WMS, and (**d**) outlet 2 using 4 × 4 WMS. The bottom of the pipeline is indicated in the outlets.

### 3.2. Flow Distribution Analysis

In this test setup, the WMS was installed at the input and at the four outlets to verify the flow distribution, according to the input slug flow pattern (Figure 10). Original time series was synchronized, as the measurement occurs in different test points. A slug flow of $J_L = 1.0$ and $J_G = 1.0$ was observed at the input (Figure 10a) and propagated through the outlets (Figure 10b–e) with the same liquid-gas profile. The flow distribution was visually symmetrical in the outlets.

A longer void fraction time-series of 30 s, using superficial gas and liquid velocities of 1.5 m/s is presented in Figure 11. In this setup, numerous gas pockets followed by liquid bulks were observed, which enabled a reasonable quantitative analysis of the flow distribution. The void fraction at the input (59%) agreed with the mean void fraction at the outlets (62.40%). The distributed void fraction between the outlets was almost the same, with a worst-case difference of approximately 3.5% between outlet 3 (63.7%) and outlet 4 (60.2%).

We also compared the mean void fraction profile in three different sets of velocities: $J_L = 0.5$ m/s and $J_G = 0.5$ m/s; $J_L = 1.0$ m/s and $J_G = 1.0$ m/s; $J_L = 1.0$ m/s and $J_G = 2.0$ m/s (Figure 12). The mean void fraction at the input increased from 44.8% to 68.8%, as the superficial gas velocity increased from 0.5 m/s to 2 m/s. An important result was the similarity between the void fraction profiles of the four outlets, even with different superficial velocities.

The complete result is summarized in Table 2, where eleven different sets of superficial gas and liquid velocities are evaluated. These results show a void fraction difference between the outlets to a worst-case difference of 8.2%. Using this flow distribution system, four smaller separators with 1/4 of the original size may be used with only an additional tolerance of 10% due to the uneven flow distribution. The difference between the outlet's decay at higher flow rates ($J_L \geq 1.0$ m/s and $J_G \geq 1.0$ m/s), allowing a tolerance of around 5%.

Despite using the same sensors in the outlets, we needed to consider the effects of the 4 × 4 WMS uncertainties. According to [21], WMS can measure the void fraction within 11%—relative to other methods—but the accuracy can be better, depending on the flow pattern. There was also a spatial resolution issue, as WMS pitch effects caused differences in the order of 1 to 4% for bubbly flows, but up to 20% peak at a low void fraction. In future work, we may use a higher resolution WMS and also evaluate the liquid flow rate after the separation process to enhance this study.

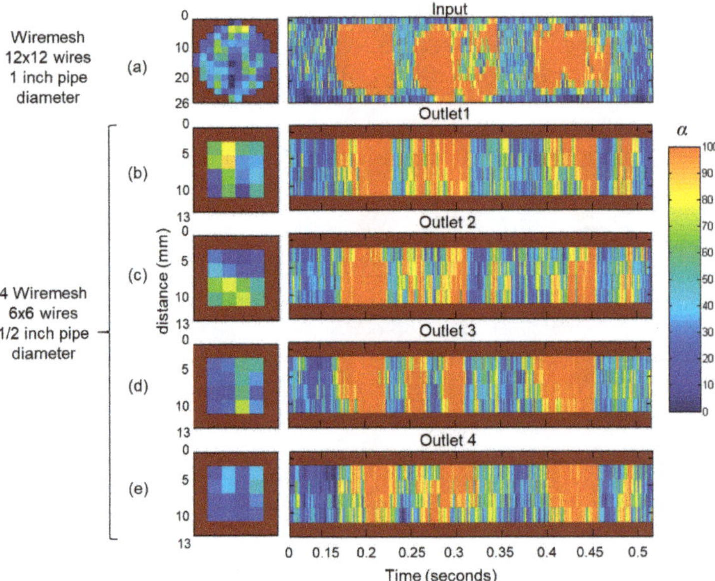


**Figure 10.** Axial and transversal view of the void fraction, using the wire-mesh sensor in two-phase flow with $J_L = 1.0$ m/s and $J_G = 1.0$ m/s. (**a**) Input and (**b**–**e**) outlets 1–4.

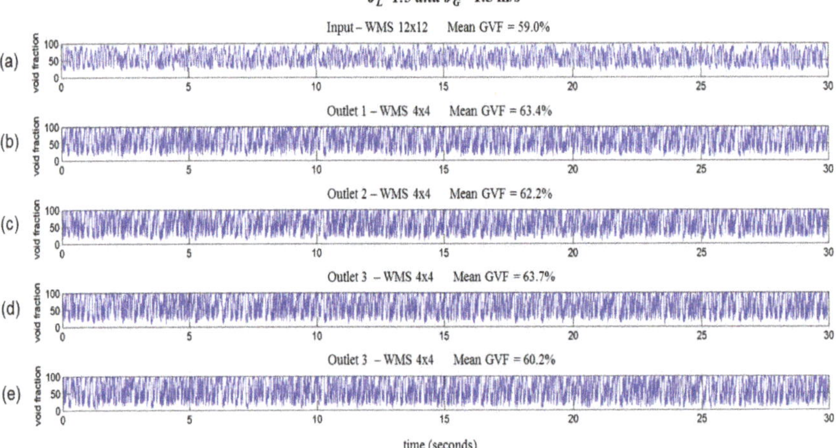


**Figure 11.** Void fraction time series (30 s. $J_L = 1.5$ m/s and $J_G = 1.5$ m/s): (**a**) Input and (**b**–**e**) outlets 1 to 4.

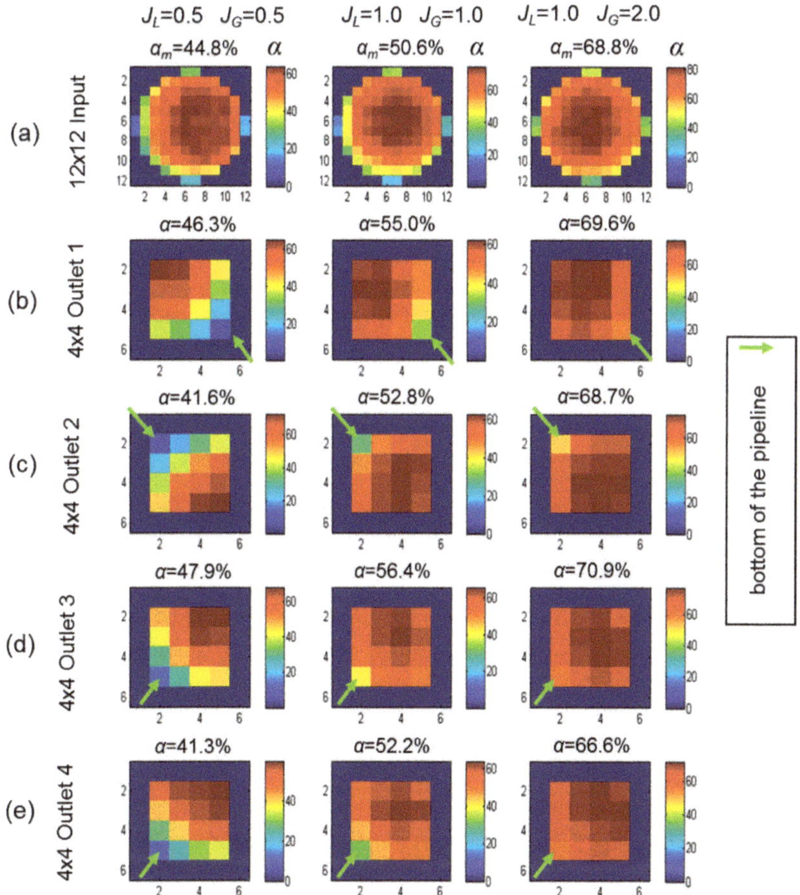



**Figure 12.** Mean axial void fraction ($\alpha_m$) measured in 30 s by WMS in three set of superficial velocities: $J_L$ = 0.5 m/s and $J_G$ = 0.5 m/s; $J_L$ = 1.0 m/s and $J_G$ = 1.0 m/s; and $J_L$ = 1.0 m/s and $J_G$ = 2.0 m/s. (**a**) Input and (**b–e**) outlets 1 to 4. WMS data at the outlets are rotated due to a different installation.

**Table 2.** Summary of mean void fraction results.

| $J_L$ (m/s) | $J_G$ (m/s) | Void Fraction (%) | | | | | | | |
| | | Input | Outlet 1 | Outlet 2 | Outlet 3 | Outlet 4 | Mean | Worst Case Difference | |
| | | $12 \times 12$ | $4 \times 4$ | $4 \times 4$ | $4 \times 4$ | $4 \times 4$ | Outlets | Value | % |
|---|---|---|---|---|---|---|---|---|---|
| 0.5 | 0.5 | 44.84 | 46.30 | 41.63 | 47.91 | 41.34 | 44.29 | 3.61 | 8.2 |
| 0.5 | 1.0 | 60.18 | 64.22 | 60.23 | 65.66 | 58.17 | 62.07 | 3.90 | 6.3 |
| 1.0 | 0.5 | 33.08 | 40.68 | 38.67 | 42.01 | 37.78 | 39.79 | 2.22 | 5.6 |
| 1.0 | 1.0 | 50.59 | 55.04 | 52.82 | 56.47 | 52.29 | 54.15 | 2.32 | 4.3 |
| 1.0 | 1.5 | 62.03 | 63.63 | 62.27 | 65.20 | 60.54 | 62.91 | 2.38 | 3.8 |
| 1.0 | 2.0 | 68.83 | 69.62 | 68.70 | 70.93 | 66.67 | 68.98 | 2.31 | 3.3 |
| 1.5 | 0.5 | 29.93 | 40.63 | 38.41 | 41.44 | 37.88 | 39.59 | 1.85 | 4.7 |
| 1.5 | 1.0 | 47.69 | 53.48 | 52.59 | 54.96 | 50.44 | 52.87 | 2.42 | 4.6 |
| 1.5 | 1.5 | 59.00 | 63.41 | 62.19 | 63.79 | 60.21 | 62.40 | 3.20 | 5.1 |
| 2.0 | 0.5 | 30.80 | 44.14 | 44.32 | 46.14 | 41.93 | 44.13 | 2.20 | 5.0 |
| 2.0 | 1.0 | 45.49 | 56.99 | 55.71 | 57.80 | 54.73 | 56.31 | 1.58 | 2.8 |

## 4. Conclusions

This work presents an experimental analysis of a flow distribution system for gas-liquid separation, using tomographic wire-mesh sensors. The study focused on a qualitative analysis of the slug flow pattern at the input, its behavior through the cyclonic chamber, and the flow distribution symmetry across the outlets. Use of wire-mesh sensors and a camera enabled the visualization of complete flow inside the cyclonic chamber. Results showed an ascendant liquid film in the cyclonic chamber with the gas phase at the center of the pipe generating a symmetrical flow. Dispersed bubbles coalesced into a gas vortex, due to the centrifugal force inside the cyclonic chamber. The behavior favored the separation of smaller bubbles from the liquid bulk, which was an important parameter for gas-liquid separator sizing. The void fraction analysis of the outlets showed an even flow distribution with less than 10% difference. At higher flow rates, the distribution rate difference decayed to 5%. A part of the observed difference could be due to the $4 \times 4$ WMS uncertainties. In future work, we may use a higher resolution WMS and also evaluate the liquid flow rate after the separation process. In summary, these results contribute to a new flow distribution system design, which is suitable for subsea gas-liquid separation problems. Hence, wire-mesh sensors were successfully applied to investigate a case study of the offshore oil and gas industry.

**Author Contributions:** Conceptualization, C.Y.O., H.K.E., E.N.d.S. and C.C.R.; Methodology, C.Y.O., H.K.E., E.N.d.S. and C.C.R.; Formal Analysis, C.Y.O., E.N.d.S., M.J.d.S.; Investigation, C.Y.O., H.K.E. and C.C.R.; Resources, P.H.D.d.S., M.J.d.S., F.N.J. and R.E.M.M.; Data Curation, C.Y.O., H.K.E.; Writing—Original Draft Preparation, C.Y.O. and H.K.E.; Writing—Review & Editing, C.Y.O., H.K.E., E.N.d.S., C.C.R., M.J.d.S. and P.H.D.S.; Visualization, C.Y.O.; Supervision, F.N.J., P.H.D.d.S. and R.E.M.M.; Project Administration, F.N.J., P.H.D.d.S., R.E.M.M. and P.V.S.R.D.; Funding Acquisition, F.N.J., R.E.M.M. and P.V.S.R.D.

**Funding:** Authors acknowledge the financial support provided by PETROBRAS. César Ofuchi thanks the support of CAPES through a PNPD post-doctoral grant.

**Conflicts of Interest:** The authors declare no conflict of interest.

## References

1. Prasser, H.M. Novel experimental measuring techniques required to provide data for CFD validation. *Nucl. Eng. Des.* **2008**, *238*, 744–770. [CrossRef]

2. Hernandez-Perez, V.; Azzopardi, B.; Kaji, R.; da Silva, M.J. Wisp-like Structures in Vertical Gas-Liquid Pipe Flow Revealed by Wire Mesh Sensor Studies. *Int. J. Multiph. Flow* **2010**, *36*, 908–915. [CrossRef]

3. Da Silva, M.J.; Hampel, U.; Arruda, L.V.R.; Amaral, C.E.F.D.; Morales, R.E.M. Experimental Investigation of Horizontal Gas-Liquid Slug Flow by Means of Wire-Mesh Sensor. *J. Braz. Soc. Mech. Sci. Eng.* **2011**, *33*, 237–242.

4. Tat, T.; Kikura, H.; Murakawa, H.; Tsuzuki, N. Measurement of Bubbly Two-Phase Flow in Vertical Pipe Using Multiwave Ultrasonic Pulsed Dopller Method and Wire Mesh Tomography. *Energy Procedia* **2015**, *71*, 337–351. [CrossRef]

5. KRagna, I.; Brito, R.; Scheicher, E.; Hampel, U. Developments for the Application of the Wire-Mesh Sensor in Industries. *Int. J. Multiph. Flow* **2016**, *85*, 86–95. [CrossRef]

6. Ragna, K.; Kryk, H.; Schleicher, E.; Gustke, M.; Hampel, U. Application of a Wire-Mesh Sensor for the Study of Chemical Species Conversion in a Bubble Column. *Chem. Eng. Technol.* **2017**, *40*, 1425–1433.

7. Jaworski Artur, J.; Meng, G. On-Line Measurement of Separation Dynamics in Primary Gas/Oil/Water Separators: Challenges and Technical Solutions—A Review. *J. Pet. Sci. Eng.* **2009**, *68*, 47–59. [CrossRef]

8. Prescott, N.; Mantha, A.; Kundu, T.; Swenson, J. Subsea Separation—Advanced Subsea Processing with Linear Pipe Separators. In Proceedings of the Offshore Technology Conference, Houston, TX, USA, 2–5 May 2016.

9. Orlowski, R. Marlim 3 Phase Subsea Separation System—Challenges and Solutions for the Subsea Separation Station to Cope with Process Requirements. In Proceedings of the Offshore Technology Conference, Houston, TX, USA, 30 April–3 May 2012.

10. Ninahuanca, H.E.M.; Stel, H.; Morales, R.E.M.; Ofuchi, C.; da Silva, M.J.; Neves, F., Jr. Characterization of the Liquid Film Flow in a Centrifugal Separator. *AIChE J.* **2016**, *62*, 2213–2226. [CrossRef]

11. Rosa, E.S.; França, F.A.; Ribeiro, G.S. The Cyclone Gas-Liquid Separator: Operation and Mechanistic Modeling. *J. Pet. Sci. Eng.* **2001**, *32*, 87–101. [CrossRef]
12. Sunday, K.; Yeung, H.; Lao, L. Study of Phase Distribution in Pipe Cyclonic Compact Separator Using Wire Mesh Sensor. *Flow Meas. Instrum.* **2017**, *53*, 154–160.
13. Kouba, G.E.; Shoham, O. A Review of Gas-Liquid Cylindrical Cyclone (GlCC) Technology. In Proceedings of the Production Separation System International Conference, Aberdeen, UK, 23–24 April 1996.
14. Kouba, G.; Wang, S.; Gomez, L.; Mohan, R. Review of the State-of-the-Art Gas/Liquid Cylindrical Cyclone (GLCC) Technology—Field Applications. In Proceedings of the 2006 SPE International Oil & Gas Conference and Exhibition, Beijing, China, 1–9 December 2006.
15. Huang, L.; Deng, S.; Chen, Z.; Guan, J.; Chen, M. Numerical Analysis of a Novel Gas-Liquid Pre-Separation Cyclone. *Sep. Purif. Technol.* **2018**, *194*, 470–479. [CrossRef]
16. Abrand, S.; Bonissel, M.; Roberto, D.S.V.; Raymond, H. Liquid Gas Separation Device and Method, in Particular for Crude Oil Liquid and Gaseous Phases. Patent US2010/0116128 A1, 2010.
17. Barnea, D. A Unified Model for Predicting Flow-Pattern Transitions for the Whole Range of Pipe Inclinations. *Int. J. Multiph. Flow* **1987**, *13*, 1–12. [CrossRef]
18. Prasser, H. A New Electrode-Mesh Tomograph for Gas–liquid Flows. *Flow Meas. Instrum.* **1998**, *9*, 111–119. [CrossRef]
19. Da Silva, M.J.; Schleicher, E.; Hampel, U. Capacitance Wire-Mesh Sensor for Fast Measurement of Phase Fraction Distributions. *Meas. Sci. Technol.* **2007**, *18*, 2245–2251. [CrossRef]
20. Ali, P.L.; Svrcek, W.Y.; Monnery, W. Computational Fluid Dynamics-Based Study of an Oilfield Separator—Part II: An Optimum Design. *Oil Gas. Facil.* **2013**, *2*. [CrossRef]
21. Casey, T.; Prasser, H.M.; Corradini, M. Wire-Mesh Sensors: A Review of Methods and Uncertainty in Multiphase Flows Relative to other Measurement Techniques. *Nucl. Eng. Des.* **2018**, *337*, 205–220. [CrossRef]

MDPI

St. Alban-Anlage 66

4052 Basel

Switzerland

Tel. +41 61 683 77 34

Fax +41 61 302 89 18

www.mdpi.com

*Sensors* Editorial Office

E-mail: sensors@mdpi.com

www.mdpi.com/journal/sensors